The Mind Has No Sex?

The Mind Has No Sex?

Women in the Origins of Modern Science

Londa Schiebinger

Harvard University Press
Cambridge, Massachusetts
London, England

Printed in the United States of America
FIFTH PRINTING, 1993

First Harvard University Press paperback edition, 1991

Library of Congress Cataloging-in-Publication Data

Schiebinger, Londa L.
The mind has no sex? : women in the origins of modern science / Londa Schiebinger.
p. cm.
Bibliography: p.
Includes index.
1. Women in science—History. 2. Sex discrimination in science—History. 3. Sex discrimination against women—History. I. Title.
Q130.S32 1989 88-34945
508.8042—dc19 CIP
ISBN 0-674-57623-3 (alk. paper) (cloth)
ISBN 0-674-57625-X (paper)

For Robert

Acknowledgments

A number of friends and critics have contributed to this project. Evelyn Fox Keller and Carolyn Merchant were generous with their time and ideas about women's place in science. Frigga Haug and Karin Hausen made my time in Berlin both pleasant and profitable. I. Bernard Cohen, Thomas Laqueur, Margaret Rossiter, Richard Kremer, Roger Hahn, Joan Scott, and Donald Fleming read and commented on various parts of the manuscript along the way.

Support from several grants gave me the much needed time and resources to pursue research for this book in libraries and archives throughout Europe. My thanks go first to the Fulbright-Hayes Graduate Scholar Program in Germany (1980–81) for support while I first framed my problem and began research; a Charlotte W. Newcombe Fellowship from the Woodrow Wilson Foundation supported early research on this topic; and a Marion and Jasper Whiting Fellowship made possible my summer in Paris in 1982. A National Endowment for the Humanities Research Fellowship and Rockefeller Foundation-Changing Gender Roles Fellowship provided me with two important years for research and writing; and a Deutscher Akademischer Austauschdienst Grant allowed me to spend a summer in Berlin to collect additional documents.

Many archivists and librarians helped me track down materials and documents. My thanks for their kind assistance to Christa Kirsten, Director, and the staff of the Zentrales Akademie-Archiv der Akademie der Wissenschaften der DDR; Gerda Utermöhlen and the staff of the Leibniz Archiv, Niedersächsische Landesbibliothek, Hanover; the staff of the Observatoire de Paris; Charlotte Wellman and the staff of Special Collections, Stanford University Libraries; Richard J. Wolfe and the staff of the Rare Book Collection of the Boston Medical Library in the Francis A. Countway Library. This book could not have been written without the extraordinary resources of the Harvard Libraries, the Preussische Staatsbibliothek, the Deutsche Staatsbibliothek, the Bibliothèque Nationale, and the British Library. My thanks, too, to the many others who helped with the arduous task of collecting illustrations. Sections of the introduction and Chapters 3, 5, and 7 include materials from articles that originally appeared in *Signs, Isis, Critical Inquiry,* and *Representations;* I thank these journals for permission to reprint those materials.

To those special friends whose wit and humor made it all worthwhile, my thanks: Judith Walkowitz, Leora Auslander, Arnold Davidson, David Kennedy, Mary Pickering, and especially Robert Proctor, who first suggested I explore the problem of women in science and to whom I dedicate this book.

Contents

Introduction

L'esprit n'a point de sexe.

—François Poullain
de la Barre, 1673

When the ardent Cartesian Poullain issued this declaration, that "the mind has no sex," he based his argument on the new science of anatomy. Women have sense organs similar to men's and brains with the same power of reason and imagination, so why, he asked, should they not be the equals of men, serving as professors, judges of the court, military officers, or ambassadors? Poullain's words reverberated across all of Europe. A woman chemist cited them in 1674 to defend her publication; a literary man invoked them in 1884 in support of women's admission to the Académie Française.

This popular refrain did not go unchallenged, however. In the days preceding the French Revolution, anatomists and medical doctors asserted that the mind does indeed have sex, and that sex extends "through more or less perceptible nuances into every part of the body"—including the brain.[1] William Whewell, in 1834, in the same paper in which he coined the term *scientist,* assured his readers that "notwithstanding all the dreams of theorists, there is a sex in minds."[2] Women, so these scholars taught, are essentially different from men, and female nature destines women ("the sex" as they were often called) for lives as mothers, confined to hearth and home.

The question of male and female equality in the sphere of intellect has proved an enduring one. Nowadays we ask, as have others before us, why are there so few women scientists? In the seventeenth century, the English natural philosopher Margaret Cavendish spoke for many when she wrote that women's brains are simply too "cold" and "soft" to sustain rigorous thought. The alleged defect in women's minds has changed over time: in the late eighteenth century, the female cranial cavity was supposed to be too small to hold powerful brains; in the late nineteenth century, the exercise of women's brains was said to shrivel their ovaries. In our own century, peculiarities in the right hemisphere supposedly make women unable to visualize spatial relations.

It seems unnecessary, however, to jump to rigidly biological explanations when one considers the obstacles that have been thrown in women's paths. For centuries, women were barred from academies and universities for no reason other than their sex. Those few who were able to succeed in science often failed to enjoy the recognition of that office: Marie Curie, the first person ever to win two Nobel prizes, was denied membership in the prestigious Académie des Sciences in 1911 because she was a woman. A woman was not elected to full membership in the academy until 1979, more than three hundred years after it first opened its doors.

Perhaps we should be asking a different question: why are there so few women scientists *that we know about?* Perhaps women have been scientists in the past but their stories have not been remembered. Or perhaps women have dominated certain fields but these fields have not been recognized as science. As long ago as 1830, the German physician Christian Harless lamented the "longstanding gap in the history of the natural sciences . . . There has been no historical and evaluative survey of all the women who, from the earliest times until our own, have distinguished themselves in the various sciences."[3] How long has the problem of women in science been a problem—how many times has the battle been fought, lost, and then forgotten? Science is not a cumulative enterprise; the history of science is as much about the loss of traditions as it is about the creation of new ones.

My purpose in this book is to explore the long-standing quarrel between science and what Western culture has defined as "femininity." What is it about being a woman that has made men of science fearful of female intrusion? What is it about science that has made it susceptible to such fears? To answer these questions I analyze the rise of modern science in Europe in the seventeenth and eighteenth centuries, focusing especially on the circumstances that led to the exclusion of women. In the seventeenth century science was a young enterprise forging new ideas and institutions. Men of science at this time can be seen as standing at a fork in the road. They could either sweep away the traditions of the medieval past and welcome women as full participants, or they could reaffirm past prejudices and continue to exclude women from science. What were the social, political, and intellectual circumstances that directed science down one road? What was the cultural cost of that journey?

The project of writing histories of women in science is not an entirely new one. As early as 1405, Christine de Pizan asked if women had made original contributions in the arts and sciences:

> I realize that you are able to cite numerous and frequent cases of women learned in the sciences and the arts. But I would then ask whether you know of any women who . . . have themselves discovered any new arts and sciences . . . which have hitherto not been discovered or known. For it is not such a great feat of mastery to study and learn some field of knowledge already discovered by someone else, as it is to discover by oneself some new and unknown thing.[4]

De Pizan's "Lady Reason" gave the answer of many modern historians of women: "rest assured, dear friend, that many great and noteworthy sciences and arts have been discovered through the understanding and subtlety of women, both in cognitive speculation, demonstrated in writing, and in the arts, manifested in manual works of labor."

De Pizan's celebration of the heritage of intellectual women was not unique. The first major histories were presented in the form of encyclopedias; from Giovanni Boccaccio's *De claris mulieribus* (1355–1359) through the eighteenth century, the names of women learned in the arts and sciences were collected largely in an attempt to prove that there were indeed more accomplished women than had been previously imagined.[5] It was not until the late eighteenth century, however, that the first encyclopedia appeared devoted exclusively to the history of women's achievements in the natural sciences. In 1786 the French astronomer Jérôme de Lalande included in his *Astronomie des dames* the first short history of women astronomers.[6] Fifty years later, in his *Verdienste der Frauen,* Christian Harless emphasized that both men and women are capable of doing science. "It is not exclusively for the man to research nature, to penetrate her creations with a keen eye and to enjoy her charms with unending passion. Sensitive women may also perceive her endless magic."[7] At the same time Harless identified what he thought were significant differences between men's and women's relationship to nature. Man, he wrote, as soon as he is moved by the spirit, searches to uncover causes underlying appearances, seeking to discover laws in life and nature. Woman, by contrast, searches nature over for expressions of love—this, he concluded, is the more natural and most beautiful way to approach the external world.

The European women's movement of the 1880s–1920s drew attention once again to the question of women's ability to contribute to the sciences. In 1888, a journal entitled *La Revue scientifique des femmes* was founded in Paris. In 1894, the Saint-Simonians in Paris held the first conference in modern times on women and science, from which grew Alphonse Rebière's book, *Les Femmes dans la science.*[8] That same year Elise Oelsner published her *Leistungen der deutschen Frau* (Achievements of the German woman), in which she paid close attention to women's scientific achievements.[9] By this time, however, the encyclopedia format employed in these books no longer served as an effective strategy for proving that women had indeed been great scientists. Antifeminists—such as Gino Loria in Italy—pointed out that even if there were enough distinguished women to fill three hundred pages, an equivalent project for men would run to three thousand pages. What woman, Loria asked, can rival Pythagoras or Archimedes, Newton or Leibniz?[10]

In response to this kind of criticism, European and American feminists turned from the strategy of emphasizing the achievements of exceptional women and began to emphasize instead the *barriers* to women's participation in science. The first detailed work of this sort was published in America in 1913 by H. J. Mozans (a pseudonym for the Catholic priest J. A. Zahm) under the title *Woman in Science.* It was an impassioned attempt to show that whatever women have achieved in science has been through "defiance of that conventional code which compelled them to confine their activities to the ordinary duties of the household." Mozans urged women to join the scientific enterprise and thereby unleash half the energies of humanity. He expected each woman to act as a Beatrice inspiring her very own Dante to achieve his full potential; in this way, man and woman would complement each other and together form a perfect *androgyne.* Only then would the world enter a new golden age, the golden age of "science and perfect womanhood."[11]

The works of de Pizan, Harless, Oelsner, Rebière, and Mozans are landmarks in the field of the history of women in science. Yet it should be noted that these authors, who wrote about outsiders, were also themselves by and large outsiders. Within the academy, as might be expected, the study of women in science was no more welcome than were women scientists. Despite scattered interest since the time of Christine de Pizan, records of women's contributions did not become part of the historical canon.

Nor was this picture to change with the emergence of the modern discipline of the history of science in the 1920s and 1930s. This new field, purporting to study the relation between science and society, did not consider the role of women in science. Even the women working in the field—Marie Boas Hall, Martha Ornstein, and Dorothy Stimson—paid little attention to women's participation in the sciences. None of the major theorists exploring the social origins of modern science—Robert Merton, Edgar Zilsel, Boris Hessen—made any mention of women. Historians studied participation in science from many important vantage points—religious affiliation, class, age, vocation—but ignored entirely questions of gender. Merton, for example, in his pioneering work on seventeenth-century English science, pointed out that 62 percent of the initial membership of the Royal Society were Puritan.[12] He did not, however, explore the implications of the even more striking fact that the early membership in the Royal Society, and indeed in all seventeenth-century academies of science, was 100 percent male.[13]

Since the 1970s, with increasing numbers of women entering both science and the historical profession, there has been a steadily growing interest in the history and philosophy of women in science. Women scientists have contributed thoughtful autobiographies giving firsthand accounts of their struggles to make a mark in science.[14] Intellectual biographies have appeared of Sophie Germain, Mary Somerville, Sofia Kovalevskaia, and Clemence Royer.[15] These books assess the contributions of women to science and address an important set of questions: What sparked their interest in science? How did they obtain access to the tools and techniques of science? How did they make their scientific discoveries? What recognition did these achievements receive in the broader community of scholars?

Much of this work fits the "history of great men" mold, with women simply substituted for men. One problem with this brand of history is that it often retains the male norm as the measure of excellence. Margaret Rossiter's analysis of women scientists in America breaks from this mold by shifting the focus from the exceptional woman to the more usual patterns of women working in science.[16] Evelyn Fox Keller, though she retains the focus on an exceptional woman in her biography of Barbara McClintock, does not simply measure McClintock against traditional male standards but uses her story (told largely in McClintock's own words) as a vehicle for evaluating current methods of experimental science.[17]

Today we understand to a large extent how women have been excluded from scientific institutions. Yet, despite efforts made through affirmative action, the problem of women's limited presence in science has not disappeared. In recent years scholars from a variety of disciplines have concentrated on the problem of explaining the deeper mechanisms of exclusion.[18] Sociologists and historians have identified structural barriers, in both society and the institutions of science, that have impeded women's professional advancement.[19] Biologists have begun to unravel the "myths of gender" embedded in the female body.[20] Philosophers and historians have begun to define gender-based distortions in the norms and practices of science and to discuss alternative epistemologies for the sciences.[21] This burgeoning literature has focused attention on how sexual differences have been cultivated by societies intent upon preserving sharp social and intellectual distinctions between the sexes. But what are the origins and implications of those differences? Should differences be celebrated or overcome? Are men and women different by nature or by nurture? And what difference does difference make?

Understanding gender differences and how they operate in the world of science today requires a reexamination of the history of women in science. What role did the "woman question" and the debate over "female nature" play in the origins of modern science? In the following pages I shall be examining the revolution in European science that took place in the seventeenth and eighteenth centuries and the place of gender in that revolution. I have analyzed the problem into four constituent parts—institutional organizations, individual biographies, scientific definitions of female nature, and cultural meanings of gender. Rather than privilege any one of these as causal agents, I see them as interdependent parts of a dynamic system. My purpose is to bring together elements of what are sometimes seen as different historical methods—each of which is crucial for understanding women's place in scientific culture.

The first part of this study looks at the *institutions* of science as agencies mediating between science and society, with a focus on how gender boundaries were negotiated in universities and scientific academies of the seventeenth century. Medieval universities were closed to all but a few exceptional women. Modern science, arising outside of and in opposition to the medieval university, was fostered in academies, princely courts, Parisian salons, and the artisanal workshop—that is, in a social landscape expansive enough to include a number of women. Chapter 1 argues that in this period it was not at all obvious that women would be excluded from the new institutions of science.

The second part concentrates on women as *historical actors* maneuvering within the gender boundaries prescribed by society. A number of women experimented in early modern Europe with the limits of convention in order to secure their place among the men of science; chapters 2, 3, and 4 tell the stories of some of them. My method here is comparative; the diversity among women should not be ignored. Women's participation in science varied greatly from country to country, class to class, and town to town. Women scientists in this period came predominantly from two social groups—the aristocracy and the crafts. Chapter 2 looks at England and France, where the greatest contributions came from women of the aristocracy. Noblewomen gained a limited access to science in the seventeenth and eighteenth centuries, much as they had access to political power and influence by virtue of their noble status. In Germany (Chapter 3), by contrast, women's scientific activities were prompted by their involvement in craft production. The strength of the artisanry in Germany may explain the remarkable fact that between 1650 and 1720 a sizable proportion (six out of forty-two) of German astronomers were women. In Chapter 4 I shift the focus to women's traditions in science. Women traditionally dominated the field of midwifery, for example, but with the scientific and social revolutions of the seventeenth and eighteenth centuries the "man-midwife" came to encroach upon that ancient monopoly. Finally, Chapter 9 traces the lives of some women in science at the close of this era; by the nineteenth century, women were effectively barred from the new institutions of science and restricted to the increasingly private sphere of the family, where they served as "invisible assistants" to brothers, husbands, or fathers.

The third part examines how the biological *sciences* have read and misread sex and gender (no real distinction was made by eighteenth-century anatomists) in women's bodies, and how these scientific readings of female nature were used to argue for or against the participation of women in science. Chapter 6 sets the stage by sketching the cosmological assumptions behind definitions of sex and gender in the *ancien régime* of science. One might expect dramatic changes in the understanding of woman's place in society and nature during the tumultuous years of the scientific revolution—a revolution which itself arose as part and parcel of larger movements toward participatory democracies. We find, however, that modern science—strident in its claims to displace the old—was curiously silent on the issue of gender. Not until deep into the eighteenth century did scientists (especially anatomists) undertake a thoroughgoing reform of definitions of sexuality, what in Chapter 7 I describe as "the scientific revolution in views of sexual difference."

The fourth part explores *cultural meanings* of femininity and masculinity and how understandings of gender became embedded in debates over women's ability to do science. One idea explored in this book is that femininity represents a set of values that has been excluded from science. Yet it is important to understand that femininity itself is profoundly historical. Chapter 5 examines how notions of gender often refer as much to the manners of a particular class or a particular nation as to those of a particular sex through two examples—the rise and fall of the feminine image of science, and the battles over intellectual style played out in the salons of Paris. Chapter 8 goes on to explore how the theory of sexual complementarity justified purging both women and what came to be defined as the feminine from the public world of science. Chapter 10, by way of conclusion, examines the self-reinforcing character of the gender system in science. Scientists helped crystallize gender roles by constructing views of men and women that bolstered emerging ideals of masculinity and femininity. Yet science and philosophy did not do so from a privileged vantage point untouched by social struggle—science was itself part of the terrain that divided the sexes. Ultimately we must ask: What were the consequences of the exclusion of women for the methods and priorities of science?

Though my focus is on women in the origins of modern science, my hope is that this book will also shed light on how gender relations have molded (and continue to mold) scholarship and knowledge more generally. The nature of science is no more fixed than the social relations of men or women: science too is shaped by social forces. One of those forces has been the persistent effort to distance science from women and the feminine. Shedding light on the origins of these efforts may help to add historical understanding to the problems of gender and science facing us today.

1

Institutional Landscapes

> It would be a pleasant thing indeed to see a lady serve as a professor, teaching rhetoric or medicine; or to see her marching in the streets, followed by officers and sergeants; or playing the part of an attorney, pleading before judges; or seated on a bench to administer justice in the supreme court; or leading an army, serving in battle; or speaking before states and princes as the head of an embassy.
>
> —François Poullain de la Barre, 1673

In 1910 the physicist Marie Curie was recommended for election to the prestigious Académie des Sciences in Paris. The following year Curie would become the first person ever—man or woman—to win a second Nobel prize. The fact that Curie was a woman provoked animated discussion in the Comité Secret and throughout the Académie. Some members felt that the question of whether to admit women to France's academy system was of sufficient import that it should be raised at a plenary session of the five academies constituting the Institut de France. Such a meeting was held, and there they read to the president of the institute a petition against Curie's election, reminding him that this was not the first time that a woman had presented herself for election to an academy: George Sand was not admitted to the Académie Française, nor Rosa Bonheur to the Académie des Beaux Arts, nor Sophie Germain to the Académie des Sciences. Curie's opponents urged that tradition be respected, despite the brilliance of the name advanced.

When Curie's name was finally put to a vote within the Académie des Sciences, she lost to Edouard Branly, a pioneer in wireless communication, by a narrow margin of two votes. But Curie's case raised the more general question whether women should be admitted to any of the great academies of France. This issue was settled by a greater margin; by a vote of 90 to 52, members of the Institut de France decided that no woman should ever be elected to its membership. Jacques Bétolaud (a lawyer and member of the Institut's Académie des Sciences Morales et Politiques) and his *confrères* had carried the day: by their vote members of the Institut held it "eminently wise to respect the immutable tradition against the election of women." Though some agreed with Henri Poincaré that merit should be rewarded wherever it may be found, others preferred not "to break the unity of this elite body."[1]

But was the tradition to which members of the institute appealed in fact immutable? Today we tend to assume that the exclusion of women from science was not an issue of debate until the late nineteenth or even the twentieth century. Women were not scientists, so what was there to debate? Evidence from the seventeenth and eighteenth centuries, however, suggests that at least some considered women's participation in science an open question. Not long after the founding of the Académie Française in 1635 (an important precursor to the Académie des Sciences), academy critic Gilles Ménage submitted the names of three celebrated women for consideration—Mademoiselle de Scudéry, Madame des Houlières, and Madame Dacier. None of these women were ever admitted, yet they were not excluded without serious challenge. Indeed, it would be a mistake to see the exclusion of women from subsequent institutions of science as a foregone conclusion. The landscape was a varied one, rolling with peaks of opportunity and valleys of disappointment. Traditions that to some twentieth-century academicians seemed inevitable had, in fact, been crafted through a process of conflict and negotiation in previous centuries. The complexities of this process may be illustrated by two related stories: the shifting institutional foundations of modern science during the revolutions that mark its origins and the changing fortunes of women in those institutions.

Natural philosophy was in many ways a new enterprise in seventeenth-century Europe, struggling for recognition within established hierarchies. Its relation to church and state and its role in the larger society were in a state of flux. Important questions about the nature of the new science—its ideals and methods, its proper limits, and who should mold these—remained to be answered. At this pivotal point, natural philosophers were attempting to free themselves from the fetters of the medieval university and to establish new institutions more responsive to their needs. What these changes would mean for women was unclear. When science emerged from a social setting—the Renaissance court, for example—where high-ranking women held power and prestige, that prestige carried them into the world of natural philosophy in much the same way it gave them (often limited) rights to rule entire lands. In a world organized on the basis of birth, well-born women simply outranked scholars. The relative prestige of the aristocrat and the scientist, however, was to change over the course of the seventeenth and eighteenth centuries. While the prestige of science waxed, that of the nobility waned; in time women of the aristocracy lost their place in scientific settings.

Modern scientific institutions have their roots in the medieval world, in particular the monasteries and universities of Europe. Precedents for women's participation in learning were set in these institutions. The Renaissance brought new social settings for science—princely courts and royal academies. By the seventeenth century a third setting was vying for the attentions of the learned: the Parisian salon, a woman's institution par excellence. The salon, I contend, offered a real alternative to the organization of intellectual life.

Monasteries and Universities

Without proper training and access to libraries, instruments, and networks of communication, it is difficult for anyone—man or woman—to make significant contributions to knowledge. Historically, women have not fared well in official institutions of learning. From the medieval to the modern university, the history of women in these institutions has been the history of their exclusion. Yet this history has not been uniform or predetermined; certain times have favored women's participation more than others. One institution, the medieval convent, provided an opportunity for women to pursue learning. Between the sixth and eleventh centuries, the church acquired a virtual monopoly on European literacy and education. Daughters of feudal lords, without land or inheritance, took vows in local monasteries, and as members of these monasteries came to wield temporal as well as spiritual power. Clerical life was a respectable avenue to power, for men and women alike. A number of women became renowned for their learning; these included the poet and teacher St. Radegund, abbess of Poitiers, and Hildegard von Bingen, the most notable medieval woman author on medicine, natural history, and cosmology.[2]

The rise of universities in the twelfth through fifteenth centuries, however, brought a decline in educational opportunities for women. In England women lost their place in spiritual and intellectual life with the closing of the convents. Henry VIII appropriated church lands and used part of the monies to expand the English university system in the sixteenth century. The revenues and lands of the nunnery of St. Radegund, an important center of learning for women, were transferred to Jesus College, Cambridge. Whereas religious houses had been centers of learning for both men and women, English universities were open only to men. This pattern was repeated in various forms throughout Europe at this time: women were displaced from centers of learning with the establishment of universities.[3]

From their beginnings European universities were, in principle, closed to women. Unlike monasteries, universities provided formal training in theology, medicine, and law, aimed at preparing young men for careers in government, teaching, law, medicine, or the church. Women, barred from these careers, were not expected to enter university.[4]

What was true of medieval universities was equally true of the reformed universities in the sixteenth century. University training became especially important as expanding bureaucracies of the new nation-states required more civil servants. As the nobility of the robe fused with the nobility of the sword, noblemen in England and France perceived that education was necessary to prepare young gentlemen to wield power in the state. In England this new conjunction between state service and education brought forth what Lawrence Stone has called a "revolution in education."[5] Enrollments expanded, including larger numbers of young men from the gentry and middle classes.[6] This revolution was not, however, to extend to women. Though women's advocates argued long and hard for their admission to universities, women of all classes remained barred from university education.

There were, of course, always exceptions. A small number of women did study and teach at universities beginning in the thirteenth century—primarily in Italy. In 1296 Bettisia Gozzadini lectured in law at the University of Bologna. Novella d'Andrea replaced her deceased father as professor of canon law at the University of Bologna in the fourteenth century, lecturing from behind a curtain (as legend has it) in order not to distract students from their studies by her very great beauty.[7] Women continued to study at Italian universities throughout the seventeenth and eighteenth centuries. In 1678 Elena Cornaro Piscopia became the first woman to receive the doctorate of philosophy (though not of theology as she originally wanted) at Padua.[8] Maria Agnesi of Milan became well known for her textbook on differential and integral calculus, *Istituzioni analitiche,* published in 1748. She is best known perhaps for her formulation of the *versiera,* the cubic curve that has come to be known as the "witch of Agnesi." In trying to persuade her to take up a chair of mathematics and natural philosophy at the University of Bologna, Pope Benedict XIV wrote to her: "from ancient times, Bologna has extended public positions to persons of your sex. It would seem appropriate to continue this honorable tradition."[9] She accepted this appointment only as an honorary one, and after her father's death in 1752 she withdrew from the scientific world in order to devote herself to religious studies and to serving the poor and aged.[10]

The most exceptional woman in this regard was Laura Bassi, professor of physics at the University of Bologna (see Figure 1). She received her doctorate in philosophy at Bologna in 1733 and shortly thereafter became the first woman to occupy a chair of phys-

---

Figure 1. Laura Bassi (1711–1788), professor of physics at the University of Bologna and member of the Academy of Science in Bologna, at work in her cabinet wearing her doctoral wreath of laurel. From Alphonse Rebière, *Les Femmes dans la science* (Paris, 1897), facing p. 28. By permission of the Schlesinger Library, Radcliffe College.

LAVRA MARIA
CATHARINA BASSIA
Vxor Ios. Verati M. D. et P.P. Bononiensis
Philosophiae Doctrix, Collegii Lectrix publica
Instituti Scientiarum Socia
nata 31 Octobr. An. MDCCXI

ics at a university. Celebrated for her work in mechanics, Bassi became a member of the Academy of Science in Bologna. Like other members she prepared and presented annual papers ("On the compression of air" [1746], "On the bubbles observed in freely flowing fluid" [1747], "On bubbles of air that escape from fluids" [1748], and so forth) and received a small stipend. She also invented various devices for her experiments with electricity. In 1776 she replaced Paolo Balbi as professor of physics at the Institute of Bologna. The Englishman Charles Burney, who met Bassi during his tour of Italy and was much impressed, found her "though learned, and a genius, not at all masculine or assuming."[11]

Like those of other scientific women, Bassi's achievements have largely been ignored by historians. H. J. Mozans, the Roman Catholic priest who wrote an early history of women in science, noted that the prolific scholar was also a prolific mother: Mozans attributed to Bassi the birth of twelve children.[12] Paul Kristeller has recently questioned whether Bassi ever taught in a public capacity; though her name appeared on the registers of the university for forty-six years (1732–1778), he assumes she taught only privately at her home.[13] We need to know much more about this pioneering woman physicist.

Italy was an exception in Europe, and little is known about why women professors were acceptable to the church and university. What little is known suggests that women were encouraged in their studies by their fathers—who were usually themselves professors. Daughters, perhaps in the absence of a son, were taken into the profession much as apprentices were. Maria Agnesi, for example, was encouraged by her father, who hired distinguished professors as her tutors and established a salon in their home where she could present and defend various theses; in another case, Anna Manzolini replaced her husband as professor of anatomy at Bologna.[14]

These exceptions did not, however, open universities to regular study for women. When Piscopia was allowed to sit for her Ph.D. exams in the 1670s, for example, university officials decided that her case was not to set a precedent; indeed, for nearly three hundred years no other woman was awarded a doctorate at Padua.[15] Laura Bassi's position at the Institute of Bologna was an extra twenty-fifth chair scheduled to fold after her tenure, making it impossible for a woman to replace her.[16]

Renaissance Courts

Despite these few exceptions, women were generally proscribed from European universities until the end of the nineteenth and in some cases until the twentieth century. The prominence of universities today should not, however, lead us to overemphasize their importance in the past. Universities have not always been at the center of intellectual life. Modern science emerged from a variety of social locations—including the workshops of artisans, informal salons, and kings' academies. Women's involvement in scientific enterprises depended on their position in the social setting from which the new science emerged. Chief among these settings were the courts of Renaissance princes. The growth of towns and the rise of an urban aristocracy fostered a revival of ancient knowledge in sixteenth-century Europe. Supported by princely patrons, arts and letters flourished as ornaments of royal courts. Powerful patrons lent their support to artists and inventors; art and culture, in turn, celebrated the magnificence of the court. Well-born women, holding positions of prominence as queens or ladies of the court, often participated in the revival of learning, either as philosophers, poets, or patrons.

In the Middle Ages formal education, a monopoly of the church, was generally confined to contemplative study in monasteries and universities. Many of the highest nobility could neither read nor write, and in fact literacy carried with it no special status. Noblemen had little time for book-learning. A contemplative life did not suit the well-born gentleman who spent his day hunting and hawking, dining and drinking. At the end of the sixteenth century, a nobleman judged that "it becomes the son of gentlemen to blow the horn nicely, to hunt skillfully, and elegantly to carry and train a hawk. The study of letters is for rustics."[17]

Beginning in the sixteenth century, learning began to vie with hunting and jousting as a pursuit appropriate to genteel life. The invention of the printing press in the fifteenth century made many excellent and rare books available to the wealthy. In this period of transition from jousting feudal lord to refined courtier, physical strength was dethroned as a leading social virtue. As the aristocracy ceased to be primarily a military class, rhetorical skills became marks of the gently born. In 1532 Agrippa von Nettesheim sarcastically suggested that "if strength alone gives the preeminence, let men give place to their horses, confess their oxen their masters, and pay homage to elephants."[18] The superior power attributed to the mind seemed to justify women's participation in intellectual culture. Indeed Baldassare Castiglione, recording supposed conversations at the court of Urbino, reported the remark that men might be stronger than women, but that women's physical frailty produced mental alertness. "There is," this anti-Aristotelian held, "no doubt that women, for being of softer flesh, are yet more mentally acute and have an intelligence better attuned to speculation than men."[19]

Learning at the Renaissance courts of Italy arose in what Werner Gundersheimer has called a "feminine" context.[20] Two distinct cultures—the professions of arms and of letters—coexisted within these courts: government and martial arts were distinctly masculine and practiced almost exclusively by men, while the quiet "play of the intellect" engaged ladies and humanists. The contrast in these two cultures can be seen in Castiglione's famous book of manners, *The Book of the Courtier.* After an ample meal the ladies and gentlemen of the court repair to the rooms of the Duchess Elisabetta Gonzaga in Urbino for "gentle discussion and innocent pleasantries."[21] The duke retires to sleep, owing to his infirmities and exertions from the day's jousting, riding, and "handling of every sort of weapon." In the play of intellect that follows, the duchess or her female deputy (Signora Emilia) sets the questions and directs the conversation, gently prodding and judging the disputations of the male humanists. In the course of the evening, Duchess Gonzaga skillfully turns the humanists' attention to the subject of women. She asks for both a consideration of the proper role of the court lady and a reevaluation of Aristotle's pronouncements on women's inferior social position.

Several factors worked together to give a lady prominence in the intellectual culture of the court. First and foremost was her rank. In the absence of her husband, Duchess Gonzaga outranked every man in her presence—even potential rulers. Certainly she and many of her ladies outranked male humanists whose distinction lay in letters, not in land. In this context the duchess was able to exchange social status for a place within the learned world to which she was otherwise a stranger, having no special training in the philosophy of the ancients. The second factor lending ladies prominence in the intellectual world was that the play of the intellect was a form of leisure. Learning at the court of Urbino was not directed toward affairs of state; instead, fine questions were posed after dinner amidst pastimes such as singing or dancing—in a context highly suited to women. The questions posed and answered entertained the court circles and were not necessarily intended to resolve serious disputes.

The ladies, seated alternately among the men, influenced the course of learned conversation, even though the number of men at such gatherings was usually greater. Seated at the head of the company, the duchess impressed her style on those assembled, each of whom sought to exhibit fine manners and virtue. These ladies selected a courtier for his talent but also for his beauty of countenance and person; each was to cultivate that certain grace that was called an "air." In this atmosphere, male humanists observed certain restrictions. When discussing the qualities and occupations of the ideal court lady, would-be courtiers kept in check antiwoman sentiments so as not to hurt the ladies' feelings or make enemies of them. The greatest censure was the ladies' laughter.[22] Learned discourse was not only a feminine pastime but one favorable to women.

Ladies thus enjoyed a certain prominence among the learned of the Renaissance court, but their power was restricted by the same limits in these circles as in society at large. Though ladies chose topics, moderated discussions, steered the development of debates, and refereed the outcomes, it was for the most part the (male) humanist who brought content to their questions.[23] It was the men whose specialized training allowed them to speak. Though the women might manipulate certain aspects of the conversation, they were more commonly reduced to the subordinate position of asking questions.

Scientific Academies

Historians of science have focused on the founding of scientific academies as a key step in the emergence of modern science. The major European academies of science were founded in the seventeenth century—the Royal Society of London in 1662, the Parisian Académie Royale des Sciences in 1666 (after 1816, the Académie des Sciences), the Societas Regia Scientiarum in Berlin in 1700 (later called the Akademie der Wissenschaften). By the end of the eighteenth century, a network of academies stretching from Saint Petersburg to Dublin, Stockholm to Palermo had consolidated Europe's intelligentsia in what one historian has called a "unified Republic of Letters."[24] As the scepter of learning passed from courtly circles to learned academies, science took a first step toward losing its amateur status and ultimately becoming a profession. These state institutions, founded or protected by kings, provided social prestige and political protection for the fledgling science.

This first legitimation of the new science also coincides with the formal exclusion of women from science. With the founding of the academy system in Europe, a general pattern for women's place in science begins to emerge: as the prestige of an activity increases, the participation of women in that activity decreases. The exclusion of women from these academies was not a foregone conclusion, however. Women had been active participants in the aristocratic learned circles which these academies recognized as their forebears. There were, in fact, a significant number of women trained in the arts and sciences (and in the crafts, as described in Chapter 3).

The exclusion of women at this particular juncture in the history of science, then, needs explanation. The seventeenth-century scientific academy had its roots in two distinct traditions—the medieval university and the Renaissance court. Insofar as academies were rooted in universities, the exclusion of women is easily explained: women were unlikely candidates for admission to institutions deriving their membership largely from the universities, which since their founding had generally proscribed women. It is also possible, however, to argue that scientific societies arose more directly from courtly traditions. Frances Yates has identified the Platonic academy founded in mid-fifteenth-century Florence under the auspices of the great prince Lorenzo de' Medici as the root of the whole academy movement.[25] If we emphasize the continuities between scientific academies and Renaissance courtly culture—where women were active participants in intellectual circles—it becomes more difficult to explain the exclusion of women from the academies.

The founding of the Académie Française—an academy devoted generally to the promotion of the French language and literature—initiated the academy system in France. Incorporated by the king in 1635, the Académie Française was the first of the modern state academies to be founded outside Italy, predating the Royal Society of London by some twenty-five years and the more specialized Académie Royale des Sciences in Paris by some thirty years. The founding of the Académie Française is a particularly important moment for our analysis of women's place in intellectual culture, for—although this was not an academy devoted to science—it was here that women were first excluded from the modern institutions of learning.

In its early years, it was not clear that women would be excluded from membership in the Académie Française. Noblewomen had been active *académiciennes* at several of the courtly academies out of which the Académie Française developed. Take the case of Henri III's Palace Academy, an important precursor of the Académie Française. Established in the 1570s to enhance his own education, Henri's academy cultivated an encyclopedic learning devoted to every kind of philosophy, science, music, poetry, geography, mathematics, and painting. The assembly, meeting twice a week in the king's cabinet, was attended by the "most learned men" and "even some ladies," all of whom spoke on problems which they had studied beforehand. From contemporary accounts we know these ladies included Claude-Catherine de Clermont, marquise de Retz and Madame de Lignerolles.[26] The presence of these ladies was not purely ornamental. They took an active part in the discussions at hand. One academy member, Seigneur de Brantôme, reported that after six months' absence from Paris he was surprised to meet a lady of rank going to the academy (which he thought had been disbanded) where she was studying philosophy and discussing the principle of perpetual motion.[27]

Women were also active in the less formal salons that sprang up between the demise of the royal Palace Academy in the 1580s and the founding of the state-financed Académie Française in 1635. Scholars today continue to debate which literary salon should be considered the one to which Richelieu offered his official protection—that of Valentin Conrart, Marie de Gournay, or Guillaume Collelet.[28] Whichever among these is ultimately judged the true forerunner of the Académie Française, it is important to recognize that Marie le Jars, demoiselle de Gournay and a central figure in the private circles from which the academy emerged, would not become a member. Already in her seventieth year when the Académie was founded, perhaps she was considered too old for membership. Yet she was not the only prominent literary woman who deserved consideration for membership in the king's academy. Indeed, several women's names were put forward for membership in the early years of the academy, proposals that, we are told, were received with some favor.[29] Little is known about these discussions. Gilles Ménage, who included women in his own Wednesday receptions (and for unclear reasons refused academy membership), left the following note:

> A short while ago in the Académie were nominated several women (Mademoiselle de Scudéry, Madame des Houlières, Madame Dacier, and several others) who, illustrious for their intelligence and their knowledge, are perfectly capable of enriching our language with handsome works and who have already produced marvelous ones. Monsieur Charpentier supported this proposal with the example of the academies of Padua, where erudite women are admitted. My treatise, *Mulierum Philosopharum,* furnished [to those involved in this debate] ancient examples of marks of distinction granted erudite women. Nevertheless, the proposal made to the Académie produced no results.[30]

In his *Historia mulierum philosopharum,* Ménage had documented the great numbers of philosophical women of the past to support his argument that women (like Madame Dacier, the classicist to whom he dedicated the book) merited admission to learned bodies.

The literary merit of the women proposed for membership was never in doubt: Madeleine de Scudéry won the academy's first prize for eloquence in 1671; Madame des Houlières won the prize for poetry in 1687.[31] What was in question was their sex. Although the statutes of the academy did not specifically exclude women, some say it was originally Richelieu who refused to admit them, others say it was the poet Jean Chapelain who refused. Reflecting upon this situation, academy member Jean de la Bruyère wrote:

> I have not forgotten, gentlemen, that one of the principal statutes of that illustrious body advocates admitting only those whom one judges the most distinguished. You will not find it therefore strange that I give my vote to Monsieur Dacier, though all the same I prefer Madame, his wife, if you would admit among you persons of her sex.[32]

Thus, though women were respected members of French literary circles, they did not become salaried members of the academy.

Women faced similar problems in the early Académie Royale des Sciences. As was the case with the Académie Française, women were an integral part of the informal *réunions,* salons, and scientific circles that grew up in opposition to the tyranny of old methods in the French university system.[33] Women gathered among the curious every Monday at Théophraste Renaudot's Bureau d'Adresse to watch his experiments.[34] Women were especially strong among the Cartesians, who sought refuge from hostile academics in the salons of Paris. Every Wednesday persons of "all ages, both sexes, and all professions" gathered at the home of Jacques Rohault to watch him attempt to give an experimental base to Descartes's physics.[35] In the years preceding the founding of the Académie Royale des Sciences, the number of women attending informal academies and salons proliferated: women attended the Palais Précieux pour les Beaux Esprits des Deux Sexes in the 1650s; Cartesians flocked to the salons of the marquise de Sévigné and the duchess of Maine. Louis de Lesclache's lessons in philosophy were so overwhelmed by women that he was later reproached as a *professeur pour dames.*[36] The numbers of women attending informal academies grew at such a rate that the renowned grammarian Pierre Richelet added the word *académicienne* to his dictionary in the 1680s, explaining that this was a new word signifying a person of the fair sex belonging to an academy of *gens de lettres,* coined on the occasion of the election of Madame des Houlières to the Académie Royale d'Arles.[37]

Despite their prominence in informal scientific circles, women were not to become members of the Académie Royale des Sciences. Why not? Certain aspects of the French academic system could have encouraged the election of gentlewomen. Seventeenth-century academies perpetuated Renaissance traditions of mixing learning with elegance, adding grace to life and beauty to the soul. The Académie Royale des Sciences retained a conviviality in its program, with rules of etiquette and a routine of dinners and musical entertainment, all of which tended to blur the boundaries that would later separate the scientific academy from the salon.[38] This was an atmosphere in which the *salonnière* would have been at home. At the same time, the Académie was monarchical and hierarchical. At the head of the Académie sat twelve honorary nobles whose presence was largely ornamental; working scientists—the new aristocracy of talent—found themselves on a lowlier rung.[39] Yet noble birth was not enough to secure women a place in the academic system. The closed and formal character of the academy discouraged the election of women. Membership in the academy was a public, salaried position with royal protection and privileges.[40] A salaried position in itself would not preclude the admission of women (Marie de Gournay, for example, received a modest *pension* from Richelieu until her death in 1645); with membership of the Académie limited to forty, however, the election of a woman would have displaced a man.

The exclusion of women from the Royal Society of London is also difficult to explain but for different reasons. At least ideologically, the Royal Society was supposed to be open to a wide range of people. Thomas Sprat, the first historian of the society, emphasized that its philosophy was not to be a parochial one restricted to the tenets of a particular religion, nation, or profession, but rather was to be "a Philosophy of *Mankind.*" According to Sprat, valuable contributions were to come from both learned and vulgar hands: "from the Shops of *Mechanicks;* from the Voyages of *Merchants;* from the Ploughs of *Husbandmen;* from the Sports, the Fishponds, the Parks, the Gardens of *Gentlemen.*" In addition, no special study or extraordinary preparations of learning were required: "Here is enough business for *Minds* of all sizes: And so boundless is the variety of these *Studies,* that here is also enough delight to recompence the Labors of them all, from the most ordinary capacities, to the highest and most searching *Wits*."[41]

In fact the Royal Society never made good its claim to welcome men of all classes. Merchants and tradesmen comprised only 4 percent of the society's membership; the vast majority of the members (at least 50 percent in the 1660s) came from the ranks of gentlemen *virtuosi*, or wellborn connoisseurs of the new science.[42] Considering that the society relied for its monies on dues paid by members, the absence of noblewomen from the ranks of enthusiastic patrons is puzzling.

One woman in particular—Margaret Cavendish, duchess of Newcastle—was a qualified candidate, having written some six books about natural philosophy, along with several other plays and books of poetry (see Chapter 2). She had long been a generous patron of Cambridge University and would have been a financial asset to the impoverished society. One should recall that fellows of noble birth bestowed prestige upon the new society; men above the ranks of baron could become members without the scrutiny given other applicants. When the duchess asked for nothing more than to be allowed to visit a working session of the society, however, her request aroused a flood of controversy. Although never invited to join the Royal Society, Cavendish was allowed to attend one session after some discussion among society fellows. The famous visit took place in 1667.[43] Robert Boyle prepared his "experiments of . . . weighing of air in an exhausted receiver; [and] . . . dissolving of flesh with a certain liquor." The duchess, accompanied by her ladies, was much impressed by the demonstrations and left (according to one observer) "full of admiration."[44]

Although no official record of the discussion of Cavendish's visit remains, Samuel Pepys tells us that there was "much debate, *pro* and *con*, it seems many being against it, and we do believe the town will be full of ballads of it." When no other ballads appeared, Royal Society member John Evelyn was moved to write one of his own.[45] From Pepys's report it seems many fellows felt that Cavendish's membership would bring ridicule rather than honor. Evelyn's wife, probably reflecting the attitudes of many in the society, described the duchess's writings as "airy, empty, whimsical and rambling . . .

aiming at science, difficulties, high notions, terminating commonly in nonsense, oaths, and obscenity."[46]

Margaret Cavendish's visit indeed appears to have set a precedent—a negative one. No woman was elected to full membership in the Royal Society until 1945.[47] For nearly three hundred years, the only permanent female presence at the Royal Society was a skeleton preserved in the society's anatomical collection.[48]

Women at the Periphery

Though none of Europe's academies had formal statutes barring women, no woman was elected to full membership in either the Royal Society of London, the Académie Royale des Sciences in Paris, or the Societas Regia Scientiarum in Berlin until the middle of the twentieth century.[49] Only the Italian academies—at Bologna, Padua, and Rome—regularly admitted women. Two French women, Madeleine de Scudéry in the seventeenth century and Emilie du Châtelet in the eighteenth, denied admission to learned societies in their own countries, were honored with membership in the Italian academies, but French academies did not reciprocate. The Italian mathematician, Maria Agnesi, who had been elected to membership in the Academy of Science at Bologna in 1747 and whose work was translated into French under the auspices of the Académie Royale des Sciences, was not invited to join the academy in Paris.[50] Academy Secretary Bernard de Fontenelle sadly noted that, despite her academic achievements and mathematical acclaim, it was impossible for Agnesi to become a member.

That women were not allowed to become members of scientific academies does not mean that they were barred from scientific work at these institutions. In France, Madeleine Basseporte served as illustrator at the Jardin Royal des Herbes Médicinales, where she sketched on vellum the rare plants grown in the garden from 1735 until her death in 1780. She followed the famous Claude Aubriet in this post and was paid a small salary of 1,000 livres yearly.[51] Bernard de Jussieu judged Basseporte's work of great value and deposited her sketches in the archives of the Bibliothèque Royale. Only a historian of the nineteenth century judged that her reputation far outstripped her merit.[52] Basseporte's official position at Paris's botanical garden can be explained in part by perceptions that work in medicine or botany was appropriate for a woman, for women had long been active in these fields. Basseporte also owed her position at the Jardin du Roy to the strength of women in the visual arts. The fine arts were considered (then as now) a more appropriate field for women than any science; we need only recall that seven women were elected to membership in the Académie Royale de la Peinture et de la Sculpture between 1663 and 1682 (though in 1706 this academy reversed its policy and ruled that no more women could be admitted, at a time when more and more women were applying for membership).[53] Scientific academies of the same period, by contrast, did not even flirt with the election of female members.

Also on the periphery of academic life was anatomist Marie-Catherine Biheron, who studied the art of illustration with Madeleine Basseporte. This pair provides a rare example (outside of midwifery) of a woman training a young woman who was not her daughter for a career in science. Because these women left no private papers, our knowledge of their collaboration is hazy. From Basseporte's hand we have only her drawings; from Biheron's only a four-page advertisement of her anatomical collection. What little we do know concerning their collaboration comes from the memoirs of contemporaries—Diderot, Baron Melchior von Grimm, the comtesse de Genlis—who knew them only at a distance.

Marie Biheron was born in 1719 the daughter of an apothecary. Like her teacher Basseporte, she never married. It was Basseporte who counseled Biheron to turn her skill to preparing anatomical models; Biheron was eventually to become one of the leading wax modelers of her day. Wax, a particularly popular medium in this period, was used to model everything from scenes in Turkish baths to the figures of European royalty on display at Madame Tussaud's famous museum.[54] The art of modeling the human body in wax was developed in the sixteenth and seventeenth centuries for use in teaching because of the shortage of cadavers for dissecting. Anatomical wax modeling—practiced by men and women alike—reached its peak in Italy and France during the eighteenth century. Many of these wax models were gynecological, showing all parts of the uterus in every state and the fetus in various states of development, for the use of students—male and female, as we are told—in midwifery and anatomy.[55]

Though men, too, worked in wax, women were particularly prominent in modeling the female body because, until the eighteenth century, knowledge and care of the female body was a woman's domain. Anna Morandi-Manzolini (1716–1774) became particularly well known for her models displayed in the anatomical museum of the Institute of Bologna (showing how the fetus is nourished in the womb). Biheron's own work was praised for its precise and delicate replication of nature. A prestigious visitor to her collection (Sir John Pringle, physician general of Britain and later president of the Royal Society) found her models so lifelike that he reportedly exclaimed: "They want nothing but the smell." Grimm was so beguiled by her work that he refused to believe that the modeling substance was wax because it did not melt in fire.[56]

Biheron presented her work on several occasions to the French king's academicians. The first such occasion came in 1759 at the invitation of the anatomist Jean Morand. In the record of the meeting, her renderings were praised for surpassing those of William Desnoues, the leading wax modeler of previous decades: "Where Desnoues showed only the position and color of the different parts of the human body, Biheron has reproduced exactly the consistency, suppleness, and weight of the brain, kidneys, intestine, and other parts of the human body." Here too her renderings fooled the onlooker into believing they were real: Biheron "imitated nature . . . with a precision and truth which no person has yet achieved."[57]

How did a woman who was not a member of the established medical community procure bodies for her studies? Though we do not have Biheron's words describing her procedures, we are told by Louis Prudhomme, writing in the 1830s, that she hired people to steal cadavers for her from the military. The business was far from pleasant. Already putrefying when she received them, she kept the cadavers in a glass cabinet in the middle of her garden.[58] In this manner Biheron was able to make a close study of the human body and achieve astonishing perfection in her models.

In 1770 Biheron returned to Paris to demonstrate to the Académie Royale des Sciences her elaborate and lifelike model of a pregnant woman. The model reproduced exactly all the stages and mechanisms of birthing, complete with a moveable coccyx, a cervix that dilated or closed on demand, and removable infants. The device was particularly useful for demonstrating to students how to cope with dangerous deliveries without doing harm to a living subject.[59] Models like Biheron's were used by the celebrated midwife Anne Le Boursier du Coudray, hired in the 1770s by the French government to teach as many as 4,000 women throughout the provinces the art of midwifery.[60] Biheron's models were cited in academy proceedings as late as 1830 as the best example of this kind of work.[61]

In 1771 Biheron demonstrated her wax anatomy once again at the Académie Royale des Sciences, this time for the pleasure of the crown prince of Sweden. Accompanying her were men such as Lavoisier, Macquer, and Morand. Women of the court were also allowed to attend this special session. Biheron's greatest honor was the acquisition of her anatomical models by Catherine the Great for the Academy of Sciences in Saint Petersburg.[62] Biheron's artificial anatomy accompanied a collection of medical instruments sent from the Parisian Académie Royale des Sciences by Morand.[63]

These successes at the academy did not, however, translate into an academy seat, nor a *pension* from the king. For thirty years Biheron made her living by opening her collection to the curious every Wednesday for the small fee of three *livres*.[64] She also taught lessons in her home: Diderot and Genlis, for example, were among her students. Diderot later wrote that her classes were especially valuable for young women.[65] Prudhomme reported, however, that she was censured by the doctors and surgeons of Paris for infringing upon their monopoly. (Established surgeons presumably objected because Biheron had been attracting significant numbers of male students from their classes.) Biheron also visited London twice looking for work and met with little success, though her wax models did influence doctors William and John Hunter to open their own museum of comparative anatomy in the 1770s.

There are other examples of women scientists who were active on the periphery of the major academies. At the Académie Royale des Sciences in Paris, Emilie du Châtelet's essay on the propagation of fire was published in the proceedings of 1738. Marie Thiroux d'Arconville's anatomical illustrations were published by the academy in 1759, under the name and protection of academy member Jean J. Sue. In London the Royal Society published Caroline Herschel's discoveries of comets (in 1787, 1789, 1792, 1794, and 1796), and her revision of Flamsteed's catalog of stars was published by the society in 1787. Women also won prizes for their contributions to science. Sophie Germain won the *grand prix* of the Parisian academy in 1816 for her work on elasticity. In 1888 Sofia Kovalevskaia won the prestigious *Prix Bordin* of the Académie des Sciences for her work in mathematics.

Women were also to be found in academies on the periphery—in provincial academies or less prestigious academies in European capital cities. Madame des Houlières was elected member of the Académie Royale d'Arles in 1680. In 1788 Nicole Lepaute became a member of the Académie des Sciences of Béziers. Caroline Herschel became a member of the Royal Irish Academy in 1838.

Parisian Salons

The focus of historians on academies in the rise of modern science has drawn attention away from the other heir of the courtly circle—the salon—as an institution of science.[66] Discussion of science was fashionable at the salons of Madame Geoffrin, Madame Helvétius, and Madame Rochefoucauld; Madame Lavoisier received academicians at her home.[67] French salons of the seventeenth and eighteenth century competed with academies for the attention of the learned. While the academies had a salon character in their early years, combining conviviality with scientific investigation, later years saw the separation of the sciences from the humanities. It was only in the salon that the *savant* continued to be gracious as well as learned (see also Chapter 5).[68]

The grand salons of Paris offer unique examples of intellectual institutions run exclusively by women. These informal gatherings played a crucial role in restructuring French elites. Neither strictly aristocratic nor bourgeois, salons celebrated the superiority of acquired nobility over inherited nobility, emphasizing the virtues of talent and refinement over noble birth, and in this way opening polite society to the rich and talented. Like the academies, salons served as a major channel of communication among newly consolidated elites.[69] There was significant overlap in (male) membership among salons, academies, and writers for the *Journal des Savants,* the major French journal of scientific and cultural news (see Table 1). Bernard de Fontenelle, for example, longtime secretary of the Académie Royale des Sciences, became *président* of Madame Lambert's salon. There was also overlap in membership from salon to salon. Members of Madame Lambert's salon (Fontenelle, the physicist and mathematician Jean-Jacques Dortous de Mairan, the medical doctor Jean Astruc, and others) joined the gatherings at Madame Tencin's salon after Lambert's death. After Madame Tencin's death, Madame Geoffrin's salon moved to center stage; regular members included Fontenelle, Mairan, and Montesquieu.[70] It is hard to know exactly what went on in salons; like many other women's arts, salon discourse was not made to last. Unlike academies, salons had no journals, proceedings, or permanent secretaries.

Socially prominent women dominated the intellectual gatherings held in the magnificent salons, or sitting rooms, of their private homes. Despite their informal and private character, salons wielded substantial influence in public matters: Madame Lambert was said to have "made" academicians; Julie de Lespinasse's salon has been called "the laboratory of the *Encyclopédie*."[71] Well-placed women served as patrons to young men anxious to make careers in science. Women served effectively as intellectual power brokers because science at this time was organized as much through highly personal-

Table 1. A sample of salon members affiliated with the Académie Française or the *Journal des Savants*

Member of the salon of Madame de Rambouillet (1608–65)	Académie Française	*Journal des Savants*
J. Chapelain	x	x
Gilles Ménage		x
T. Corneille	x	
J. L. Balzac	x	
Madame de Sablé		x
V. Conrart	x	
P. Pellisson*	x	
Mlle Scudéry	Prize winner	

*Pellisson was also a member of the Académie Royale des Sciences.

ized patronage systems as through formal institutions.[72] Salons acted as social filters, identifying young men of talent and turning them into protégés.

The power of salon women to make or break public careers should not be underestimated. At the same time, it is important to recognize that *salonnières* (women of the salons) experienced the same limits to their power as the most royal women of the land: they served as powers behind the throne but could not themselves sit on the throne. While women maneuvered to ensure the election of their candidate to the Académie Royale des Sciences, they were powerless to bring about their own election. Throughout the seventeenth and eighteenth centuries, *salonnières* served as patrons to young men, not to young women.

Women's Academies

With royal academies closed to women, we should not be surprised to learn that women proposed to establish their own institutions of learning (see Figure 2). Alternatives to state academies were explored particularly in England, where salons were never as strong as in France. Consider the following example of academies imagined by women in self-conscious recognition of their exclusion from orthodox institutions.

The English natural philosopher Margaret Cavendish never recorded her impression of her visit to the Royal Society. In 1662, however, five years before her visit, she had written a play entitled "The Female Academy," portraying a conflict between an exclusively female academy and a male academy.[73] In Cavendish's play, noblewomen resolve to create an academy for the education of their daughters. The matron of this imaginary academy asks: are not women as capable as men of wit and wisdom? The matron portrays Lady Wit as the beautifully feminine mother of the nine muses; she surrounds herself in her court with poets—men of all nations and

Figure 2. A ladies' academy meets while Madame Dacier and Sappho look on. From Eliza Haywood, *The Female Spectator* (1744), frontispiece. By permission of Houghton Library, Harvard University.

R. Parr Sculp.

qualities—who serve as Platonic lovers to her virgin daughters. Wisdom, by contrast, is the son of the Gods. In his manly government, his chief counselors are reason, understanding, observation, experience, and judgment. His domestic servants are "appetite" and "passions." To end her discourse, the matron judges women to be more capable of wit than of wisdom, for wit is of the "female gender, while wisdom is severe and strict, serious and sober."

As the play continues, the men of the city grow angry that the women have sequestered themselves in an academy. The men set up a male academy in an adjoining hall and drill a small hole into the lecture room of the female academy, through which they spy on the ladies' activities. The women, the men say, are ungrateful. By "incloystering" themselves they are ungrateful to nature, which intended them for breeding, and they are ungrateful to men, who protect and maintain them. As the gentlemen listen to further discourses on truth, friendship, theater, vanity, vice, and wickedness, they become more and more dissatisfied. "The Academical Ladies," the men wail, "take no notice of the Academy of Men . . . they neither mention the Men, nor their Discoursings, or Arguments, or Academy, as if there were no such Men." At this the men become so angry that they resolve to sound trumpets so loudly that the women will not be able to hear themselves speak, and in this way, "draw them out of their Cloyster, as they swarm Bees."[74] And so the men blast with brazen trumpets, hoping to disperse that "swarm of Academical Ladies." With this trumpet blast, a matron of the female academy emerges to entreat the men to stop blowing their horns. The academy is not a cloister, she explains, but a school for the education of good wives.

There the play ends. With her choice of trumpet as noisemaker, Cavendish seems to be alluding to John Knox's *The First Blast of the Trumpet against the Monstrous Regiment of Women*. Margaret Cavendish seems to be suggesting that wherever women seek to usurp masculine power and authority, whether in government or in learning, men will resist and reassert their own authority.

Cavendish never told her readers whether her "Academy of Men" represented the Royal Society. Some years later, Mary Astell drew a more sober outline for a community of educated women differing greatly in its aims and goals from the Royal Society. In her *Serious*

Proposal to the Ladies (1694), Astell proposed in essence a Protestant convent dedicated to the education of girls from noble families—a retreat where they could learn philosophy and return to the world virtuous and Christian. Astell sought, above all, "to join the sweetness of Humanity to the strictness of Philosophy." Like members of the Royal Society, Astell sought royal support for her society, petitioning Queen Anne for funds. Sympathetic to the proposal, the queen set aside £10,000 for Astell's monastic retreat but withdrew her support when advised that the establishment of Protestant convents might be misconstrued by her enemies as centers of papal influence.[75]

We also have a report of an academy for women in France dating from 1772. What distinguishes this academy from a simple school for girls is that it was conceived as standing midstream in an academic tradition stretching back to Plato—the same heritage shared by the seventeenth-century academy system of Richelieu and Colbert. Whether the academy was real or imaginary, we do not know. What has survived are twenty-seven lectures given by the academy's *grand dame*, Philothée, and designed to treat the major branches of learning.[76] Philothée's academy had an unnamed patroness who supplied room for their meetings in her garden.

Unlike members of the state academies, for whom the title *membre de l'Académie* was prized, the women of Philothée's academy had to defend their right to learn at all. "Honest and upright people" reproached them for "devoting themselves too much to the sciences." This required Philothée to dedicate large portions of her first lecture to defending women's right to learn. It should not be imagined that Philothée's academy was attacked (by "upright" citizens and "sullen" women) because it was a hotbed of feminism. On the contrary, in concert with prescribed roles for women, Philothée taught that nature had ordained "women's science" to be that of religion and the management of the home, embroidering and knitting as well as public charity. Women's loving nature, according to Philothée, made them unfit to act in the state as lawyers, statesmen, or doctors. Nonetheless, women were to receive a thorough liberal education.[77]

Philothée's conception of knowledge differed radically from that espoused by the state academies. Whereas members of both the

Royal Society and Académie Royale des Sciences vowed never to discuss the mysteries of religion or affairs of state, religion and morals are central fields of study in Philothée's curriculum. To Philothée's mind, knowledge itself is an instrument to be applied to a virtuous life. She quotes Seneca, asking "why do we teach our sons the liberal arts?" Where professional science of the seventeenth and eighteenth centuries sought to disassociate fact from value, Philothée made a plea for knowledge in the service of virtue. The greatest care was to be taken to imprint true virtue on young hearts.[78]

The unformed state of intellectual culture in the seventeenth century left much room for innovation. Yet women did not fare well in the new institutions of science established in seventeenth-century Europe. Where science emerged as part of the public domain in established academies, women might be found on the periphery, as anatomical demonstrators or prizewinners, but they were not to be found participating in mainstream academic life as recognized members.

Exclusion from the academies, while it distanced women from the centers of scientific endeavor, did not end their participation in science. In the seventeenth and eighteenth centuries, as we shall see, there were a number of women working in natural history and natural philosophy, as well as the experimental sciences. Though few in number, women made real contributions. It is important to understand how these and other women, though barred from universities and scientific societies, could nonetheless acquire the training required for work in the sciences.

2

Noble Networks

> Being a Woman [I] Cannot . . . Publickly . . . Preach, Teach, Declare or Explane [my works] by Words of Mouth, as most of the Famous Philosophers have done, who thereby made their Philosophical Opinions more Famous, than I fear Mine will ever be . . .
>
> —Margaret Cavendish, duchess of Newcastle, 1663

In the early years of the scientific revolution, women of high rank were encouraged to know something about science. Along with gentlemen *virtuosi,* gentlewomen peered at the heavens through telescopes, inspecting the moon and stars; they looked through microscopes, analyzing insects and tapeworms. Many a young lady was able to calculate "both a Solar and Lunar Eclipse."[1] As late as 1788, Joseph Sigaud de Lafond reported that women continued to outnumber men in his classes on experimental physics.[2]

Science became fashionable in the middle decades of the seventeenth century. If we are to believe Fontenelle, it was not unusual to see people in the street carrying around dried anatomical preparations. Fontenelle also tells us that Etienne-François Geoffroy's thesis that humans generate from "worms" (spermatozoa, discovered some years earlier by Leeuwenhoek) so piqued the curiosity of ladies of high rank that it was necessary to translate his work into French.[3] Especially in Paris, wealthy women were ready consumers of scientific curiosities, collecting everything from conches, stalactites, and petrified wood, to insects, fossils, and agates to make their natural history cabinets "the epitome of the universe."[4]

During this period popular science written for the ladies became a major industry.[5] One of the earliest popularizations of science written expressly for women—in this case, in the form of science fiction—came from Margaret Cavendish, duchess of Newcastle. Her *Description of a New World, called the Blazing World,* addressed "To all Noble and Worthy Ladies," offered an introduction to her own brand of natural philosophy under the guise of a romance.[6] In this scientific utopia a young lady, abducted by an unsolicited suitor, is saved by the warmth of her great beauty from disaster and carried off to a New and Blazing World attached to the earth at the north pole. This new world is a world of peace and tranquility, wit and honesty, where there is "no difference of the sexes." The creatures of this Blazing World find the abducted lady so beautiful that they make her their empress. She chooses the wisest of them to instruct her: the bear-men become her experimental philosophers, the bird-men her astronomers, the fly-, worm-, and fish-men her natural philosophers, the spider-man her mathematician, the lice-men her geometricians, the magpie-, parrot-, and jackdaw-men her orators and logicians.

Longing for a scribe to record her new philosophy, the young lady instructs her counselors to recommend some worthy philosopher. Her advisers reject the "Soul of some ancient famous Writer, either Aristotle, Pythagoras, [or] Plato," because they were too wedded to their own opinions. Nor were Galileo, Gassendi, Descartes, Helmont, or Hobbes appropriate, given that they were "so self-conceited, that they would scorn to be Scribes to a Woman." The advisers mention to the empress one final possibility, a woman whom they have heard to be very learned, Margaret Cavendish, the duchess of Newcastle. The duchess is appointed to the position and serves the empress well; they become "Platonick Lovers, although they were both Females," and together they create a new world of philosophy.

Cavendish's utopia was the first but by no means the most popular. Twenty years later, Bernard Le Bovier de Fontenelle's introduction to Descartes's cosmology became a popular science text for ladies, appearing in translations and editions throughout Europe even after the eclipse of Cartesian thought. Fontenelle's *Entretiens sur la pluralité des mondes* sets forth the ideas of the "new philosophy"—infinite space and time, innumerable worlds, the existence of other unknown living creatures—in terms palatable to gentlemen, scholars, and ladies alike. In his book a marquise and a philosopher, walking together in the garden of her château on a moonlit night, enter into conversation and, turning their attention starward, contemplate Cartesian vortices and the possibility of new worlds (see Figure 3). Fontenelle explicitly introduced a woman into these conversations in order to encourage "gentlewomen by the example of one of their own sex" to study the laws of matter and motion.[7] Some women took exception to his depiction of the scientific lady; Aphra Behn, for example, the first to translate Fontenelle's work into English, found it patronizing in its approach to women.[8] Fontenelle, she complained, "introduceth a Woman of Quality . . . whom he feigns never to have heard of any such thing as Philosophy before, [and] makes her say a great many silly things." Despite her objections, Behn translated the book unamended.

Science for ladies remained popular throughout Europe in the eighteenth century. The Italian poet Francesco Algarotti published an introduction to Newtonian physics *per le dame* in 1737. In Germany Johanna Charlotte Unzer published her *Outline of Philosophy for Ladies* in 1761; and from his post at the Academy of Science in Saint Petersburg, Leonhard Euler wrote his *Letters to a German Princess on Diverse Points of Physics and Philosophy* in 1768.[9] Why were women considered an audience worthy of cultivation?

For one thing, enthusiasm among the wellborn facilitated the rapid diffusion and acceptance of the new science. At the same time, studying science was not thought to threaten the traditional virtues of a lady. Since the Renaissance learning had been thought to lead to moral virtue, and natural philosophy continued in this tradition. Bacon considered nature a great book from which men could read the power and wisdom of God, the "Author" of all things. Thus the cardinal virtues of ladies—modesty and religious reverence—were thought to be promoted by the study of natural philosophy. Furthermore, science at this time was by and large a leisure activity, and for that reason it was seen as an appropriate pastime for gentlemen and women. Leibniz even argued that, because of their abundant leisure, women could better cultivate knowledge than men:

> I have often thought that women of elevated mind advance knowledge more properly than do men. Men, taken up by their affairs, often care no more than necessary about knowledge;

Figure 3. Illustration from Fontenelle's *Plurality of Worlds,* showing the solar system as revealed by the telescope. From Bernard Le Bovier de Fontenelle's *Entretiens sur la pluralité des mondes.* In his *Oeuvres diverses* (The Hague, 1728), vol. 1. By permission of Houghton Library, Harvard University.

> women, whose condition puts them above troublesome and laborious cares, are more detached and therefore more capable of contemplating the good and beautiful.[10]

Others claimed that women have a slight edge over men in the study of philosophy because women have curiosity ("the parent of philosophy") and because their sedentary and sometimes solitary life suits them to study.[11]

For these reasons, then, scholars encouraged women to pursue natural philosophy. Abbé Nollet's *Leçons de physique expérimentale* and his *Essai sur l'électricité des corps* contained illustrations of women actively engaged in the pursuit of knowledge: a woman and young girl (perhaps mother and daughter) are shown using microscopes, and women conduct experiments on electricity (see Figures 4A and 4B). Such images provided what would today be called role models for women's participation in science, although even that participation was limited in scope and extent.

The Curious Matter of Math

Perhaps the most surprising aspect of this popular tradition was the encouragement women were given to do mathematics. In view of today's much touted "math anxiety," it is remarkable that in the early years of the eighteenth century women were commonly encouraged to sharpen their mathematical skills. The English *Ladies' Diary,* published from 1704 to 1841, was designed to teach, as the title page advertised, "Writing, Arithmetick, Geometry, Trigonometry, the Doctrine of the Sphere, Astronomy, Algebra, with their Dependants, viz. Surveying, Gauging, Dialling, Navigation, and all other Mathematical Sciences."[12] Early issues of the journal presented a variety of articles, including a chronology of famous women from Eve to Queen Anne, Robert Boyle's remedy for colic, methods for preserving apples and pears, and much advice on marriage. In the fifth volume of the journal (1709), however, editor John Tripper announced that since ladies seemed to prefer mathematics to cookery, the *Diary* would dedicate itself exclusively to "enigmas and arithmetical questions."[13]

Despite the limited formal education available to them, women were able to solve problems of considerable difficulty. In 1718 the

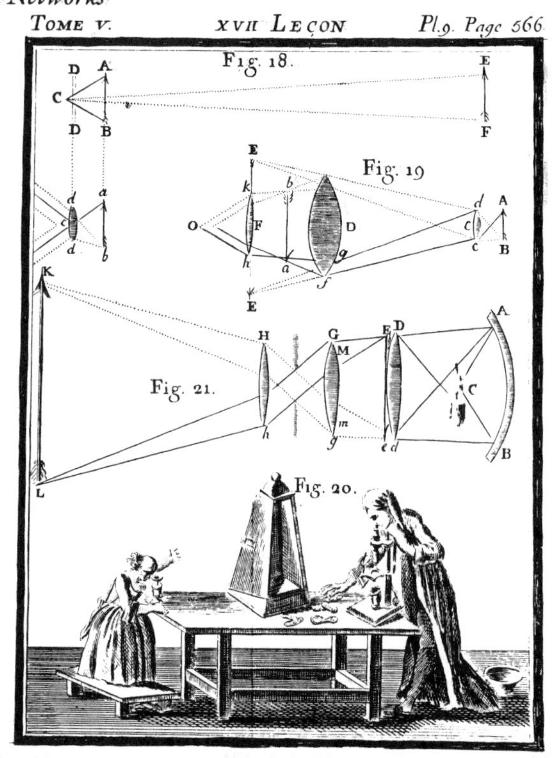

Figure 4A. Illustration from the Abbé Nollet's *Leçons de physique expérimentale* of 1743–48, providing what might today be called a role model for women scientists. A woman and young girl are shown peering into microscopes. By permission of Houghton Library, Harvard University.

Ladies' Diary editor Henry Beighton drew a positive picture of women's mathematical abilities. Women have, he wrote, "as clear Judgments, as sprightly quick wit, penetrating Genius, and as discerning and sagacious Faculties as ours, and to my Knowledge do, and can, carry through the most difficult Problems." This sparked in Beighton a note of national pride. Praising women mathematicians as "the *Amazons* of our Nation" he asserted that foreigners "would be amazed when I shew them no less than 4 or 5 Hundred several Letters from so many several Women, with Solutions *Geometrical, Arithmetical, Algebraic, Astronomical,* and *Philosophical.*"[14] Though the *Diary* was conceived as a winter amusement, it was also considered important for the development of mathematics in England. In his new edition of the mathematical problems of the *Diary* published in 1817, Thomas Leybourn, professor at the Royal Military College, praised the journal for the influence it had exerted upon the mathematical sciences in England.[15]

In fact, among women renowned for their scientific ability in this

Figure 4B. An experiment on electricity, again featuring women as participants. From Abbé Nollet, *Essai sur l'électricité des corps* (1746). By permission of Houghton Library, Harvard University.

period, a majority were mathematicians or active in math-oriented fields such as physics or astronomy. Among astronomers we find Maria Winkelmann, Maria Eimmart, Maria Cunitz, and Nicole Lepaute (see Chapter 3). Among mathematicians are Maria Agnesi and Sophie Germain. Physicists include Laura Bassi and Emilie du Châtelet. One might point to several reasons why mathematics was thought appropriate to women in the seventeenth and eighteenth centuries. For one thing, knowledge of accounts was important to the wife of a businessman: Englishmen never tired of attributing the brilliance of Dutch commerce to the mathematical skills of the Dutch wife. Mathematics was also accessible to women; the study of mathematics required neither a closet full of equipment nor a large library. Though this momentary rapprochement between science and women was not to last (see Chapter 8), for a time science and mathematics were thought to enhance a woman's life and character, and women took advantage of this opportunity.

Noblewomen in Scientific Networks

Historians tend to take the case of women as authors of and audience for popular science as the paradigmatic example of women's participation in modern science. Yet, as we shall see, relegating women to the status of amateur diminishes the contributions that women like Margaret Cavendish or Emilie du Châtelet made to science. Popular science was not sharply divided from professional science as it is today. Though today it would be difficult for anyone barred from university education to work in science, this was not the case in the seventeenth or eighteenth centuries, when few men or women were full-time or salaried scientists. Some, like Galileo, were resident astronomers at a princely court; Bacon and Leibniz were government ministers, as well as men of letters. At the end of his life Descartes was in the pay of Christina, queen of Sweden, as tutor in natural philosophy and mathematics. Emilie du Châtelet was a person of private means. This looser organization of science was one factor allowing those barred from universities and academies to find their way into scientific circles.

In the absence of clearly established prerequisites of education or certification, participation in science was regulated by informal networks. Entrée into the scientific network rested on birth and/or talent. The leisure and resources which came with noble birth gave access to learning, while the added prestige of erudition served to bolster a declining aristocracy.[16] In Paris, the title *membre de l'Académie* was prized as if it were a title of nobility; in England, it was said that he who is both nobly born and a scholar is doubly honored.[17] This worked to the advantage of noblewomen, whose high social standing gave them access to science as it did to other forms of social power and influence. In informal scientific networks, noblewomen were often able to exchange social prestige for access to scientific knowledge.

Before the advent of the scientific journal, scholars and enthusiasts exchanged news of discoveries and observations through informal networks—meetings in private homes, writing letters, working for a common patron. Marin Mersenne conducted an international correspondence with intimate friends, such as Descartes, and organized a clearinghouse of information for the learned and *grand amateur* alike, passing on ideas and putting scholars in touch with one another. His home was a kind of informal academy where friends dropped in for philosophical discussion.[18]

Royal women formed crucial links in these noble networks. With family alliances connecting European courts, queens served as ambassadors preparing the way for both cultural and philosophical exchange.[19] A prime example of this was the famed Leibniz-Clarke debate, sparked by the ascent of a German princess to the English throne.[20] Although the controversy between Newton and Leibniz had been under way for some time, the written debate began when Princess Caroline of Ansbach, one of Leibniz's pupils, moved to England upon the succession of her father-in-law (George I) to the throne. An ardent Leibnizian, Caroline's views were challenged when she entered Newtonian England. The first letter of what was to become the Leibniz-Clarke debate was written to her. In this letter, Leibniz strongly criticized Newton's philosophical views. Caroline passed the letter on to Clarke, whom she had met while looking for a translator for Leibniz's *Theodicy*. Caroline herself debated the question of the nature of the soul with Clarke (whom she considered too much of a Newtonian); such debates lasted on occasion from six until ten o'clock in the evening.

As Princess of Wales Caroline appointed herself mediator of the debate. In 1715 she wrote to Leibniz, "I should like [Sir Isaac Newton] to be reconciled with you . . . it would be a great pity if two such great men as you and he were to be estranged by misunderstandings." In a subsequent letter Caroline chided Leibniz for his dispute with Newton: "the public would profit immensely if this [reconciliation] could be brought about, but great men are like women, who never give up their lovers except with the utmost chagrin and mortal anger. And that, gentlemen, is where your opinions have got you."[21]

Women of lesser rank were also a part of these noble networks. The privileges of rank allowed Elizabeth of Bohemia to attract the attention of Descartes. Elizabeth was introduced to the French philosopher sometime after 1640 by the marquis de Dohna. The lengthy correspondence between Elizabeth and Descartes shows Elizabeth to have been a woman of serious intellectual talent. In approaching Descartes she was searching for more than learned refinements; she was looking for, as she put it, "a physician of the soul."[22] She did not hesitate, however, to voice objections to his philosophical views. Descartes's central conception of the relation between mind and matter, for example, was simply unacceptable to her. "I hope," she wrote, "you can excuse my stupidity of not being able to comprehend the idea of how the soul (without extension and immaterial) can move the body . . . it seems easier to me to concede material extension to the soul than the ability of an immaterial soul to move the body."[23] Descartes took Elizabeth's objections seriously. Her questions and objections led him to elaborate his views in his 1644 *Principles of Philosophy,* where he publicly acknowledged his regard for Elizabeth's talents.

Noblewomen thus exploited their social standing to gain access to learned circles. Yet, the dynamic of the exchange between Elizabeth and Descartes is noteworthy for what it reveals about rank and gender. Despite her high social standing, Elizabeth assumed the subordinate posture of a pupil, playing the role of a modest, self-effacing woman. For his part, Descartes—who was not unhappy to make his way into the world of royalty—played the role of a courtier, responding with the praise due to one of her rank.[24] He acknowledged his lower social standing, while she affected intellectual subordination. In matters of the intellect, the privileges of rank did not outweigh the liabilities of gender.

Even the highest rank could not entirely insulate women from ridicule. In 1650 Descartes was commissioned by the audacious Queen Christina of Sweden to draw up regulations for her scientific academy. Many blamed Christina and the rigors of her philosophical schedule for Descartes's death.[25] For her philosophical prowess the queen was often called a hermaphrodite.[26]

Noblewomen continued to participate in these informal scientific networks until late in the eighteenth century.[27] Nobility won for certain women the attention of men of lower social rank but of significant intellectual standing. This was true across Europe, as we can see from the lives and fortunes of two prominent natural philosophers of their day, Margaret Cavendish in England and Emilie du Châtelet in France.

Margaret Cavendish, Natural Philosopher

Entry into European philosophical networks gave Margaret Cavendish the background necessary for her work on natural philosophy. Cavendish (1623–1673), one of the few women in seventeenth-century England to write boldly and prolifically about natural philosophy, intended to make her mark on philosophy.[28] Though often addressed to women, Cavendish's natural philosophy was not written as a simplification or popularization for the "weaker sex." Rather she participated in discussions central to her life and times, taking up debates about matter and motion, the existence of the vacuum, the nature of magnetism, life and generation, color and fire, perception and knowledge, free will and God. Cavendish also entered into (usually one-way) correspondence with key philosophers on these issues. Her *Philosophical Letters* presented a point by point critique of Hobbes's *Leviathan,* while her *Elements of Philosophy* attacked Descartes and his vortices, Henry More's proof of God, and Helmont's "odd and strange Art of Chemistry." Her philosophical boldness remained long unmatched by any other woman.

Cavendish was born Margaret Lucas, daughter of Thomas Lucas

of the lesser gentry of Colchester. As she recorded in her autobiography, she had little formal education, and what education she had was that suited to a lady—singing, dancing, reading, and the like.[29] Though women were not "suffer'd to be instructed in Schools and Universities," as she never tired of reminding her readers, this did not dampen her appetite for ideas, for (as she wrote some years later) "thoughts are free, [and we women] may as well read in our Closets, as Men in their Colleges."[30]

Margaret Lucas recognized that women's greatest access to knowledge at this time was through educated men. "Most Scholars," she wrote, "marry, and their heads are so full of their School Lectures, that they preach them over to their Wives when they come home, so that they [the wives] know as well what was spoke, as if they had been there."[31] Consequently, Margaret married with care William Cavendish, duke of Newcastle, in the 1640s. Through marriage she became a member of what Robert Kargon has identified as the Newcastle circle, consisting of William and Charles Cavendish, Thomas Hobbes, Kenelm Digby, Mersenne, Gassendi, and (while exiled in France in the 1640s and '50s) Descartes and Roberval.[32] Without this private philosophical network, Margaret Lucas Cavendish could not have become a natural philosopher.

Advantages of rank did not, however, outweigh disadvantages of sex. Though the duchess found a place in the philosophical world, her ties to learned men remained tenuous. The intellectual status of noblewomen was not unlike the legal status of women of all ranks. Married women were legally *femmes couvertes*, literally under the "cover" of their husbands. Intellectually too, women were under the cover of male mentors. Though part of the Newcastle circle, Cavendish suffered from isolation. Her contacts with other philosophers (all male) produced few intellectual rewards. Her relations with Descartes, for example, remained indirect—perhaps more through difficulties of language, however, than anything else. She sent philosophical queries to him through the pen of her husband, but she did not correspond with him herself. When Descartes dined at the Newcastle table, the dinners passed largely in silence. As Margaret Cavendish reported, "he spake no English, and I understand no other language, and those times I saw him, which was twice at dinner with my Lord in Paris, he did appear to me a man of the fewest words I ever heard."[33]

Cavendish's isolation was not voluntary. The breeding grounds of the new science, such as the newly founded Royal Society of London, did not allow women to become members. Though Cavendish visited the Royal Society (see Chapter 1), this single encounter with the men of science could hardly have been satisfying; she never even mentioned this visit in her memoirs.

Nor did Cavendish benefit from intellectual companionship with other women. Salons did not flourish in England as they did in France, and as a result intellectual women in England suffered from isolation. Apart from a brief membership in Katherine Philips's "Society of Friendship," Cavendish cultivated few intellectual friends among women. (Indeed, she often chided the ladies of her day for playing cards and not being serious philosophers.[34]) She did not know Anne Conway, for example, a philosophical woman with whom Cavendish might have found much in common.[35]

Cavendish's chief intellectual companions were within her own family. A frontispiece reproduced in several of her philosophical works shows a "semy-circle" of ladies and gentlemen seated amicably around a table, the caption bears the title "the Duke and Dutchess of Newcastle and their Family" (see Figure 5). Cavendish learned a great deal from her brother, Lord John Lucas, one of the original Fellows of the Royal Society; she also claimed to have profited from discussions with Sir Charles Cavendish, brother of William, who had a real interest in science and mathematics and kept abreast of all the latest developments. Yet it was her husband, William, whom Margaret claimed as her "Wit's Patron."[36]

A man thirty years her senior, William Cavendish was himself a respectable *virtuoso,* reputed by William Petty to be a great patron to Gassendi, Descartes, and Hobbes. While exiled in France and Holland from 1644 to 1660, William Cavendish collected seven telescopes—four made by Estacio Divino, two by Torricelli, and one by Fontanus. William was, however, neither deeply scholarly nor profound. His greatest love was the "noble and heroick Art of *Horsemanship* and *Weapons.*"[37] If William was not the critic Margaret required, he served her well with his moral and financial sup-

Thus in this Semy-Circle, wher they Sitt,
Telling of Tales of pleasure & of witt,
Heer you may read without a Sinn or Crime,
And how more innocently pass your tyme.
Abr. à Diepenbeke delineavit.
Pet. Clouwe sculp.

port. Nearly all of her philosophical works include a laudatory verse from William. In addition, William financed the repeated private publication of her voluminous works.

Apart from literary orations, plays, and poems, Margaret Cavendish wrote a number of philosophical works, including *The Philosophical and Physical Opinions* (1655), *Natures Pictures drawn by Fancies Pencil to the Life* (1656), the fascinating *Observations upon Experimental Philosophy* (1666), to which she added *The Description of a New World, called The Blazing World,* and *Grounds of Natural Philosophy* (1668). Her declared purpose in writing was to achieve fame. She was, as she put it, "as ambitious as ever any of my Sex was, is, or can be." Cavendish identified three avenues to fame in the England of her day: leadership in government, military conquest, and innovation in philosophy. As government and military service were both closed to her by law, she took up natural philosophy. This was not, in her view, such a poor alternative, given that men "hold Books as their Crowne . . . by which they rule and governe."[38] Cavendish hoped her books would bring her similar glories.

Margaret Cavendish set out her natural philosophy most clearly in three major works: her *Philosophical Letters,* her *Observations upon Experimental Philosophy,* and her *Grounds of Natural Philosophy.*[39] Cavendish was a thoroughgoing materialist. Yet, she did not contribute to what Carolyn Merchant has described as the "death of nature," the process by which master mechanists of the scientific revolution came to think of nature as a system of dead, inert particles moved by external forces.[40] Central to her conception of nature was that matter is intelligent. For her, nature is composed of an infinite number of "intelligent" atoms, each with self-knowledge

---

Figure 5. "The Duke and Dutchess of Newcastle and their Family." The Duke and Duchess of Newcastle sit at the head of the table, crowned with laurel. In the sketch for this frontispiece, Margaret Cavendish holds up her hand for attention; in the printed version here, Margaret has relinquished the role of narrator to her husband. Frontispiece to *Natures Pictures Drawn by Fancies Pencil to the Life* (London, 1656). By permission of Houghton Library, Harvard University.

and self-propulsion so that "there is not any Creature or part of Nature without . . . Life and Soul."[41] Matter is not dead material, devoid of spirit; rather, corporeal nature is both subject and agent. Objecting to Hobbes's distinction between intelligent subject and inanimate object, Cavendish stated that "all things, and therefore outward objects as well as sensitive organs, have both sense and reason." Criticizing Descartes's radical distinction between mind and body, Cavendish held that a fundamental unity pervaded the world, that nature was composed of one material yet "self-moving" and "self-knowing" body.[42]

Cavendish's rejection of the mind/body dualism led her into the atheist camp. For her, only matter exists. This matter being itself "intelligent," there is no need for a first cause. "Self-moving matter, which is sensitive and rational," she wrote, is "the only cause and principle of all natural effects." On this basis she objected to Descartes's notion of vortices set in motion by God. "I cannot," she wrote, "well apprehend what Des Cartes means, by Matter being at first set a moving by a strong and lively action [God]."[43] Margaret Cavendish was never strident in her atheism. Much like the members of the Royal Society, she too struck a compromise with established religion by separating philosophy from theology and banishing things spiritual to a sphere beyond experimental science. In one of her earlier works, she conceded that knowledge of God might be innate. Unlike Descartes, however, Cavendish made knowledge of God a part of both inanimate and animate matter. "All parts of Nature," she wrote, "even the inanimate, have an innate and fixt self-knowledge, it is probable that they may also have an interior self-knowledge of the existency of the Eternal and Omnipotent God, as the Author of Nature."[44]

Margaret Cavendish's rejection of a sharp distinction between animate and inanimate nature led her to reject the Cartesian imperative that man through science should become master and possessor of nature. Such a view Cavendish held to be impossible. "We have," she insisted, "no power at all over natural causes and effects." Man is merely one part of nature. The whole (nature itself) can know the parts, but the parts (men) cannot know the whole. Consequently, since he is not above nature, man must be content with things as nature has ordered them, "for man is but a small part, . . .

his powers are but particular actions of Nature, and he cannot have a supreme and absolute power."[45]

Nor was Cavendish as quick as Descartes or Henry More to proclaim man the greatest of nature's creatures. Cavendish thought that man was in no position to judge such an issue, since he was himself author of the debate. She thus found man "partial" in this matter where other creatures were given no voice. She argued moreover that "Elemental Creatures" (that is, nonhumans) are as "excellent and Wise" as man, for what man, she asked, is as clever as a bee and able to build a honeycomb? The much-praised man, she deemed, is not so useful to his fellow creatures as his fellow creatures are to him, for men are less profitable and more apt to make spoil.[46]

Within two years of her critique of the rationalists Hobbes and Descartes, Cavendish penned an equally sharp critique of the experimentalists (though unnamed, most probably) Robert Boyle and Robert Hooke.[47] Cavendish judged a philosophy based on the human senses unreliable, for eyes, ears, and noses are prone to error and cannot serve as a sure foundation for philosophy. The new telescopes and microscopes she found even more unreliable: with glass often cracked, concave, or convex they distorted the figure, presenting a "hermaphroditical" view of things—partly artificial, partly natural—so that a louse appears like a lobster. More important, these impure images go no further than reason in providing true knowledge—what she called "the interior natural motions of any part or creature of Nature." Cavendish also criticized experimental philosophy for being impractical: does the inspection of a bee through a microscope, she asked, bring more honey?[48]

Cavendish's bold attack on rationalist and empiricist, ancient and modern, was sharply censured by Joseph Glanvill, one of the leading figures in the Royal Society. In explicit reference to her work, Glanvill warned that "he is a bold Man" who dares to attack "the Physics of Aristotle himself, or of Democritus . . . or Descartes, or Mr. Hobbs."[49] The duchess of Newcastle did not, however, take such criticism lying down. She made it clear that her want of learning—for which she repeatedly apologized—was not peculiar to her but was a liability of her sex: "[That] I am not versed in [learning], no body, I hope, will blame me for it, since it is sufficiently known, that our Sex is not bred up to it, as being not suffer'd to be instructed in Schools and Universities."[50] She also recognized some criticism for what it was—the prejudice of "spiteful tongues."[51] Though Cavendish reported "censorious" criticisms of her work, we know very little about the source of that criticism or how it was communicated to her. In fact, her work suffered that worst censure of all—neglect. Unlike the work of Maria Merian or Emilie du Châtelet, Cavendish's was not reviewed in major European journals.

By her own account, Margaret Cavendish remained, for most of her life, cloistered in her study working in her own world of philosophy. This she attributed, in part, to a bashful nature and, in part, to her condition as a woman. Excluded by custom and temperament from public life, the duchess tried to make contact with the learned world through her books. These she dedicated to the "most famously learned" men of the universities and sent copies to Europe's leading libraries. She sent each of her beautifully published volumes to Oxford and to Cambridge, where her husband and two brothers had been educated, and a complete set of her philosophical works to Christian Huygens at the University of Leiden.[52] In return, she received letters of thanks and vapid praise of the type a courtier would offer a lady. Only Joseph Glanvill and Huygens engaged her in serious correspondence: Glanvill discussed with her his work on witchcraft; Huygens discussed with her "Rupert's exploding drops."[53]

Cavendish, a Feminist?

The duchess of Newcastle was frustrated by the limitations placed upon her because of her sex. And indeed, she had a great deal to say about women—not all of it favorable. In her early work Cavendish uncritically accepted the long-standing pronouncements of the ancients on women. Along with Aristotle, she judged the masculine spirit superior to the feminine. It is not, she wrote, so great a fault in nature for a woman to be masculine as for a man to be effeminate: "for it is a Defect in Nature to decline, as to see Men like Women, but to see a Masculine Woman, is but onely as if Nature had mistook, and had placed a Mans Spirit in a Womans

Body."[54] Cavendish also accepted the ancient view that women's brains are cold and soft. The softness of the female brain (not, surprisingly, women's lack of education, which she mentioned time and again) explained, for Cavendish, women's poor showing in philosophy:

> This [the softness of the female brain] is the Reason why we are not Mathematicians, Arithmeticians, Logicians, Geometricians, Cosmographers, and the like; This is the Reason we are not Witty Poets, Eloquent Orators, Subtill Schoolmen, Subtracting Chimists, Rare Musicians, [and the like] . . . What Woman was ever so wise as Solomon, or Aristotle . . . so Eloquent as Tully? so demonstrative as Euclid? It was not a Woman that found out the Card, the Needle, and the use of the Loadstone, it was not a Woman that invented Perspective-Glasses to pierce into the Moon; it was not a Woman that found out the invention of writing Letters, and the Art of Printing; it was not a Woman that found out the invention of Gunpowder, and the art of Gunns . . . what ever did we do but like Apes, by Imitation?[55]

Like so many of her day, Cavendish judged the supposed inferiority in women's physical and intellectual character consistent with their social disfranchisement. Men were right to close government to women, because the wisest woman is never so wise as the wisest man. Yet she left a caveat to explain her own achievement. "Some women," she wrote, "are wiser than some men." In her view, women of the educated classes were superior in learning to "Rustick and Rude-bred men."

Cavendish never renounced these views borrowed, for the most part, from the ancients. In her "Femal Oration" (1662), however, Cavendish seemed less certain about them. The "Oration" is composed of five voices, each presenting a different view of women's character and social condition. The first voice strongly opposes the "Tyrannical Government" of men:

> Ladies, Gentlewomen, and other Inferiours, but not less Worthy, I have been Industrious to Assemble you together, and wish I were so Fortunate, as to persuade you to make a Frequentation, Association, and Combination amongst our Sex, that we may Unite in Prudent Counsels, to make our Selves as Free, Happy, and Famous as Men . . . Men are so Unconscionable and Cruel

> against us, as they Indeavor to Barr us of all Sorts or Kinds of Liberty, as not to Suffer us Freely to Associate amongst our Sex, but would fain Bury us in their Houses or Beds, as in a Grave; the truth is, we Live like Bats or Owls, Labour like Beasts, and Dye like Worms.[56]

Though Cavendish energetically recorded this view, it was never her own. A second voice counters with the argument that nature, and not man, has made women inferior by making them less ingenious, witty, and wise. Voice number three—probably Cavendish's own—pleads for a nurturing of women's bodies and minds in order to develop within them a "masculine" strength:

> For Strength is Increased by Exercise, Wit is Lost for want of Conversation . . . let us Hawk, Hunt, Race, and do like Exercises as Men have, and let us Converse in Camps, Courts, and Cities, in Schools, Colleges, and Courts of Judicature, in Taverns, Brothels [!], and Gaming Houses, all which will make our Strength and Wit known, both to Men, and to our own Selves, for we are as Ignorant of our Selves, as Men are of us . . . Wherefore, my Advice is, we should Imitate Men, so will our Bodies and Minds appear more Masculine, and our Power will Increase by our Actions.[57]

Yet another voice (perhaps an undecided Cavendish) warns that *nurture* cannot contradict *nature*. To educate women and to extend to them liberties would be like grafting peach branches onto an apple tree that might then bear the wrong fruit. To contradict nature's will, in other words, is dangerous. A feminine manner becomes a female body; to attach masculine virtues to a feminine body would be unnatural and unwise:

> The former Oration was to Persuade us to Change the Custom of our Sex, which is a Strange and Unwise Persuasion, since we cannot make our selves Men; and to have Femal Bodies, and yet to Act Masculine Parts, will be very Preposterous and Unnatural; . . . Wherefore, let me Persuade you, since we cannot Alter the Nature of our Persons, not to Alter the Course of our Lives.[58]

Above all, this fourth speaker warns women against becoming "hermaphroditical," corrupt and imperfect. The hermaphrodite—the "womanish man" or "manly woman"—provoked uneasy feelings in the Europe of her day, and Cavendish used the term degradingly to refer to anything of a mixed nature—as, for example, impure alloys of tin or brass. If metals were to be censured for ambiguous identity (being, as she pointed out, half natural and half artificial), how much more serious was the charge of ambiguous sexual identity. The speaker in this oration (like Cavendish herself) admonishes women to follow their own nature and remain properly "feminine," "huswifely," "cleanly," and "of few words."

A fifth and final voice closes the oration with the neo-Platonic view that women are different from, and indeed superior to, men:

> Why should we Desire to be Masculine, since our Own Sex and Condition is far the Better? for if Men have more Courage, they have more Danger; and if Men have more Strength, they have more Labour . . . ; if Men are more Eloquent in Speech, Women are more Harmonious in Voice; if Men be more Active, Women are more Gracefull . . . Wherefore, Women have no Reason to Complain against Nature, or the God of Nature.[59]

In this "Femal Oration" Cavendish left unresolved the source of women's subordination—tyrannical men, nature, or nurture. She also left unresolved the relative value of masculinity and femininity. Are the strengths and liberties of masculinity the preferred traits, and thus to be cultivated in women as well as in men? Or, are the sexes to strike a bargain, wherein each perfects its own virtues? Or, alternatively, are the beauty and grace of femininity, in fact, culturally superior qualities? As Cavendish later remarked, she spoke freely in these orations—*pro* and *con*—but did not take sides.[60]

After this essay, only occasional remarks on women appeared in prefaces to Cavendish's works and in her letters. She abandoned the woman question in her philosophical works; she did not set old questions concerning women on new philosophical foundations or integrate women and their distinctive concerns into mainstream philosophical discourse. In this, Cavendish followed the example of the men she critiqued; like Descartes, she, too, spoke only of generic "man."[61]

Was there a potential (however unfulfilled) within Cavendish's philosophy for a liberal posture toward women? In her later works, consistency alone demanded that Cavendish abandon her earliest notions that feminine weakness—the small arms and tender feet, the soft and moist brain—adequately explained women's subordination, for in these works there is no possibility of stronger or weaker minds: rational matter is homogeneous. Rational matter, being all of the same quality, can have no differences in strength. Cavendish portrayed a kind of democracy among the infinite bits of matter. Harmony in nature required that each bit of matter follow its own inner logos. If the strong dominated the weak, the methodical and regular workings of the weaker parts would be violated and there would be no harmony. Nature's laws would be violated.[62] Though her views on matter might conceivably have been extrapolated to human relations, Cavendish left that potential unfulfilled. A staunch royalist, the good duchess was uncomfortable with anything that threatened ancient privilege.[63] Though philosophically a modern, she frowned upon those "unconscionable Men in Civil Wars" who endeavor to pull down ancient learning along with the hereditary mansions of the aristocracy.[64] Cavendish did not advocate changes which might have threatened the privilege she herself enjoyed over men of lower rank.

Cavendish's hesitant approach to the woman question was never consistent with her own ambitions. She had refused from her earliest years to follow a traditionally female path. In her youth, she took up the pen and not the needle. In her maturity, she took up philosophy and not housewifery. "I cannot for my Life be so good a Huswife, as to quit Writing . . . the truth is, I have somewhat Err'd from good Huswifry, to write nature's Philosophy."[65] Every part of her project—her voluminous publication, her visit to the Royal Society, her autobiography (which she later suppressed), her early atheism, her criticism of "learned men"—overstepped the bounds of convention.[66]

How are we to explain Cavendish's departure from the English custom of quiet, pious feminine deportment? Cavendish never revealed the source of her feminism. She was not only unaware of her intellectual predecessors—women such as Christine de Pizan or Anna van Schurman—but she dreaded hearing of such women. "I have not read much in History to inform me of the past Ages, . . . for I fear I should meet with such of my Sex, that have out done all the glory I can aime at."[67] The upheaval of the civil war, bringing with it a certain tolerance for public voices for women, may have influenced Cavendish in her youth. Perhaps during her years of exile she took in the feminist air of the Continent. The Restoration, in any case, along with the return of the duke and duchess to England in 1660, served to silence her philosophical voice. Two years after her critique of experimental philosophy and shortly after her visit to the Royal Society in 1667, Margaret Cavendish published a more modest edition of *Grounds of Natural Philosophy* (taking back many of her earlier claims) as her last philosophical work. She died soon thereafter and was buried in Westminster Abbey, where she still lies today.

Emilie du Châtelet and Physics

After Margaret Cavendish died, no woman in England wrote as boldly on natural philosophy. Anne Conway kept a philosophical notebook in the 1660s and '70s, but, unlike Cavendish, she never intended it for publication. Her friends, Henry More and Franciscus van Helmont, readied it for publication after her death. Perhaps the ridicule which Cavendish experienced discouraged other women from similar efforts. Perhaps the sharp decline of the nobility in England brought an end to the philosophical networks which had emboldened Cavendish. The noble network survived, however, and was at its best in France. Throughout the eighteenth century, this network spawned a number of scientific women. Madame du Pierry, Nicole Lepaute, and Madame Le Français de Lalande practiced astronomy.[68] Madame Lavoisier and Madame Thiroux d'Arconville translated chemistry texts from English into French;[69] Sophie Germain was a prizewinning mathematician.[70] The best known of these, however, was Madame du Châtelet, the physicist. Much has been written about du Châtelet in both our times and her own; her association with Voltaire for more than sixteen years catapulted her into the limelight.[71]

Gabrielle-Emilie Le Tonnelier de Breteuil, marquise du Châtelet (1706–1749), was able to exercise a greater freedom in choosing intellectual companions than was Margaret Cavendish. Though she married out of considerations of rank, she chose her mentor according to her own intellectual tastes and needs. At age nineteen she married Forent-Claude, marquis du Châtelet and count of Lomont, a military officer of an illustrious but rather impoverished family from Lorraine. After the conventional marriage expected of a woman of her standing and three children, Emilie du Châtelet's interests turned increasingly toward science. She also met Voltaire, already a celebrated poet. To Voltaire, du Châtelet offered an exchange. For his part, Voltaire received sanctuary at Cirey-sur-Blaise, du Châtelet's country estate, and her influence at court. Having been exiled for his publication of his *Lettres philosophiques,* Voltaire found du Châtelet's connections at court invaluable. For her part, du Châtelet received entrée into intellectual circles through the prestige of Voltaire's literary reputation. She envisioned her retreat at Cirey as an intellectual center, with Voltaire as the main attraction.

Emilie du Châtelet met Voltaire in Paris in 1733 and, while married and on good terms with her husband, developed an increasingly intimate relationship with Voltaire that was to last the rest of her life. Aristocratic women at this time enjoyed certain freedoms in matters of love; while extramarital relations were not encouraged, they were in fact tolerated. In 1734 Voltaire's secretively published *Lettres philosophiques* were seized, denounced, and publicly burned. Threatened with arrest, he and du Châtelet withdrew to her château at Cirey. As Madame du Châtelet described their retreat, "everyone stays in his or her own room until nine o'clock in the evening . . . sometimes several days together will pass without anyone seeing me."[72] There were few of the amusements—card playing or gambling—which she associated with high society in Paris, save for the theater du Châtelet had built to stage Voltaire's plays, in which she often played the leading role.

At Cirey, du Châtelet began her systematic philosophical studies (see Figure 6). Voltaire, who had developed a taste for Newtonian physics during his stay in England in 1728, introduced her to France's ardent Newtonians—Pierre Maupertuis, who wrote the first French work devoted to the Newtonian world system (*Discours sur la figure des astres;* 1732), and Alexis-Claude Clairaut. Until 1737 she, like Voltaire, supported the French Newtonians in their struggle against the Cartesians of the Académie Royale des Sciences.[73] Under the tutelage of Maupertuis, who agreed to give her lessons in algebra as a personal favor in 1734, she soon outstripped Voltaire in mathematics and physics. When Voltaire began his *Elé-*

Figure 6. Madame du Châtelet with compass and celestial globe.

ments de la philosophie de Newton in 1736, Châtelet supplied the mathematical expertise he lacked. This period marked the height of purely Newtonian enthusiasm at Cirey. The young Italian poet and fellow Newtonian, Francesco Algarotti, prepared his popular version of Newtonian optics, *Il Newtonianismo per le dame*, while a guest at Cirey. Madame du Châtelet for a while intended to translate Algarotti's work into French.

In 1737 the Académie Royale des Sciences announced a prize competition on the nature of fire and heat. Voltaire, determined to enter, had installed at Cirey a fine *cabinet de physique,* which was set up with instruments sent by the Abbé Nollet. Though initially collaborating with Voltaire on his essay as she had on earlier projects, du Châtelet decided at the last minute to write and submit her own essay, where she argued against many of his ideas. This she did without his knowledge. Afraid of angering Voltaire, she hid her work from him, writing secretly at night, keeping herself awake by dipping her hands in iced water. "I did not tell Monsieur de Voltaire anything about it, because I did not want to feel ashamed of an enterprise . . . which I was afraid would displease him."[74] Her work, unlike his, was done outside the laboratory. Not wanting to arouse his suspicion, she did not go back into the laboratory to test her hypotheses. Only when she learned that neither of them had won the prize did she reveal her authorship to him. Though neither had won, Voltaire arranged nonetheless to have their memoirs included with those of the winners in the academy's publication.[75]

With the prize competition, Emilie du Châtelet began parting ways intellectually with Voltaire. She had come to distrust his radical antimetaphysical stance, which he had adopted from Locke and Newton. In the belief that natural science was incomplete without some kind of metaphysical foundation, she began her work on her *Institutions de physique,* initially conceived as a book on the principles of Newtonian physics for the instruction of her son, Louis-Marie. Again she worked in secret. She had arranged for one of her close friends, Madame de Chambonin, to take the work to the censor and printer in order to preserve her anonymity.[76] Though working secretly, du Châtelet felt the need for comments on her work and chanced showing it to Maupertuis, the only one of her friends she felt qualified to read it. In order to keep her secret even from Maupertuis, she presented it to him not as proposed chapters of a book but simply as exercises.

Maupertuis's criticism of her mathematics led her to begin looking for a tutor who could give her more time. She hired Samuel König, a disciple of the German Christian Wolff, who introduced her to Leibnizian metaphysics still virtually unknown in France. In

Leibniz's doctrine of *force vive* and the principle of sufficient reason, du Châtelet found the metaphysical framework for which she had been searching. In her *Institutions,* she sided neither with Leibniz nor with Clarke in their great debate but attempted to synthesize Newtonian physics with German metaphysics.[77]

Her tutor König, from whom her project had also been kept secret, unwittingly helped her rewrite the entire manuscript until—in November of 1739—he chanced to see some proof sheets sent from the printer. Feeling that the office of tutor was beneath his dignity, König revealed her authorship on the eve of publication, putting out the story that he himself was the true author—that she had simply copied out his notes and sent them to the printer as her own work.[78] After König left du Châtelet was unable to find another tutor whom she trusted and so finished her book herself, publishing it anonymously at the end of 1740 (a revised edition appeared in Amsterdam in 1742).

After the König fiasco, she returned again to the project of introducing the works of Newton to a French audience, the work for which she would be most remembered. Her translation of Newton's *Principia mathematica* with a commentary, published after her death, remains today the standard French translation of that work.[79] In 1749 she went to Paris to finish her commentary and theoretical supplement in collaboration with Clairaut. She soon discovered, however, that at age forty-two she was pregnant (by the poet, marquis de Saint-Lambert, whom she had met at the court of Stanislas, ex-king of Poland, at Lunéville). She died a few days after the birth of her daughter of childbed fever (the child also died). Before her death she had entrusted the manuscript of her annotated translation of the *Principia* to the librarian of the Bibliothèque du Roi in Paris. It appeared in 1759 (one of the few works to bear her name) and remains the sole French translation of that work.

Even a woman of Emilie du Châtelet's stature suffered from the restrictions placed on women. As one biographer put it, Madame du Châtelet was privileged, but not free.[80] Excluded from membership in the Académie Royale des Sciences and the free exchange of ideas that membership offered, her intellectual life—like that of Margaret Cavendish—was mediated through men such as Voltaire or Maupertuis. Maupertuis might have corrected her work as a personal favor but—busy with other students and his Arctic expedition—he never gave her the time she required. Dependent for intellectual guidance on those who came as guests or tutors to her estate, she could not develop her interests as she would have liked.

These restrictions limited the scope of du Châtelet's work. Her greatest contribution was to spread the ideas of Newton and of Leibniz into France. Her work was by and large synthetic; even in her *Institutions de physique,* she did not develop her own foundation for physics but chose instead to present a clear and faithful translation of Wolff and Leibniz's metaphysics. She had a profound sense of her own limitations and was uncertain about her ability—even perhaps her right—to make original contributions to science.[81] Thus she limited her work to translation, feeling that "it is better to do a good translation of an esteemed English or Italian book than to write a bad book in French."[82] She added her own comments in quotation marks so that the reader could distinguish them from the original.

Like Margaret Cavendish, Emilie du Châtelet felt oppressed by "the prejudice which excludes us [women] so universally from all the sciences."[83] "Why," she asked, "for so many centuries has there never been a good tragedy, a good poem, a valued history, a beautiful painting, a good physics book produced by a woman?" Unlike Margaret Cavendish, du Châtelet did not accept the explanation that there is something in the physical nature of women that prevents them from exercising the same reason as men. On the contrary, she believed that women's limited contribution came from their limited education. In the preface to her translation of Mandeville's *Fable of the Bees* (first published in the 1940s), she wrote, "I leave to the naturalists the search for a physical reason, but until one is found, women have the right to complain about their education." She then imagined a scientific experiment she would perform if she were king: "I would reform an abuse which cuts off, so to speak, half the human race. I would make women participate in all the rights of humankind, and above all in those of the intellect."[84] She believed that everyone would benefit from education for women: women by gaining a new appreciation of their own talents, men by interacting with these talented women.

Emilie du Châtelet is probably the best-known woman scientist of the eighteenth century. Known to her contemporaries as "Emilie," a name popularized by Voltaire, her reputation rests as much on her *liaison* with him as on her own scientific achievement. As one contemporary remarked, "women are . . . like conquered nations . . . whatever originality, greatness, and sometimes genius they possess is considered only as a reflection of the spirit of the famous man they loved."[85]

In the seventeenth and eighteenth centuries, natural philosophy remained a part of elite literary culture. Noblewomen were able to insinuate themselves into networks of learned men by exchanging patronage or public recognition for tutoring from men of lesser rank but of intellectual stature. The privileges of her rank allowed Christina of Sweden to attract Descartes to her court as tutor and philosophical adviser. Rank and patronage introduced Margaret Cavendish, duchess of Newcastle, into social circles where Hobbes and Descartes appeared as guests at the dinner table. Madame du Châtelet's estate at Cirey allowed her to shelter Voltaire and his coterie. Yet, as we have seen, there were limits to this exchange. In the same way that privilege gave women only limited access to political power and the throne, nobility gave them only limited access to the world of learning. Because women were barred from the centers of scientific culture—the Royal Society of London or the Académie Royale des Sciences of Paris—their relationship to knowledge was inevitably mediated by a man, whether that man was their husband, companion, or tutor.

But not all women working in science were aristocratic. At a time when participation in science was regulated to a large extent by social standing, women working in science came from two distinct social groups—the aristocracy and the artisan class. Social origins shaped to a certain degree the kind of science they did. In France women working in science came overwhelmingly from the aristocracy and engaged primarily in theoretical work. But in Germany women's scientific activities centered around craft production, where the emphasis was on observational science, especially astronomy and entomology. Women's participation in the household economy and craft production gave them a surprisingly strong position in early modern science.

3

Scientific Women in the Craft Tradition

> If one considers the reputations of Madame Kirch [Maria Winkelmann] and Mlle Cunitz, one must admit that there is no branch of science . . . in which women are not capable of achievement, and that in astronomy, in particular, Germany takes the prize above all other states in Europe.
>
> —Alphonse des Vignoles, 1721

It may be surprising that between 1650 and 1710 a significant proportion—some 14 percent—of all German astronomers were women.[1] These women came not from the aristocracy but from the workaday world of the artisanal workshop, where women as well as men were active in family businesses. Craft traditions, central to working life in early modern Europe, also contributed to the development of modern science. This route to science was more open for women in Germany, where craft traditions remained especially strong. To be sure, Germany had its outstanding royal women—Caroline of Ansbach, Princess Elizabeth, and Sophie Charlotte, the founder of the Academy of Sciences in Berlin—but it was working women who made steady contributions to the empirical base of science. As Alphonse des Vignoles, vice-president of the Berlin academy, observed, there were more women astronomers in Germany at the turn of the eighteenth century than in any other European country.

Edgar Zilsel was among the first historians to point to the importance of craft skills for the development of modern science in the West.[2] Zilsel located the origin of modern science in the fusion of three traditions: the tradition of letters provided by the literary humanists; the tradition of logic and mathematics provided by the

Aristotelian scholastics; and the tradition of practical experiment and application provided by the artist-engineers.

What Zilsel does not point out, however, is that the new value attached to the traditional skills of the artisan also allowed for the participation of women in the sciences. Of the various institutional homes of the sciences, only the artisanal workshop welcomed women. Women were not newcomers to the workshop: it was in craft traditions that the fifteenth-century writer, Christine de Pizan, had located women's greatest innovations in the arts and sciences—the spinning of wool, silk, linen, and "creating the general means of civilized existence."[3] In the workshop, women's (like men's) contributions depended less on book learning and more on practical innovations in illustrating, calculating, or observing.

Women's position in the crafts was stronger than has generally been appreciated. In fifteenth-century Nuremberg and Cologne, for example, craftswomen were active in nearly all areas of production: of the thirty-eight guilds that Margret Wensky has described in her study of working women in Cologne (a city where women's economic position was especially strong), women were full members of more than twenty of those guilds.[4] Women's membership in these guilds conferred on them limited civic rights—they could buy and sell and be represented in a court of law, for example, but they could not hold city office.

Astronomers and entomologists were never, of course, officially organized into guilds. Yet craft traditions were very much alive in the practice of these sciences. This was especially true in Germany, where stirrings of industrialization came late. Whereas in England and Holland guilds declined after the midseventeenth century, in Germany they remained an important economic and cultural force well into the nineteenth century.[5]

Women's work in household workshops differed widely from trade to trade, from town to town. Yet it is possible to sketch general patterns.[6] Women participated in craft production as: (1) daughters and apprentices; (2) wives who assisted their husbands as paid or unpaid artisans; (3) independent artisans; or (4) widows who inherited the family business. As we shall see, these categories were also important for defining women's place in scientific production.

Maria Sibylla Merian and the Business of Bugs

Maria Sibylla Merian was a leading entomologist of the eighteenth century (see Figure 7). At a time when travel was difficult for women (as for men), she sailed to the Dutch colony of Surinam, where she undertook a series of studies that broadened significantly the empirical base of European entomology. In the seventeenth and early eighteenth centuries, the apprentice system was the key to women's training in science. Maria Sibylla Merian was born in Frankfurt am Main in 1647, daughter of the well-known artist and engraver, Matthäus Merian the elder.[7] In her father's workshop she learned the techniques of illustrating—drawing, mixing paints, etching copperplates. From the age of thirteen, Maria Merian served an informal apprenticeship with her stepfather, guild painter Jacob Marell (her own father died when she was three), and with her stepfather's apprentice, Abraham Mignon. A contemporary, Joachim von Sandrart, confirmed that "in her home, Merian received good training in sketching and in painting (both oil and water-color) all manner of flowers, fruit, and birds, and in particular . . . worms, flies, mosquitos, and spiders."[8]

Interestingly, it was this training in art that gave Merian her entrée to science; the primary value of her studies of insects derived from her ability to capture in fine detail what she observed. In early modern science women commonly served as observers and illustrators. A woman's success as an illustrator rested, in part, on her ability to adapt to a new field skills in which women excelled (nuns had long illuminated manuscripts; other women were active members of painters' guilds).[9] The recognized need for exact observation in astronomy, botany, zoology, and anatomy in this period made the work of accomplished illustrators particularly valuable.

Merian's education followed the pattern typical of a guild master's daughter—namely, the daughter trained as an apprentice in her own home. Young women did not (like young men) travel over the course of several years serving as journeymen with various masters. Merian's future husband, Johann Graff, for example, studied for two years with a local master in Frankfurt, then traveled to Rome to study for four years at the academy of art.[10] Merian, in contrast, did not travel from workshop to workshop. The step-

Figure 7. Maria Merian shown with exotic specimens brought from Surinam and displayed at the Stadthaus in Amsterdam. To offset the cost of her research, Merian sold specimens to the curious for three florins each. Anonymous Dutch engraving from the first half of the eighteenth century. By permission of the Öffentliche Kunstsammlung, Kupferstichkabinett Basel.

daughter of a prominent master, however, Merian had the advantage of his training and, when he was away from home for five years at a time, she trained with the master who took his place.

In 1665, Merian married Graff, one of her stepfather's apprentices, and the couple moved to Nuremberg. Though both Merian and Graff were painters, Merian did not work (as was common) as a partner in her husband's business but established one of her own—selling fine silks, satins, and linens that had been painted with flowers of her own design. In both Frankfurt and Nuremberg Merian gathered around her a group of women students (her *Jungfern Combanny,* as she called them) who served both as assistants and apprentices. Most of her pupils came from the homes of Nuremberg painters—Magdalena Fürst was herself to become a famous painter of flowers; Dorothea Maria Auer helped Merian run her business in painters' colors.[11] During this period Merian also began experimenting with technique. Testing different ways to make her fabrics both beautiful and durable, she eventually developed a type of watercolor that withstood multiple washings.

Merian began her scientific career with the publication of her *Wonderful Metamorphosis and Special Nourishment of Caterpillars* in 1679, a book that captured in pictures the transformation of caterpillars.[12] As Merian told later in life, this study emerged from years of observation and research:

> Since my youth, I have studied insects. In my place of birth, Frankfurt am Main, I began studying silkworms. When I realized that butterflies and moths develop more quickly than other caterpillars, I collected all the caterpillars that I could find, in order to observe their metamorphosis. Thus, I withdrew from human society and engaged exclusively in these investigations. In addition, I learned the art of drawing so that I could draw and describe them as they were in nature. I gathered all the insects I could find in the vicinity of Frankfurt and Nuremberg and painted them . . . very exactly on parchment.[13]

In fifty copperplates she drew the life cycle of each insect—from egg to caterpillar to cocoon to butterfly—attempting to capture each change of skin and hair and the whole of their life "as much as possible in black and white" (see Figure 8). Merian undertook her study of caterpillars in an attempt to find other varieties that, like

Figure 8. A page from one of Merian's works on the life cycle of caterpillars, showing metamorphosis from larva to butterfly along with the kinds of plants the organisms feed on at each stage. Merian undertook her study of caterpillars hoping to find another variety as economically profitable as the silkworm. From Maria Sibylla Merian, *De Europische Insecten* (Amsterdam, 1730), plate 5.

the silkworm, could be used to produce fine thread. Others in Germany shared Merian's interest in the silk business. Leibniz, as president of the Berlin Academy of Sciences, imported mulberry trees from China for the new academy. Though the king had granted the academy a monopoly on silk making in 1700, the trees did not flourish and silk was not as profitable as Leibniz had hoped. Maria Merian continued her research, over a period of five years searching out and collecting various caterpillars along with enough of their particular food to sustain them for the days or months of observation and drawing. After much earnest and wearisome study, she found many caterpillars that change into moths or flies, but none that spin a useful thread similar to that of the silkworm.

Merian's second book, *Neues Blumenbuch,* was published in "magical" (as she called it) color in 1680.[14] This book of flowers, drawn from life, provided guild artists with designs for painting and embroidery. Merian hoped to capitalize on the flower craze then sweeping Europe; as Merian reported, one tulip bulb could fetch 2,000 Dutch florin—a shocking sum—and a garden of tulips could cost as much as 70,000 florin. In order to capture the living beauty of the flower, Merian developed a new printing technique. Following established procedures, she first drew the flower on parchment, then etched it onto a copperplate and printed it. She then sent the freshly inked print back through the press, printing a reverse image. The artistic advantage of the reprinted copy was that it did not bear the harsh outline of the copperplate, and it was not a reverse image but a true rendering of the original sketch.[15] The business advantage of this method was that each print yielded two copies; the first was colored by one of her daughters or apprentices, the second she colored herself. A reviewer found her colors so beautiful that they were "more like painting than illustration."[16]

In Maria Merian we find a confident and independent woman directing her own business interests, training young women in her trade, experimenting with technique, and following her own scientific interests. In the prefaces to her publications she never apologized for her achievements (as did so many women of this period) nor spoke, as did Margaret Cavendish, of the "softness" of the female brain. Yet even Merian found it necessary to profess a certain modesty. She had been persuaded, she wrote, to publish her work by "learned and well respected people." This she did "not for my own glory, but for the glory of God alone, who created such wonders."[17] Craftswomen were also required to keep well-regulated households. Joachim von Sandrart made a point of saying that Merian's business did not interfere with her household duties.[18]

After fourteen years in Nuremberg, Merian returned to Frankfurt in 1682 to care for her recently widowed mother. Up to this point Merian (or "Gräffin," as she called herself) had lived with her husband. In 1685 or 1686 she left him and reclaimed her maiden name. Newspapers of the time reported that Merian left her husband after twenty years of marriage because of his "shameful vices" (we are not told what these were).[19] A later issue of this same paper withdrew this report, insisting instead that Merian was to blame for the separation. According to this and later reports, Merian left Graff, a respected citizen of Nuremberg, out of a certain "caprice" and moved with her two daughters to the experimental religious Labadist community.[20] Graff reportedly went to the Labadist colony in an attempt to bring Merian and their two daughters back with him to Nuremberg. Merian, however, refused, and Graff turned the matter over to the Nuremberg authorities. Merian was publicly censured, and when she did not respond Graff was given freedom to marry once again. Elisabeth Rücker has found the divorce notice in the Nuremberg archives: "Johann Andreas Graffen, painter, applies to be . . . separated completely from his wife [Weib] who left him seven years ago to join the Labadists."[21]

Contemporaries gave conflicting explanations for why Merian reclaimed her father's name. Some claimed that her father, who died when she was three, recognized his talent in her and told her to keep his name always. Others claimed that Merian changed her name in order to distance herself from her husband's scandalous reputation, even though she had had several children by him.[22] Merian's divorce was not as uncommon in early eighteenth-century Germany as we might think. Marriages in this period were often disrupted. Many spouses died; widows and widowers often remarried. Couples also separated. Georg Gsell was separated from his first wife before marrying Merian's daughter, Dorothea Maria. Merian's other daughter, Johanna Helena, later separated from her husband.[23]

The Labadist colony that Merian joined was an experimental religious community at the Walta castle in West Friesland owned by the Sommelsdijk family. Merian was attracted to the community for the protection from her husband it offered (Labadists regarded null and void marriages with anyone outside the community) and because her half-brother lived there. In addition, the Labadists were sympathetic to independent and accomplished women; Anna van Schurman (the famed "learned maid" of Utrecht) had been a follower of Jean de Labadie and had helped found the Walta community some years earlier. Merian left no record of her ten-year sojourn with the Labadists. She was no doubt active in their self-sufficient economy—baking bread, weaving cloth, and printing books—and she also sharpened her scientific skills, learning Latin and studying the flora and fauna sent to her from the Labadist colony in Surinam.

The Labadist community began to dissolve in 1688, and in 1691, after her mother's death, Merian gave up her civic rights (*Bürgerrechte*) in Frankfurt and moved to Amsterdam, a city rich with "many rarities from the East and West Indies."[24] In Amsterdam Merian supported herself and her two daughters by doing the same kind of work she had done in Nuremberg—selling her colored fabrics and preparing and selling paints for artists. At the same time she continued her work in scientific illustration, preparing, for example, 127 illustrations for a French translation of Joannes Goedaert's *Metamorphosis et historia naturalis insectorum.*[25] More importantly, she met Caspar Commelin, director of the botanical gardens, and had the opportunity to study the many rich collections of natural history that Amsterdam had to offer.

> In Holland I saw beautiful animals from the East and West Indies . . . I had the honor of seeing the beautiful collections of Dr. Nicolaas Witsen, mayor of Amsterdam and director of the East India Company, and the collection of Jonas Witsen, secretary of Amsterdam. I also saw the collection of Fredericus Ruysch, doctor of anatomy and professor of botany, and the one of Levinus Vincent and many others.[26]

Merian was disappointed, however, that these collections presented only a static view of the life of insects. What interested her were the processes by which caterpillars spin cocoons and turn into butterflies. Thus Merian set out to do her own research. "This all resolved me to undertake a great and expensive trip to Surinam (a hot and humid land) where these gentlemen had obtained these insects, so that I could continue my observations."

In 1699, at the age of fifty-two, Merian and her daughter Dorothea set sail for the Dutch colony of Surinam to continue her insect research—an undertaking unusual for either a woman or a man. For two years Merian collected, studied, and drew insects and plants of the region, gathering specimens early in the cool of the day and preparing them in the evening. As she reported to Johan

Georg Volckamer of Nuremberg, "in Surinam I gathered worms and caterpillars, feeding them daily, and observing them as they went through their transformations. These and the plants they fed upon I painted and described." She also described how she prepared her specimens:

> The snakes and similar animals I put in glasses with ordinary brandy, and seal the glass with perforated paper . . . With butterflies, I place the tip of a needle in a flame until it is hot or glowing, and stick the needle into the butterfly. The butterfly dies quickly and is not damaged.[27]

Merian lived for a time at the Labadist mission on the rubber tree plantation of Cornelis van Sommelsdijk, the Dutch governor of Surinam.[28] Her writings from this period reveal an undercurrent of conflict between her and the European planters. Conflicts arose not because she was a woman but because she was a scientist. In the commentary to her Surinam book, Merian told how the planters "mock me, because I am interested in something other than sugar." In turn, Merian criticized the planters for failing to explore other plants of the region, such as cherries or plums, that could be cultivated for sale. She was particularly critical of the treatment the Indians received at the hand of the colonists. In her description of a plant used by the natives to induce abortion (*Flos pavonis*) she noted:

> the seeds of this plant are used by women who have labor pains, and who must continue to work, despite their pain. The Indians, who are not treated well by the Dutch, use the seeds to abort their children, so that their children will not become slaves like they are. The black slaves from Guinea and Angola have demanded to be treated well, threatening to refuse to have children. In fact, they commit suicide because they are treated so badly, and because they believe that they will be born again, free and living in their own land. They told me this themselves.[29]

The rigors of the climate, not the uninviting planters, forced Merian to return to Amsterdam in 1701, sooner than she had intended. Overcome with malaria, Merian reported that to obtain her insects, "I nearly paid with my life."[30] Her trip, however, was a great success for both her science and her business. Merian brought with her from Surinam exotic specimens that the mayor put on display at the town hall. Among her brandy-preserved treasures were a crocodile (called by Réaumur a "fierce insect"), many types of snakes, and other animals—including twenty jars of butterflies, bugs, fireflies, and iguanas. Several of these specimens she sold for three florins each. She also sold one crocodile, two large and eighteen small snakes, turtles, and "other insects" for twenty florins.[31] Her illustrations fetched a higher price, selling for as much as forty-five florins each. Through these and other sales Merian hoped to recoup the price of her passage.[32]

Upon her return to Amsterdam, Maria Merian began work on her major scientific work, the *Metamorphosis insectorum Surinamensium.* In sixty illustrations, Merian detailed the life cycle of various caterpillars, worms and maggots, moths, butterflies, beetles, bees, and flies—important empirical work, given that it had been only thirty years since Francesco Redi had recognized that insects hatch from eggs and do not generate spontaneously from excrement as Aristotle had thought.[33] As well as showing the reproduction and development of insects, Merian's illustrations revealed to Europeans "plants never before described or drawn." Enthusiasts judged Merian's *Metamorphosis* the "first and strangest work painted in America." As she wrote, "this work is rare and will remain rare . . . since the trip is costly and the heat makes living [in Surinam] extremely difficult."[34]

In addition to broadening the empirical base of entomology, Merian sought to integrate her work into the learned world by citing from works of Thomas Moufet, Joannes Goedaert, Jan Swammerdam, and others. Nonetheless Merian felt constrained by that world, and in the preface to her *Metamorphosis* she wrote, "I could have given a much fuller text, but because the world today is very sensitive, and because the views of the learned differ so greatly, I present only my observations."[35] Merian did, however, depart from standard practices of the learned world by retaining the names given plants by native Americans (Latin names were added to her text by Caspar Commelin). She also incorporated into her commentary practical aspects of the fruits she drew. In addition to describing a plant and its history, Merian provided recipes for its use. In her description of the pineapple, for example, Merian pointed out that "one eats it raw and cooked, one can make wine and brandy from it." To her description of the cassava root Merian added a recipe for cassava bread, eaten by both Indians and Europeans in America. "If the root is eaten raw, one dies of its poisons; if prepared correctly, it makes a tasty bread similar to Dutch Zwieback."[36] Recent biographers have attributed her dinner recipes to her "household interest" but, as we shall see in Chapter 4, natural histories of this period commonly included medicinal or culinary recipes.[37]

For most of her life Merian financed her own research and scientific projects.[38] The printing of her large Surinam volume with its many copperplates was extremely expensive—each copy costing forty-five florins—fifteen for printing, thirty for illustration. She spared no expense in this work, employing, as she reported, "the most famous engravers and the best paper so that the connoisseur of art as well as the lover of insects could study it with pleasure and joy." To cover production costs she sold subscriptions or advance orders. Merian did not intend to profit from the publication of her work. Rather, she wrote, "I was satisfied to recover my costs."[39]

Merian left her mark on entomology. Six plants, nine butterflies, and two beetles are named for her.[40] Merian's *Metamorphosis* met with great success. It was praised in Germany's *Acta eruditorum* and was well received by the learned world.[41] Christoph Arnold (1627–1685) wrote that "what Gesner, Wotton, Penn and Muset have neglected to do has come to life in Germany through the hands of a clever woman."[42] Her work was also admired by *virtuosi* of natural history. Between 1675 and 1771 her three books appeared in a total of nineteen editions, and her *Metamorphosis* became a standard fixture in drawing rooms and natural history libraries.[43] In her own time Merian's work was much admired by Peter I of Russia. The czar hung Merian's portrait in his study and purchased two volumes of her work in 1717 (the year of her death) for 3,000 florins. The portrait and several of her illustrations were placed on display at the "Kikin Palace," where Peter I opened his scientific collection to the public.

Merian's training and skills did not die with her but were carried on by her daughters, who completed the third volume of her Surinam book. In 1717 Dorothea moved to Saint Petersburg, where she and her husband, Georg Gsell, became court painters. Their daughter (Merian's granddaughter) eventually married Leonhard Euler.[44]

The independent Merian, who wrote a great deal about her life and times, gave few apologies for her sex, and she encountered little criticism of the sort women scientists often faced—at least during her lifetime. Her work remained popular throughout the eighteenth century and well into the nineteenth. Goethe marveled at Merian's paintings for the way they moved between art and science. In his 1840 edition of *The Cabinet Cyclopedia,* William Swainson rightly claimed that Merian was one of the first ever to publish on insects.[45]

It was not until the nineteenth century that Merian's work suffered severe criticism. In a review of her work published in 1834 in the *Magazine of Natural History,* Reverend Lansdown Guilding praised this female "votary" of the sciences for having quit the comforts of her home to seek for two years the gratification of her curiosity in an unwholesome climate. Yet he found her *Metamorphosis* full of errors and her drawings "rude" and "worthless." The greatest defect, to Guilding's mind, was what he called her "anthropological flair"—the attention she gave to recording the knowledge of indigenous peoples. Guilding accused Merian of being beguiled by "some cunning negroes" and called her reports of traditional remedies and beliefs "idle stories." That Negroes never kill a particular type of spider because they believe it brings bad luck Guilding found an "absurd superstition"—one that had "served to protect a useful creature," but for the wrong reasons. Merian's criticism of the abuses of black and Indian slaves by Dutch plantation owners Guilding ignored. Concerning her discussion of the uses of *Flos pavonis* for abortion, he only remarked that this plant, used by the "Creole doctresses," formed a "pretty hedge." The tone of his attack suggests that more than the question of scientific exactitude was at stake. The Reverend Guilding missed no chance to remind his reader that Merian was of the "fair sex." Any "*boy* entomologist," he claimed, would not make such simple errors. In particular Guilding blamed Merian for drawing several types of *Lepidoptera* in such a way that had led Linnaeus to misname them (he failed to blame Linnaeus, however, for perpetuating the error).[46]

In 1854 the German naturalist Hermann Burmeister launched an equally sharp attack against Merian's work. In an address to the Société Impériale des Naturalistes de Moscou, Burmeister asked if Merian's great popularity was due to the content of her work or to its "showy" format.[47] In recent years, however, Merian's work has experienced a renaissance. Exquisite new editions of her major works have been republished in Leipzig, and the Academy of Sciences in Leningrad, where her papers were carried by her daughter, has published for the first time her notebooks and working papers.[48]

It would be a mistake to think that Maria Merian was merely an exceptional woman who, defying convention, made her mark on science. Merian's life and career may have been exceptional but it was not unusual; Merian did not forge a new path for women as much as take advantage of routes already open to women. She emerged from the artisanal workshop, where it was not uncommon for women to engage in various aspects of production, and her ties to craft traditions facilitated her contribution to science. Few women followed Merian's lead into the science of entomology. The more usual craft-based science for women in this period was a different science altogether: astronomy.

Women Astronomers in Germany

The late sixteenth and early seventeenth centuries saw the birth of modern astronomy. Copernicus published his *De revolutionibus orbium coelestium* in 1543; Galileo first turned his telescope to the sky in 1609. Astronomers in this period served in a variety of social roles—that of academician, servant of the court, or amateur enthusiast.[49] It is also possible to argue that the German astronomer of the late seventeenth century bore a close resemblance to the guild master or apprentice, and that the craft organization of astronomy gave women a prominence in the field. Between 1650 and 1710 a surprisingly large number of women—Maria Cunitz, Elisabetha Hevelius, Maria Eimmart, Maria Winkelmann and her daughters Christine Kirch and Margaretha—worked in German astronomy. All these women worked in family observatories—Johannes Hevelius built his private observatory across the roofs of three adjoining houses in 1640; Georg Christoph Eimmart built his on the Nuremberg city wall in 1678. Of this group only Maria Cunitz was not the daughter or wife of an astronomer who, in guildlike fashion, assisted a master in his trade.

It is perhaps unfair to include the example of Maria Cunitz (1610–1664) among women working within the crafts tradition, for her father was a landowner. Nonetheless, her education too depended on training given her by her father, the medical doctor Heinrich Cunitz, who owned several estates in Silesia. Sometimes called the "second Hypatia," Maria learned from her father six languages—Hebrew, Greek, Latin, Italian, French, and Polish—as well as history, medicine, mathematics, painting, poetry, and music.[50] Her principal occupation, however, was astronomy. In 1630 she married Eliae von Lowen, a medical doctor and amateur astronomer. During the Thirty Years' War her family took refuge in Poland, where she prepared astronomical tables published in 1650 as *Urania propitia*. The main purpose of this work was to simplify Kepler's Rudolphine Tables, used for calculating the position of the planets. Maria Cunitz was not merely a calculator. Her book also treated the art and theory of astronomy.

Though Cunitz published *Urania propitia* under her maiden name, few believed that the work was her own. Her husband found it necessary to add a preface to later editions asserting that he had taken no part in the work.[51] In the preface Cunitz assured her readers that her astronomy was reliable, though done by "a person of the female sex." Cunitz emphasized that her diligence in spending both "days and nights gathering knowledge from one or the other sciences or arts" had sharpened her understanding or—as she wrote—"at least what understanding is possible in a woman's body."[52] But Cunitz's diligence was not rewarded. In 1706, just forty years after her death, Johann Eberti judged that Cunitz had sacrificed her womanly duties to her astronomy:

> She was so deeply engaged in astronomical speculation that she neglected her household. The daylight hours she spent, for the most part, in bed (concerning which all manner of ridiculous events have been reported) because she had tired herself from watching the stars at night.[53]

This story was repeated throughout the eighteenth century in an effort to discredit her.

Maria Eimmart (1676–1707), though less well known, also practiced astronomy. From her father, Georg Christoph Eimmart, astronomer and director of the Nuremberg Academy of Art from 1699 to 1704, Maria Eimmart learned French, Latin, drawing, and mathematics. As a young girl she also learned the art of astronomy at her father's observatory, where she worked alongside his other students. Like Maria Merian, Maria Eimmart owed her place in astronomy largely to the strong position of women in the arts. Much of Eimmart's scientific achievement derived from her ability to make exact sketches of the sun and moon. Between 1693 and 1698 she prepared 250 drawings of phases of the moon in a continuous series that laid the groundwork for a new lunar map. She also made two drawings of the total eclipse of 1706.[54] A few sources claim that in 1701 Eimmart published a work on the sun, *Ichnographia nova contemplationum de sole,* under her father's name, but there is no evidence that this was her work.[55] Apart from her astronomical sketches, Maria Eimmart was known for her many drawings of flowers, birds, ancient statues, and, interestingly, of ancient women. These drawings have all been lost.

After training as an apprentice to her father, the scientifically minded Eimmart secured her position at the observatory by marrying Johann Heinrich Müller in 1706. Müller was a physics teacher at a Nuremberg Gymnasium and since 1705 director of her father's observatory. Müller also benefited from the marriage. Through the principle of daughter's rights, the Eimmart observatory became part of the daughter's inheritance, passing through the daughter to her husband.[56] Maria Eimmart-Müller's astronomical career was cut short when she died in childbirth in 1707.

Elisabetha Koopman (later Hevelius, 1647–1693) of Danzig also took care to ensure her career in astronomy. In 1663 she married a leading astronomer, Johannes Hevelius, a man thirty-six years her senior. Hevelius, a brewer by trade, took over the lucrative family beer business in 1641. His first wife, Catherina Rebeschke, had managed the household brewery, leaving Hevelius free to serve in city government and to pursue his avocation, astronomy. When

Elisabetha Koopman, who had been interested in astronomy for many years, married the widowed Hevelius, she served, in appropriate guild fashion, as chief assistant to her husband, both in the family business and in the family observatory.

Margaret Rossiter has described "women's work" in nineteenth- and twentieth-century science (and especially in astronomy) as typically involving tedious computation, lifelong service as an assistant, and the like—all of which are a legacy of the guild wife.[57] The role of the guild wife, however, cannot be collapsed into that of a mere assistant; wives were of such import to production that every guild master was required by law to have one.[58] The very different structure of the workplace—in the seventeenth century the observatory was in the home, not part of a university—allowed the wife a more comprehensive role. For twenty-seven years Elisabetha Hevelius collaborated with her husband, observing the heavens in the cold of night by his side (see Figure 9).[59] After his death she edited and published their joint work, *Prodromus astronomiae,* a catalogue of 1,888 stars and their positions.[60]

Maria Winkelmann at the Berlin Academy of Sciences

Of all the women astronomers in Germany, Maria Winkelmann was the most outstanding. In 1710 Winkelmann petitioned one of the newly founded learned societies, the Academy of Sciences in Berlin, for an appointment as assistant astronomer. Already a respected astronomer when her husband, Gottfried Kirch, died, Winkelmann asked to be appointed in her husband's stead. In so doing, she invoked a principle well established in the organized crafts that recognized the right of a widow to carry on the family business after her husband's death. The perpetuation of craft traditions had allowed women of the seventeenth century access to the secrets and tools of the astronomical trade, but were these traditions—part and parcel as they were of an older order—to secure a place for women in the new institutions of science?

Maria Margaretha Winkelmann was born in 1670 at Panitzsch (near Leipzig), the daughter of a Lutheran minister. She was educated privately by her father and, after his death, by her uncle. The

Figure 9. Like Gottfried Kirch and Maria Winkelmann, Elisabetha and Johannes Hevelius collaborated in astronomical work. This illustration from Hevelius's *Machinae coelestis* shows them working together with the sextant (Danzig, 1673, facing p. 222). By permission of Houghton Library, Harvard University.

young Winkelmann made great progress in the arts and letters, and from an early age she received advanced training in astronomy from the farmer and self-taught astronomer Christoph Arnold, who lived in the neighboring town of Sommerfeld. Had Maria Winkelmann been male, she would probably have continued her studies at the nearby universities of Leipzig or Jena. Though women's exclusion from universities set limits to their participation in astronomy, it did not exclude them entirely. Debates over the nature of the universe filled university halls, yet the practice of astronomy—the actual work of observing the heavens—took place largely outside the universities. In the seventeenth century the art of observation was commonly learned under the watchful eye of a master. Gottfried Kirch, for example, studied at Hevelius's private observatory in Danzig; this was as important for his astronomical career as his study of mathematics with Erhard Weigal at the University of Jena.

It was at the astronomer Christoph Arnold's house that Maria Winkelmann met Kirch, Germany's leading astronomer. Though Winkelmann's uncle wanted her to marry a young Lutheran minister, he consented to her marriage to Kirch, a man some thirty years older than Winkelmann. Knowing she would have no opportunity to practice astronomy as an independent woman, Winkelmann moved, through her marriage, from being an assistant to Arnold to being an assistant to Kirch. And Kirch found in Winkelmann a much-needed second wife who could care for his domestic affairs, as well as a much-needed astronomical assistant who could help with calculations, observations, and the making of calendars.[61]

In 1700 Kirch and Winkelmann took up residence in Berlin, the newly expanding cultural center of Brandenburg. The move represented an advance in social standing for both husband and wife. A university education at Jena and apprenticeship to the well-known astronomer Hevelius afforded Kirch the opportunity to move from the household of a tailor in the small town of Guben to the position of astronomer at the Societas Regia Scientiarum.[62] Maria Winkelmann's mobility, by contrast, came not through education but through marriage. Though coming via different routes, both served at the Berlin academy: Gottfried as astronomer, Maria as an unofficial but recognized assistant to her husband.

During her first decade at the Berlin academy, Maria Winkelmann's scientific accomplishments were many and varied.[63] Every evening, as was her habit, she observed the heavens beginning at nine o'clock.[64] During the course of an evening's observations in 1702, she discovered a previously unknown comet—a discovery that should have secured her position in the astronomical community. (Her husband's position at the academy rested partly on his discovery of the comet of 1680.) There is no question about Winkelmann's priority in the discovery. In the 1930s F. H. Weiss published her original report of the sighting of the comet (see Figure 10).[65] Kirch also recorded in his notes from that night that his wife found the comet while he slept:

> Early in the morning (about 2:00 A.M.) the sky was clear and starry. Some nights before, I had observed a variable star, and my wife (as I slept) wanted to find and see it for herself. In so doing, she found a comet in the sky. At which time she woke me, and I found that it was indeed a comet . . . I was surprised that I had not seen it the night before.[66]

News of the comet, the first "scientific" achievement of the young academy, was sent immediately to the king. The report, however, bore Kirch's, not Winkelmann's, name.[67] Published accounts of the comet also bore Kirch's name, which unfortunately led many historians to attribute the discovery to him alone.[68]

Why did Winkelmann let this happen? Surely she knew that recognition for her achievements could be important to her future career. Nor was she hesitant about publishing; she was to publish three tracts under her own name between 1709 and 1711. Her inability to claim recognition for her discovery hinged, in part, on her lack of training in Latin—the shared scientific language in early eighteenth-century Germany—which made it difficult for her to publish in the *Acta eruditorum,* then Germany's only scientific journal. Her own publications were all in German.

More important to the problem of credit for the initial sighting of the comet, however, was the fact that Maria and Gottfried worked closely together. The labor of husband and wife did not divide along modern lines: he was not fully professional, working in an observatory outside the home; she was not fully a housewife, confined to hearth and home. Nor were they independent profes-

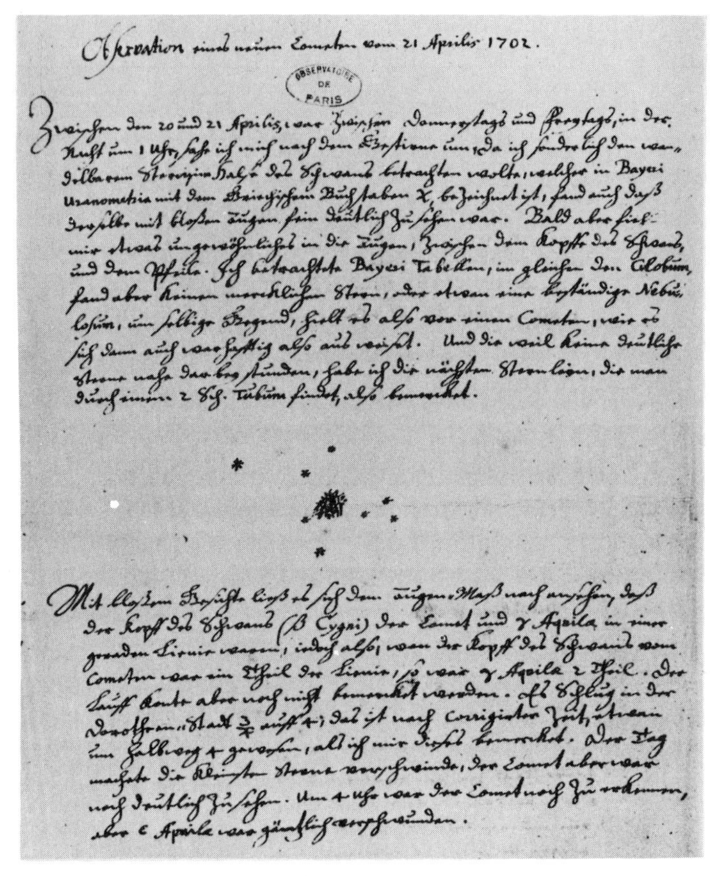
Observation eines neuen Cometen vom 21 Aprilis 1702.

Figure 10. Winkelmann's report of her discovery of the comet of 1702. Reproduced with permission of the Observatoire de Paris.

sionals, each holding a chair of astronomy. Instead, they worked very much as a team and on common problems. As Vignoles put it, they took turns observing so that their observations followed night after night without interruption. At other times they observed together, dividing the work (he observing to the north, she to the south) so that they could make observations that a single person could not make accurately.[69] After Winkelmann's sighting of the comet on the twenty-first of April, both Kirch and Winkelmann followed its course until the fifth of May.

Though Gottfried Kirch published the report under his own name and as if he alone had made the discovery, it would be an oversimplification to fault him for "expropriating" his wife's achievement. According to Vignoles, a family friend, Kirch was timid about acknowledging his wife's contributions to their common work and so published the first report of the comet without mentioning her. Later, however, someone (we do not know who) told him "that he could feel free to acknowledge her contributions." Thus when the report of the comet was reprinted eight years later, in the first volume of the journal of the Berlin academy, *Miscellanea Berolinensia,* Kirch mentioned Winkelmann's part in the discovery. This report, published in 1710, opens with the words: "my wife . . . beheld an unexpected comet."[70]

In addition to their scientific work, Kirch and Winkelmann took an active interest in the development of astronomical facilities at the academy. The Academy of Sciences in Berlin had been founded primarily to promote astronomy. In 1696 Sophie Charlotte, electress of Brandenburg, later queen of Prussia, had directed her minister, Johann Theodor Jablonski, to build an observatory, a project that took a decade to complete.[71] The Kirch family struggled long and hard, squeezing money from academy and royal purses, to create the conditions necessary for good astronomical observations. Winkelmann took an active part in these efforts. On November 4, 1707, she wrote to Leibniz (adviser to Sophie Charlotte and president of the academy), describing her reported sighting of the northern lights ("the likes of which my husband has never seen"), yet her real motive in writing was to secure housing for the astronomers more convenient to the observatory. She asked Leibniz's intervention.[72]

During the years of their acquaintance at the Berlin academy Leibniz had expressed a high regard for Winkelmann's scientific abilities. Though his letters to her have not been preserved, her letters to him reveal his interest in her scientific observations.[73] In 1709 Leibniz presented her to the Prussian court, where Winkelmann was to explain her sighting of sunspots. In a letter of introduction Leibniz wrote:

> There is [in Berlin] a most learned woman who could pass as a rarity. Her achievement is not in literature or rhetoric but in the most profound doctrines of astronomy . . . I do not believe that this woman easily finds her equal in the science in which she

> excells . . . She favors the Copernican system (the idea that the sun is at rest) like all the learned astronomers of our time. And it is a pleasure to hear her defend that system through the Holy Scripture in which she is also very learned. She observes with the best observers, she knows how to handle marvelously the quadrant and the telescope [*grandes lunettes d'approche*].[74]

He added that if only she had been sent to the Cape of Good Hope instead of Peter Kolb, the apprentice given the job, the academy would have received more reliable observations.[75]

Maria Winkelmann apparently made a good impression at the court of Frederick I. The ambassador of Denmark, on a visit to the Royal Observatory, praised her for the aid and assistance she offered her husband in his astronomical work. While at court Winkelmann distributed copies of her astrological pamphlet, *Vorstellung des Himmels bey der Zusammenkunfft dreyer Grossmächtigsten Könige*.[76] Leibniz remarked that Winkelmann's tract was "an astrological note that on the second of that month the sun, Saturn and Venus would be in a straight line. One supposes that there is significance in this."[77]

Maria Winkelmann's three pamphlets published between 1709 and 1711 were all astrological. In his 1721 eulogy to Winkelmann, Vignoles tried to explain away her interest in astrology. "Madame Kirch," as he called her, "prepared horoscopes at the request of her friends, but always against her will and in order not to be unkind to her patrons."[78] Perhaps Winkelmann's interest in astrology was purely financial, as Vignoles suggested, but her correspondence with Leibniz reveals a belief in nature as something more than matter in motion. In her description of the extraordinary northern lights of November 4, 1707, she wrote to Leibniz, "I am not sure what nature was trying to tell us."[79] Another of Winkelmann's pamphlets, *Die Vorbereitung zur grossen Opposition*, predicting the appearance of a new comet, was reviewed favorably in the *Acta eruditorum* in 1712.[80] The reviewer praised her talents, ranking her skill in observation and astronomical calculation as equal to that of her husband. Even though Winkelmann made "concessions" to the art of astrology, the reviewer judged her work valuable. The review closed with a lavish tribute to this woman who understood matters incomprehensible without "the force of intelligence and the zeal of hard work." Several months after the pamphlet appeared in 1711, Jablonski reported favorably that Winkelmann was becoming famous. Nowhere is there a hint that the Berlin academy objected to her astrological work.

Winkelmann mixed astrology and astronomy in calendar making, a project of both scientific and monetary interest for her and the academy. Unlike many major European courts, the Prussian court did not yet have its own calendar. In 1700 the Reichstag at Regensburg ruled that an improved calendar similar to the Gregorian calendar was to be used in German lands.[81] Thus the production of an astronomically accurate calendar became a major project for the Berlin Academy of Sciences, founded in the same year. In addition to fixing the days and months, each calendar predicted the position of the sun, moon, and planets (calculated on the basis of the Rudolphine tables); the phases of the moon; eclipses of the sun or moon to the hour; and the rising and setting of the sun within a quarter of an hour for each day of the year.

The monopoly of the sale of calendars was one of two monopolies granted to the academy by the king (the other was silk). Throughout the eighteenth century, the Berlin Academy of Sciences derived a large part of its revenues from the sale of various forms of calendars. This income (some 2,500 talers per year in the early 1700s) made the position of astronomer particularly important. Calendars—which Leibniz called "libraries for the common man"—had been issued since at least the fourteenth century and drew much of their popular appeal from astrology. Until 1768 there was little distinction between academy calendars and farmer's almanacs; each predicted the best times for haircutting, bloodletting, conceiving children, planting seeds, and felling timber.[82]

Weather prediction, another valuable function of the academy calendars, was an important part of the duties of the academy astronomer. Between 1697 and 1774 different members of the Kirch family kept a daily record of weather. Winkelmann's observations, as was common at that time, were made with the aid of a "weatherglass," a term used for both the barometer and thermometer. Daily observation, she noted, sharpens prediction and can be of great usefulness in many areas of life, especially in agriculture and navigation. It was Winkelmann's hope that "weather can be more accurately forecast, if more diligence is applied."[83]

The Attempt to Become Academy Astronomer

Gottfried Kirch died in 1710. It fell to the executive council of the academy—President Leibniz, Secretary Jablonski, his brother and court pastor, D. E. Jablonski, and the librarian—to appoint a new astronomer. The council needed to make the appointment quickly, as the academy depended on the yearly revenues from the calendar; but apart from one in-house candidate, Jablonski could think of no one qualified for the position.[84] Ten years earlier the council had settled on Gottfried Kirch, who despite his advanced age (sixty-one) was the best in the field.[85] Though there were few candidates, Maria Winkelmann's name was not even considered in 1710. This is even more surprising when one considers that her qualifications were not that different from her husband's when he was appointed. They both had long years of experience preparing calendars (before coming to the Berlin Academy of Sciences, Kirch had earned his living by selling Christian, Jewish, and Turkish calendars); they had both discovered comets—Kirch in 1680, Winkelmann in 1702; and both had prepared ephemerides and recorded numerous observations. What Winkelmann did not have, which nearly every member of the academy did, was a university degree.

Kirch died in July; Winkelmann made her move in August. Since her name had not come up in discussions about the appointment, Winkelmann submitted it herself, along with her credentials. In a letter to Secretary Jablonski, she asked that she and her son be appointed assistant astronomers in charge of preparing the academy calendar (see Figure 11).[86] Winkelmann made it clear that she was asking only for a position as *assistant* calendar maker. "I would not," she wrote, be so "bold as to suggest that I take over completely the office [of astronomer]." Her argument for her candidacy was twofold. First, she argued, she was well qualified, since she had been instructed in astronomical calculation and observation by her husband. Second, and more important, she had been engaged in astronomical work since her marriage and had, de facto, been working

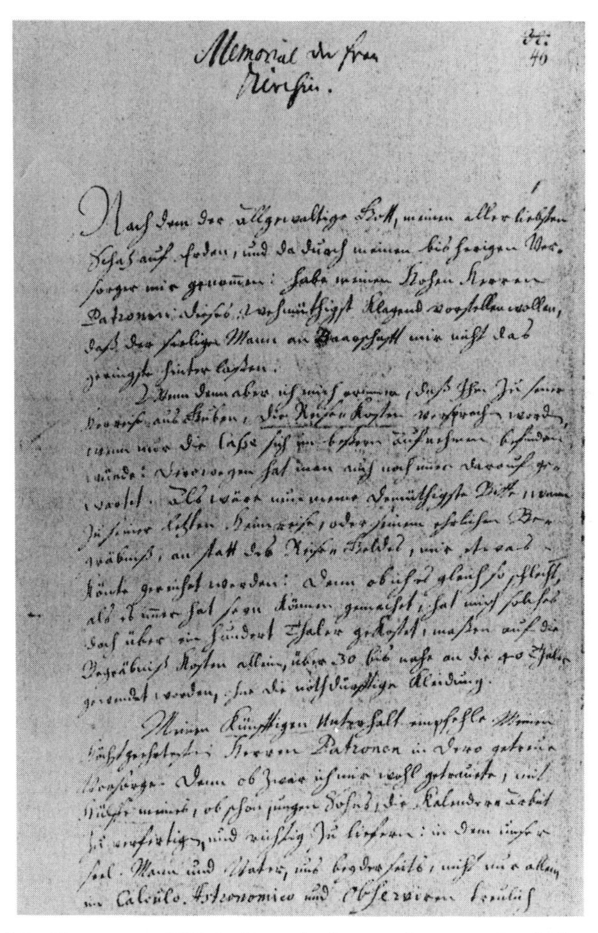

Memorial der frau Kirchin.

Figure 11. The first page of Winkelmann's six-page letter to the Berlin Academy of Sciences asking to be appointed assistant astronomer. Reproduced with permission of the Zentrales Akademie-Archiv, Akademie der Wissenschaften der DDR.

for the academy since her husband's appointment ten years earlier. Indeed, she reported, "for some time, while my dear departed husband was weak and ill, I prepared the calendar from his calculations and published it under his name." She also reminded Jablonski that he himself had remarked on how she lent a helping hand to her husband's astronomical work—work for which she was paid a wage—and asked to be allowed to stay in the astronomer's quarters. For Winkelmann, a position at the Berlin academy was not just an honor, it was a way to support herself and her four children. Her husband, she reported, had left her with no means of support.

Jablonski was aware that the academy's handling of the Winkelmann case would set important precedents for the place of women in Germany's leading scientific body. In September 1710 he cautioned Leibniz: "You should be aware that this approaching decision could serve as a precedent. We are tentatively of the opinion that this case must be judged not only on its present merits but also as it could be judged for all time, for what we concede to her could serve as an example in the future."[87] The effect on the academy's reputation of hiring a woman was a matter of some concern. Again Jablonski wrote to Leibniz:

> That she be kept on in an official capacity to work on the calendar or to continue with observations simply will not do. Already during her husband's lifetime the society was burdened with ridicule because its calendar was prepared by a woman. If she were now to be kept on in such a capacity, mouths would gape even wider.[88]

By rejecting Winkelmann's candidacy, the academy ensured that the stigma attached to women would not further tarnish its already dull reputation.[89]

Leibniz was one of the few at the academy who supported Winkelmann. In the council meeting of March 18, 1711 (one of the last meetings at which he presided before leaving Berlin), Leibniz argued that the academy, considered as either a religious or an academic body, should provide a widow with housing and salary for six months as was customary. At Leibniz's urging the academy granted Winkelmann the right to stay in its housing a while longer; the proposal that she be paid a salary, however, was defeated. Instead the council paid her forty talers for her husband's observation notebooks. Later that year the academy showed some goodwill toward Winkelmann by presenting her with a medal.[90]

After Leibniz left Berlin Winkelmann took her case to the king. With Leibniz gone, however, the council became even more adamant in denying her requests. In 1712, after one and a half years of active petitioning, Winkelmann received a final rejection. The council found her request unseemly (*ungereimt*) and inadmissible (*unzulässig*). "We must," the minutes read, "try and persuade her to be content and to withdraw of her own accord; otherwise we must definitely say no."[91]

The academy never spelled out its reasons for refusing to appoint her to an official position, but Winkelmann traced her misfortunes to her sex. In a poignant passage she recounted her husband's assurance that God would show his grace through influential patrons. This, she wrote, does not hold true for the "female sex." Her disappointment was deep: "Now I go through a severe desert, and because . . . water is scarce . . . the taste is bitter." It was about this time that Winkelmann felt compelled to defend women's intellectual abilities in the preface to one of her scientific works. Citing biblical authority, she argued that the "female sex as well as the male possesses talents of mind and spirit." With experience and diligent study, she wrote, a woman could become "as skilled as a man at observing and understanding the skies."[92]

Although Winkelmann had been involved in preparing the calendar for ten years and knew the work well, the position of academy astronomer was awarded to Johann Heinrich Hoffmann. Hoffmann had been a member of the academy since its founding in 1700 and had long hoped to be appointed academy astronomer. Yet his tenure as astronomer was not a happy one. By December 1711 he was already behind in his work. Jablonski wrote to Leibniz complaining that Hoffmann was guilty of neglecting his work. Jablonski suggested that perhaps Hoffmann needed an assistant; ironically he suggested "Frau Kirchin, for example, who would spur him on a bit." In 1712 Jablonski again had occasion to complain to Leibniz about Hoffmann's performance. Hoffmann had not completed the yearly observations as he should have, and his work for the calendar was incomplete. Hoffmann was officially censured by the academy for his poor performance. While Hoffmann was being reprimanded, Winkelmann was becoming, as Jablonski reported, "rather well known" for her pamphlet on the conjunction of Saturn and Jupiter.[93]

During this period, conflict arose between Winkelmann and Hoffmann, each of whom considered the other a competitor at the observatory. Jablonski reported to Leibniz that Winkelmann had complained that "Hoffmann uses her help secretly, yet denounces her publicly, and never lets her use the observatory." Unemployed and unappreciated for her scientific skills, Winkelmann moved across Berlin in October 1712 to the private observatory of Baron Bernhard Frederick von Krosigk. This did not end Hoffmann's problems with the academy. In 1715 Jablonski complained once again to Leibniz that Hoffmann was neglecting his duties.[94]

The Clash between Guild Traditions and Professional Science

Did Winkelmann have a legitimate claim to the post of assistant astronomer? How was it possible in 1700 for a woman to hold a semiofficial position (as Winkelmann did) as assistant astronomer to her husband at the Berlin academy? Winkelmann owed her position at the academy to the perpetuation of guild traditions. Wolfram Fischer has argued that the relation of apprentice-journeyman-master provided a model for many German institutions. Fischer gives the example of the masons; W. V. Farrar has developed the example of the universities. According to Farrar, the guild character of the university system survived longer in Germany than elsewhere.[95]

But while retaining vestiges of the guild system, the Berlin academy incorporated other traditions. We should distinguish two levels of participation in the academy. At the top was a tier of university-educated, internationally renowned scientists and philosophers. This part of the organization had nothing in common with the guilds; rather, class standing and scientific distinction were important for membership at this level. Like members of the Royal Society in London and the Académie Royale des Sciences in Paris, many "gentlemen" members of the Berlin academy were of high social standing. It was its financial structure that set the Berlin academy apart from its counterparts in Paris or London and nearer craft traditions. Members of the Académie Royale des Sciences in Paris drew pensions directly from the king's purse in order to distance themselves from traditional trades and professions, considered "mere occupations."[96] The Berlin academy, in contrast, drew much of its revenues from two trades—calendar making and silk making—and hired artisans, the second tier of participants in its activities, to carry out the tasks required.

The academy astronomer was caught between the two tiers of the hierarchy: as a university-educated mathematician, he was a distinguished gentleman; as calendar maker, he was an artisan engaged for the services he could provide. The "gentlemen" of the academy (except the president and secretary) were not paid, nor did they pay for their membership. The astronomer, however, like the other artisans of the academy, was paid a living (500 talers per year) from its coffers. It should be noted that though Maria Winkelmann asked to continue as calendar maker, she never asked to become a member of the academy (nor was she granted membership).[97]

As the wife of an artisan-astronomer Winkelmann enjoyed a modest measure of respect at the Academy. When she petitioned the council to continue as assistant calendar maker, she was invoking (although not explicitly) principles well established in the organized crafts. In most cases guild regulations gave a widow the right to run the family business after the death of her husband.[98] In her study of thirty-eight Cologne guilds in the late Middle Ages, Margret Wensky found that eighteen of those guilds allowed a widow to continue the family business after her husband's death.[99] The rights of widows followed three general patterns. In some guilds the widow was allowed to serve as an independent master as long as she lived. In others she was allowed to continue the family business but only with the help of journeymen or apprentices. In still others she filled in for one or two years to provide continuity until her eldest son came of age.[100] Within lower echelons of the academy, widows were allowed to continue in their husband's position. A woman whom we know only by the name of "Pont," widow of the keeper of the academy's mulberry trees, was allowed to complete the last four years of her husband's six-year contract.[101] This is what Maria Winkelmann tried to do. After the death of her husband she tried to carry on the "family" business of calendar making as an independent master. Yet, as we have seen, she found that traditions that had once secured women a (limited) role in science were not to apply in the new institutions.

Though the academy retained vestiges of an older order, it also contained the seeds of a new. The founding of the academy in 1700 was a first step in the professionalization of astronomy in Germany. Earlier observatories—those of Hevelius in Danzig and Eimmart in Nuremberg—had been private. The academy's observatory, however, was a public ornament of the Prussian state. Astronomers were no longer owners and directors of their own observatories but employees of the academy, selected by a patron on the basis of personal merit rather than family tradition. This shift of the character of scientific institutions from private to public had dramatic implications for women's work in science. As astronomy moved more and more out of private observatories and into the public world, women lost their toehold in modern science.

A Brief Return to the Academy

Although Winkelmann could not remain at the Berlin academy, she did continue her astronomical work. At Baron von Krosigk's private observatory in Berlin, where she and Gottfried Kirch had worked while the academy observatory was under construction, Winkelmann reached the height of her career. With her husband dead and her son away at university, she enjoyed the rank of a "master" astronomer. She continued her daily observations and—now the master—had two students to assist her. The published reports of their joint observations bear her name.[102] During this period she also supported herself and her daughters by preparing calendars for Breslau and Nuremberg. When Krosigk died in 1714, Maria Winkelmann left his observatory for a position in Danzig as an assistant to a professor of mathematics.[103] This part of her life remains sketchy. When this position fell through, Winkelmann again found a patron. The family of Johannes Hevelius (Gottfried Kirch's teacher) invited her and her son, Christfried, now a student in Leipzig, to reorganize the deceased astronomer's observatory and to use it to continue their own observations.

In 1716 the Winkelmann-Kirch family received an invitation from Peter the Great of Russia to become astronomers in Moscow. The family decided instead to return to Berlin when Christfried was appointed observer for the academy following the death of Hoffmann. Academy officials expressed grave reservations about the abilities of their newly appointed astronomers—Christfried Kirch was not well grounded in astronomical theory and could not express himself decently in Latin or his native German; J. W. Wagner was weak in astronomical calculation. Academy funds, however, were insufficient to support the appointment of a "celebrated" astronomer, who would require a higher salary, better housing, and a number of assistants. Under these circumstances a factor weighing in Kirch's favor was that, along with him, the academy received an extra astronomical hand—Winkelmann—with skills very similar to those of the two astronomers under consideration. Thus, Winkelmann returned once again to the work of observation and calendar making for the academy, this time as assistant to her son.[104]

But all was not well. The opinion was still afoot that women should not do astronomy, at least not in a public capacity.[105] In 1717 Winkelmann was reprimanded by the academy council for talking too much to visitors at the observatory. The council cautioned her to "retire to the background and leave the talking to Wagner and her son." A month later the academy again reported that "Frau Kirch meddles too much with Society matters and is too visible at the observatory when strangers visit." Again the council warned Winkelmann "to let herself be seen at the observatory as little as possible, especially on *public* occasions."[106] Maria Winkelmann was forced to make a choice. She could either continue to badger the academy for a position of her own, or, in the interest of her son's reputation, she could retire, as the academy requested, to the background. Vignoles reports that she chose the latter option. Academy records show, however, that the choice was not hers. On October 21, 1717, the academy resolved to remove Winkelmann—who apparently had paid little heed to their warnings—from academy grounds. She was forced to leave her house and the observatory. The academy did not, however, seem to want her to give up her duties as mother; officials expressed the hope that Winkelmann "could find a house nearby so that Herr Kirch could continue to eat at her table."[107]

In 1717 Winkelmann quit the academy's observatory and continued her observations only at home, as was thought appropriate, "behind closed doors"—a move which Vignoles judged detrimental to the progress she might have made in astronomy. With few scientific instruments at her disposal, she was forced to quit astronomical science. Maria Winkelmann died of fever in 1720. In Vignoles's opinion "she merited a fate better than the one she received."[108]

Invisible Assistants

Maria Winkelmann was not the only woman present at the founding of the Berlin Academy of Sciences. Sophie Charlotte, queen of Prussia, was important as an ambassador of scientific ideas at the court in Berlin. Working closely with Leibniz and her ministers, Sophie Charlotte carried forth plans and negotiations for the founding of the Berlin academy with such vigor that, as we have seen, Leibniz claimed that women of elevated mind should be the ones to cultivate knowledge.[109] Frederick II, her grandson, credited her with establishing the Academy of Sciences. He wrote that "she founded the royal Academy and brought Leibniz and many other learned men to Berlin. She wanted always to know the first principle of things." Since she died shortly after its founding, it remains unclear whether Sophie Charlotte intended to take an active part in the academy or to serve merely as patron.[110]

The founding statutes of the Berlin Academy of Sciences did not bar women from membership. In fact, Leibniz thought women should benefit from participation. In his sketch of academy regulations of 1700, he wrote that a scientific academy would foster good taste, solid understanding, and an appreciation of God's handiwork, not only among German nobility, "but among other people of high standing (as well as among women)."[111] Yet despite his intentions women were not admitted. Perhaps the decision to use the scientific societies of London and Paris as models for the academy in Berlin reinforced the exclusion of women. Although neither the London nor the Paris society had regulations excluding women, neither society admitted them.

The fate of Winkelmann's daughters—Christine and Margaretha—reveals a process of privatization of women within the academy. Trained in astronomy from the age of ten, both Kirch daughters worked for the academy as assistants to their brother Christfried. According to Vignoles, "Margaretha, the younger sister, usually took a telescope; Christine, the older, most often took the pendulum in order to mark exactly the time of each individual observation." Christine also did calculations for her brother; she and Christfried checked each other's calculations for accuracy. Yet having witnessed the lost battles of their mother, Christine and Margaretha did not ask (as Winkelmann had) for official positions. Nor did they exude the fire of their mother, badgering the academy for housing or greeting foreign (male) visitors. Rather they molded their behavior to fit academy prescriptions, becoming "invisible helpers" to their brother. Again, Vignoles described the sisters' situation: "They helped their brother carry out his professional duties; . . . nonetheless they remained very private and spoke with no one but their close friends. By the same modesty, they avoided going to the observatory when there was to be an eclipse or other observation that might attract strangers."[112]

When Christfried died in 1740, the Kirch sisters lost their male protector and were forced to observe more often at home. Although they watched the heavens daily, their situation made serious work almost impossible. When Christine sent their observations of the comets of 1742 and 1743 to Joseph-Nicolas Delisle, director of the Paris Observatory, she complained that "we observed daily [the course of the comet] as well as we could . . . but our observations were done under very bad conditions and with inferior instruments, namely with a two foot [*zwei Schühe*] telescope . . . We could not use a larger telescope because our house had no window large enough to accommodate it."[113] Though Christine and Margaretha Kirch had little opportunity to go to the observatory after their brother's death, Christine continued to prepare the academy calendar—silently and behind the scenes—from at least 1720 until her death in 1782. This is not surprising; by the 1740s calendar making was no longer on the cutting edge of astronomical science but tedious and time-consuming work. Never married, Christine supported herself through her calendar work, for which she received a small pension of 400 talers per year.[114]

After Christine Kirch retired, there were no other women doing scientific work for the Berlin Academy of Sciences until well into the twentieth century. During the eighteenth century the academy did, however, grant honorary membership to some women of the noble classes. The first to be granted honorary membership at the (then) Académie Royale des Sciences et Belles-Lettres was one of the most powerful persons in Europe at the time, Catherine the Great of Russia. Rank still spoke loudly in Prussia, and the prestige of rank outweighed the liabilities of sex. Catherine's position in the academy was wholly honorary.[115] After Frederick the Great's tenure as president, few women were elected. One exception was poet and writer Duchess Juliane Giovane, who was awarded honorary membership in 1794. No other woman was elected for 106 years, and even then it was for purely nonscientific reasons: In 1900 Maria Wentzel was awarded honorary membership for her gift of 1,500,000 marks.[116]

It is clear that before 1949 only women of the very highest social standing were admitted to the Berlin Academy of Sciences. Though Catherine the Great and Juliane Giovane were women of intellectual stature, they were also women of social rank. Maria Winkelmann, however, was a tradeswoman who dirtied her hands in the actual work of astronomy (she was referred to by academy officials as a *Weib,* not a *Frauenzimmer*). The election of a woman purely on scientific merit had to wait until 1949, when the physicist Lise Meitner was elected—but only as a corresponding member. Meitner was followed by the chemist Irène Joliot-Curie, daughter of Marie Curie, and then by the medical doctor Cécilie Vogt in 1950. The first woman to be awarded full membership was the historian Liselotte Welskopf, in 1964. Since the founding of the Academy of Sciences in Berlin in 1700, only fourteen of its 2,900 members have been women. Of those fourteen, only four have enjoyed full membership.[117] As of 1983 no woman had ever served in a position of leadership as academy president, vice-president, general secretary, or head of any of the various scientific sections.

In seventeenth- and eighteenth-century Europe, craft traditions gave women limited access to the tools of science. Science at this time was a new enterprise forging new ideals and institutions. With respect to the problem of women, science may be seen as standing at a fork in the road: it could either affirm and broaden practices inherited from craft traditions and welcome women as full participants, or it could reaffirm university traditions and continue to exclude them. As the case of Maria Winkelmann demonstrates, the Berlin Academy of Sciences chose to follow the latter path.

The poor representation of women in the Berlin Academy of Sciences cannot be traced simply to an absence of women qualified in science. Instead, the exclusion of women resulted from policies consciously implemented at an early period in the academy's history. These decisions, made in the early eighteenth century, held serious consequences for women's later participation.

4

Women's Traditions

All the arts have been invented by man, not by woman.

—Voltaire, 1764

Voltaire, like others of his day, claimed that women lack the fire of imagination and strength of perseverance required for inventive genius—to create "things mechanical, gunpowder, printing, and the clock."[1] Voltaire was not without evidence for his claim: women in the eighteenth century were active in astronomy, physics, and entomology and yet, even under the best conditions, their participation in fields developed by men remained marginal. But have women been leaders in other sciences—sciences developed primarily by women? Voltaire, like many subsequent historians of science, ignored women's contributions to the arts and sciences by focusing primarily on developments of importance to industry, the state, or the military. If Voltaire had broadened his list to include innovation in spheres he considered principal occupations of the human species—lodging, nourishment, and clothing—he might have better appreciated women's inventive genius.

This is precisely the point Christine de Pizan made in her *Book of the City of Ladies,* published in 1405. De Pizan argued that it was women's arts, not men's, that have contributed most to civilizing the world; arts developed by women, she claimed, have been more valuable to humankind than the works of the most profound philosophers.[2] Among the ladies eclipsing Aristotle's glory de Pizan named Minerva, Isis, Ceres, and Arachne. It was Isis, she wrote, who discovered the art of constructing gardens and of planting grain. Ceres taught humankind how to grind grain and make bread. Arachne invented the art of dyeing wool and of making tapestries, flax, and linen. The eighteenth-century feminist Mademoiselle Archambault expanded de Pizan's list, numbering among women's accomplishments the invention of medicine and the embalming of bodies.[3]

How did men's sciences come to overshadow whatever arts or sciences may once have been cultivated by women? As early as 1779 Dr. William Alexander, author of a history of women, suggested that women's arts have been devalued for the simple reason that it is men who write history. Alexander argued that in writing history, "partiality and self-love" bias men to value those activities in which they alone are involved. Thus men set the greatest value on martial arts and literary skills while hardly ever considering what women do. He also pointed out that women's skills—particularly their skills in raising and caring for children—are essential to humankind and should be accorded the same value as men's. "Are the female arts less useful than the desolating arts of war, or even the speculations of the statesman and improvements of the philosophers?"[4] Alexander was not alone in claiming that women's inventions were as valuable as men's. François Poullain de la Barre claimed that it took as much skill to embroider a tapestry—variegating the color, watching the proportions of the figures, and distributing the silk across the canvas—as it did to engage in men's sciences, where there was "nothing to do but to observe the uniform laws of nature."[5] Both Alexander and Poullain called for a reappraisal of women's arts.

One of the goals of recent women's history has been to recover women's traditions and innovations in an attempt to discover if, when women controlled some area of life, things were done in a different manner. To those, like Voltaire, who say that women have developed no sciences, it is important to point out that women did develop at least three science-related arts: midwifery, nursing, and the science of home economics. These fields have generally been ignored by historians of science, perhaps because they have been considered second rate. Yet such a view may rest more on the fact that midwifery and nursing have been practiced by women (not always of the most refined classes) than on the actual value of the service rendered or the skill involved. The science and technology of birthing—an essential aspect of the whole field of women's health care—was developed and carried out exclusively by midwives for at least two thousand years.

In this chapter I use midwifery and medical cookery as examples of arts developed and practiced by women most often for the benefit of other women. Unlike midwifery, medical cookery was not developed exclusively by women, though this was a field to which they made substantial contributions. Affirming the common belief that food is medicine, medical cookbooks prescribed a regimen of preventive medicine designed to keep healthy people healthy and return the sick to a state of well being. These "poor-folks' pharmaceuticals" explained in easy terms how to prepare and administer remedies for common ailments. An important aspect of the story of both midwifery and medical cookery, however, is the story of their demise: these arts did not survive the scientific revolution intact but declined dramatically with the rise of new and exclusively male professional groups. In the course of the sixteenth century, the "man-midwife" began encroaching on women's ancient monopoly on female health care, taking over the more scientific (and lucrative) parts of birthing; for two centuries midwives and medical men were locked in a struggle over professional turf and who should properly treat women. This story has been told before, but I sketch it here because it serves as an important backdrop to developments in scientific views of sexuality which we look at in subsequent chapters.[6] Medical cookery also changed dramatically. During the 1700s it was transformed into the academic fields of nutrition, botany, and pharmacy, sciences increasingly staffed by men. Only the everyday activity of preparing the family dinner remained in women's hands.

Midwifery

Midwives—wise women or *sages-femmes*—held a monopoly on birthing until the seventeenth century.[7] Even queens entrusted their precious heirs to the experienced hands of the midwife. The midwife was typically a widow or an older woman.[8] Though paid a fee for service, she was never particularly well-off. Except in the unusual circumstance where the delivery of a royal heir might bring a tidy sum—upwards of one hundred pounds in early eighteenth-century England—the midwife's wage was similar to that of other tradespeople. Like her medical peers—the barber-surgeon or the spicier-apothecary—the midwife was generally semiliterate (able to sign her name and, after 1500, perhaps able to read manuals on childbirth) but she learned her trade either on her own or through a three- or four-year apprenticeship.[9]

Before 1550 medical men and midwives coexisted peaceably enough. Tensions increased, however, in the course of the seventeenth and eighteenth centuries, when traditional crafts were upgraded and eventually professionalized. The consolidation of the (predominantly male) medical profession took place in several steps. In the sixteenth century, for example, barber-surgeons in England organized themselves into guilds, improving their status markedly and limiting all practice to members of their ranks. In the eighteenth century surgeons once again improved their status by cutting their ties with barbers (in France in 1743 and in England in 1745) and founding societies such as the Company of Surgeons in England. In France surgeons had been upgraded sufficiently to be initiated into academies—the Académie Royale de Chirurgie was founded in France in 1731; the College of Surgeons in London in 1800. Surgeons also began taking university degrees and eventually became the equals of university-trained physicians.

A similar story can be told for apothecaries who severed association with spiciers, dentists who cut their ties with toothpullers, and veterinarians who broke with blacksmiths; each of these areas of medical practice became professions in the course of the eighteenth century. One traditional craft, however—midwifery—did not follow the path of other crafts. When the professions of obstetrics and gynecology grew up around birthing, these professions were practiced by men, not women; midwifery, by contrast, remained a traditional art practiced by women primarily for the benefit of the poor.[10]

Midwives did not incorporate into self-governing corporations—though not for want of trying. In 1616 the Huguenot Peter Chamberlen, member of a family of successful man-midwives living in England, petitioned the king on behalf of the midwives of London for permission to form a self-regulating body. Chamberlen asked for improved training for midwives, especially in the new science of anatomy. Chamberlen's petition, however, was defeated by the College of Physicians, joined in this case by a large number of midwives who felt that Chamberlen alone stood to profit from securing authority to instruct, approve, and license midwives. In 1634 midwives again petitioned the king, this time on their own behalf, to form a corporation. Again their petition was rejected by the College of Physicians.

Denied the right to regulate their own craft, midwives remained under the control of different male agents. In Paris midwives were trained by one of the city's four *matrones jurées,* women supervised by the king's chief barber-surgeon, throughout the sixteenth and seventeenth centuries. In the eighteenth century, midwives were more often licensed by city fathers, who were as interested in the midwife's role in regulating illegitimate births as in her technical competence. In England they were licensed by the Church, interested primarily in certifying that the midwife was of high moral character. After 1642 regulation in England passed to the surgeons since by that time midwifery had come to be seen as a surgical procedure. In the 1660s licensing returned once again to ecclesiastical authorities.

But midwives faced a far greater challenge in this period. In the second half of the sixteenth century men began practicing midwifery. In France, it was the aristocracy that was the first routinely to engage men in attending births. The middle classes soon followed suit, and by the 1760s the practice was spreading to the tradesman and artisan class.[11] The *accoucheur,* or so-called man-midwife, was no longer called in only to attend abnormal labor; he was engaged in routine cases, thus coming into direct competition with the midwife (see Figure 12).

Male incursion into a female domain brought with it a certain amount of linguistic confusion. The unlikely term *man-midwife* entered the English language in 1625.[12] Edmund Chapman, a surgeon and man-midwife practicing in London in the 1730s, queried, "how can a man be a wife without being a hermaphrodite?" To whom does the "wife" in the term *midwife* refer? Is it the wife about to be delivered of child and pain, or is it the woman assisting the

Figure 12. A man-midwife, depicted over the caption "a newly discovered animal not known in Buffon's time." Behind the man-midwife are his surgical tools—forceps, boring scissors, and a blunt hook; he holds a lever. On the shelf beneath are vials of aphrodisiacs, his "Love Water, Cantharides, and Cream of Violets," labeled "for my own use." The midwife, by contrast, stands in a kitchen, where she boils water. Frontispiece to [S. W. Fores], *Man-Midwifery Dissected* (London, 1793). Courtesy of the U.S. National Library of Medicine.

birth? In either case, Chapman pointed out, the woman might not be a wife at all, but a maid or, indeed, a widow. In view of these uncertainties, Chapman suggested calling a man assisting a birth a "mid-man" and the woman in attendance a "mid-woman."[13] Another man-midwife suggested that these newcomers to the field be called "Andro-Beothogynists," or "Man-Helpers of Women."[14] By the 1820s, with the consolidation of male control of the profession, the confusion was resolved and the man-midwife, now with university training, became known as an obstetrician.

One should not underestimate the rivalry between medical men and female midwives in the sixteenth through the eighteenth century. Men began entering midwifery for a number of reasons. With the rise of the new science of anatomy, surgeons such as Ambroise Paré (surgeon to the king of France) began making advances in understanding the mechanism of labor. Paré reintroduced "podalic version," a method (known to the ancients) of turning the child in the womb and delivering feet first. The development of forceps also speeded delivery, forestalling in some cases the death of either mother or child (or both). As surgeons and learned physicians developed new techniques—knowledge they were usually unwilling to share with midwives—the man-midwife soon appeared more competent than the midwife.[15]

Midwives were caught in a double bind: they were ignorant of new methods and practices because they could not attend university or establish their own medical colleges, but they could not do so simply because they were women. Throughout the seventeenth century midwives struggled to found their own corporations to remedy the situation. Only in France did midwives succeed in establishing an institutional base. After the 1630s a prospective midwife could enroll in a three-month course of study offered by midwives at the great public hospital Hôtel Dieu near Notre Dame in Paris.[16] Yet training at the Hôtel Dieu did not go beyond traditional birthing techniques, and midwives continued to demand additional instruction from the Faculty of Medicine. This, however, was denied.

Some fifty years after the founding of the Hôtel Dieu in Paris, Elizabeth Cellier in London tried to establish an institution for English midwives similar to the College of Physicians. In 1687 she petitioned James II "to unite the whole Number of skillful Midwives . . . into a Corporation."[17] Midwives were to be instructed through lectures and discussions in the "most perfect Rules of Skill." Cellier developed her proposal in some detail. The number admitted to the college was not to exceed one thousand of the ablest midwives, each of whom would pay the sum of five pounds annually—funds which would be used to build a hospital or home for abandoned children. In this way Cellier sought to solve two pressing problems: improved instruction for midwives and shelter for foundlings. Like Chamberlen before her, Cellier may have hoped for personal profit from this scheme (she proposed that she be paid a salary to administer the project), yet this seems less important than the fact that her plan—and others like it—failed. In the days crucial to the founding of modern institutions of medicine, midwives were not allowed to found a self-regulating body. When professional instruction for midwives was finally implemented in eighteenth-century lying-in hospitals, these programs were initiated by (male) physicians who trained midwives to become only competent medical assistants.

Throughout the eighteenth century midwives continued to try to secure their ancient privileges, citing every authority from the Bible to the new ethnology to show that among all peoples it was women who assisted in birthing. Increasingly midwives argued for woman's natural propriety in this sphere.[18] The English midwife Elizabeth Nihell, who wrote an influential polemic against male practice in the 1760s, argued that her art was a gift innate to women. The fashionable preference for the man-midwife she held to be a fatal inversion of the "natural order of things." Attempting to profit from the new theory of sexual complementarity, Nihell argued that there are certain employments and vocations more proper for one sex than for another: a woman who sets up an academy for fencing and riding aims at something above her sex, but a man sinks beneath his sex when he interferes in the female province. "It is not with quite so good a grace as a woman that he would spin, make beds, pickle and preserve, or officiate as a midwife." This division of labor, Nihell argued, was natural and thus did not "impeach" the superiority of men.[19] Nihell's argument that the practice of midwifery lay beneath the dignity of men went unanswered. By the end of the eighteenth century the midwife had lost her traditional monopoly on women's health care and had become a mere auxiliary to the art she had once dominated.

The decline of the midwife had consequences beyond the loss of jobs for women. For Elizabeth Nihell, the intrusion of men fundamentally changed the practice of midwifery. The woman midwife had traditionally assisted all women—rich and poor—and not simply those best able to pay.[20] Men entered midwifery, she claimed, only because it was lucrative. Nihell reported that she had delivered over nine hundred patients free of charge, even though she had trained at the best school in Europe, the Hôtel Dieu in Paris, for two years. She challenged her male counterparts to say the same.[21]

Some of Nihell's accusations proved true. Although the midwife was soon supplanted by the obstetrician within the upper and middling strata, no one challenged the midwife's right to treat the poor. After 1700 state pensions in France for deliveries were increasingly given the man-midwife, yet magistrates did not withdraw pensions for midwives to aid and assist the poor.

Elizabeth Nihell charged not only that men practiced midwifery for their own gain, but that their incursion into the field changed the art in other significant ways. Nihell found women midwives gentler, more patient and sympathetic to the suffering of one of their own sex, for most had themselves borne a child. The man-midwife, Nihell charged, is rash and tends to be overly eager to utilize his instruments—the forceps. Nihell thought the new instruments were irrelevant to the claim that the man-midwife saved lives. Men go to work with their instruments, she claimed, where the skill and management of a good midwife would have probably prevented the difficulty in the first place. Nihell claimed that the primary function of this new technology was to exclude women from midwifery. First, she claimed, men forge the phantom of incapacity in women, and next insist on the necessity of "murderous" instruments.[22]

Nihell was not moderate in her criticism; many of the best doctors had already expressed their opposition to indiscriminate use of forceps. Though these instruments often terrified a young mother, improved forceps often allowed surgeons to assist difficult deliveries without mutilating or dismembering the child. Nonetheless, surgeons guarded the secret of their instruments from midwives (since the thirteenth century, the right to use surgical instruments belonged exclusively to surgeons) and the introduction of forceps coincided with the decline of the midwife in Europe.[23]

Gunnar Heinsohn and Otto Steiger have recently argued that the knowledge of fertility control was suppressed along with the midwife.[24] In the ancient world, knowledge of contraception was part of established medicine. As late as 1600 two hundred contraceptive and abortion methods, both medicinal and mechanical in nature, were commonly used.[25] There is also evidence that coitus interruptus, abortion, and infanticide regulated population in the seventeenth century. Elizabeth Cellier claimed that in the space of twenty years (between 1660 and 1680) thirteen thousand children had been aborted.[26] According to Heinsohn and Steiger, modern medicine did not achieve a knowledge of fertility control comparable to that practiced by women in early modern Europe until the last third of the nineteenth century.

Heinsohn and Steiger claim that the midwife was not put out of her profession by surgeons and physicians alone. Though the immediate conflict was a professional one, the need to supplant the midwife emerged as part and parcel of state policies to increase population. Medieval families were small (some 2.44 children per household in central Italy) not because of high mortality rates but because peasants limited offspring to the number their landholdings would support. After the Black Death and the catastrophic decline in population in the fourteenth century, however, state and church officials turned their attention to fostering population growth. Midwives became prime targets of these pronatalist, mercantilistic policies, for without a ready knowledge of fertility control women would conceive and raise more children. The demographic transition that is so key to recent centuries, Heinsohn and Steiger argue, may be traced not to decreased mortality rates but to the increased number of births that prevailed despite relatively poor economic conditions. It was in this larger context, Heinsohn and Steiger claim, that the role of the midwife changed. Where she had earlier been both a doctor and a teacher of other women, she now became an agent of church and state whose license did not guarantee professional competence but moral character. As agents of the church, midwives baptized infants near death; as agents of the state, they certified virginity and, more important, registered illegitimate births, making sure that mothers did not kill or conceal their offspring. With the demise of the midwife and rise of the male expert, women lost control over their own bodies.[27] Women without midwives had more children and understood less about what was coming to be known as gynecology.

Cookbooks for the Health and Pleasure of Mankind

A similar division of labor between the expert and the everyday arose in the field of medical cookery. From ancient times until well into the eighteenth century the art of cooking was an essential part of medicine.[28] It was a common saying that "cooking is the best doctor, and he who is well fed, does not need a doctor or apothecary."[29] Medical cookery was not a field like midwifery where women dominated, but it was a field to which they made substantial contributions. In an age when chemists prescribed sometimes harmful or expensive medicines and physicians did not treat the poor, books of medical cookery fulfilled a distinct need. Written for private families and "publick-spirited gentlewomen," the cookbook rendered women and the poor self-sufficient in medical care.[30] It provided the lady of the manor, who served as unofficial doctor of the village, with a ready knowledge of medicines and salves for common diseases.

Medical cookery was the poor cousin of a number of medical fields—especially botany and chemistry.[31] In seventeenth-century universities, professors of medicine were also professors of botany—many taught anatomy during the winter months, when the cold would preserve bodies for dissection, and botany during the warm, summer months. As late as 1771 Albrecht von Haller, in a bibliography of botanical works, classified cookery as part of botany, including in this category the works of Anna Weckerin, Maria Sophia Conring, and the duchess of Jägerndorf, along with Maria Merian's natural history of plants and Elizabeth Blackwell's herbal.[32] The good cook was also considered an empirical chemist.[33] In his *Culinary Chemist* of 1821, the chemist Frederick Accum held to the notion (by then outdated) that the art of preparing good food should remain a branch of chemistry. The kitchen, as he saw it, is a "chemical laboratory," the broilers, stewpans, and cradle spit of the cook rivaling the digestors, evaporating basin, and crucibles of the chemist.[34] Baking, broiling, conserving, and pickling, according to Accum, are all founded in the principles of chemistry.

The association of chemistry with the kitchen gave women a certain confidence to publish in the field. In 1666 Marie Meurdrac published her *La Chymie charitable et facile, en faveur des dames*—a work widely hailed as the first treatise on chemistry written by a woman.[35] Within the context of medical cookery, however, Meurdrac's work is not exceptional. Meurdrac understood *chemistry* to mean the distillation and mixing of substances. She divided her book into six parts—an explanation of chemical equipment (vessels, furnaces, fires, and weights), an outline of the properties of the basic elements (salt, sulfur, and mercury), the chemical use of metals and minerals, medical remedies, and cosmetic recipes. Though apprehensive about publishing a book inappropriately displaying her knowledge (women, she said, should remain silent), she decided that because "minds have no sex" her work would not be inferior to a man's. Meurdrac's is one of a number of books to include cosmetic recipes—rare secrets, as she called them, for the ladies. Her recipe for "water of the queen of Hungary" she claimed to be a true copy of the one written by the queen shortly before her death. The water, Meurdrac said, gave the queen a youthful air even in her seventy-second year.[36]

As a form of popular medicine cookery (or *kitchen physick* as it was sometimes called) emphasized preventing disease, rather than curing a disease already in progress.[37] A cure from the kitchen was much preferred to one from the pharmacy. In his *Kitchin-Physick* Thomas Cocke prescribed directions for preventing sickness and curing diseases by diet and "such things as are daily sold in the Market."[38] In 1769 the physician William Buchan wrote a *Domestic Medicine* aimed at rendering the medical arts more generally useful by showing people what is in their own power both with respect to the prevention and cure of diseases.[39] Buchan gave special attention to diet, drink, air, and other elements of a healthy regimen.

Because cookbooks were written for a broad audience, it is often said that they were merely practical texts without theoretical underpinnings.[40] These underpinnings, however, were assumed. Belief in the healing power of food was most commonly based in Galenic medicine, according to which the human body is composed of four humors—blood (hot and moist, warms and nourishes the fleshy parts of the body), phlegm (cold and moist, nourishes the brain and tempers the heat of the blood), yellow bile (hot and dry, provokes expulsion), and black bile (cold and dry, stirs up appetite and nourishes the spleen). A healthy life required a balance among these elements. Should heat and dryness gain the upper hand, balance should be restored through a cool and moist therapy.[41] Health resulted from the balance between the individual (the microcosm) and his or her environment (the macrocosm).

Herbals written by academic physicians provided the philosophical framework for cooking. Foods, including herbs and spices, were thought to have certain "virtues" that might restore the balance between the organism and its environment. For a hot disease one might take a peach, which is both cooling and nourishing.[42] John Gerard's influential sixteenth-century herbal was typical in describing the virtue and temperature of an herb or food. Saffron, for example, was considered hot in the second degree and dry in the first (according to Galen). Among its "virtues" (in moderate use) was the ability to quicken and enliven the senses and strengthen the heart. An immoderate use brought on headaches and harmed the brain. For infections Gerard prescribed the use of saffron in a mixture of walnuts and figs.[43] In books revealing the "secrets" of the arts and sciences food was catalogued as either simple or medicinal; the first nourishing and restoring, the second nourishing but, at the same time, altering the disposition of the body.[44] As one author wrote, "I count bread and everything we eat and drink Physick."[45]

Temperament was also affected by diet. Certain foods were considered appropriate for certain conditions and professions. As a cookbook from Leipzig pointed out, the spicy food appropriate for a newly married couple does not belong in a cloister for monks and nuns.[46]

Allegiance to the theory of humors also lay behind the claim (found in several cookbooks) that the first inhabitants of the earth—including Adam and Eve—were vegetarians, and for this reason were healthy because plants are easily digested and produce temperate humors.[47] According to this story, it was only in 2000 B.C. that humans changed from an exclusively vegetable to an animal diet—a diet which produced an overabundance of melancholy. *Adam's Luxury, and Eve's Cookery,* the vegetarian cookbook published in 1747, recommended vegetables for their medical virtues. Artichokes and asparagus were good diuretics, lettuce cooled the body and induced sleep, and spinach stopped coughing.[48]

Though cookbooks followed Galenic teaching, they also served to record practical experience. Anna Weckerin, widow of the *Stadtarzt* (town physician) of Colmar, published her cookbook from notes kept while accompanying her husband on house calls. Weckerin's husband, who emphasized that food and drink were an important part of any cure, found his wife particularly adept at nourishing the ill. He encouraged her to visit the sick along with him and to keep a notebook of her observations, which she published so that her knowledge would not go to the grave with her but would be readily available to others.[49]

Cookbooks were written by a diverse group, including master cooks, doctor's wives, noblewomen, and members of scientific academies. In one sample of such books published in England between 1701 and 1800, eighteen were authored by women, six by men, and eleven anonymously.[50] And even though a proportionately greater number of cookbooks were written by men in the seventeenth century, the cookery tradition was closely associated with the housewife. Women's knowledge—long held secret—was to be revealed, as many titles suggested: Thomas Dawson's *The Widdowes Treasure: Plentifully furnished with sundry precious and approved secrets in Physicke and Chirurgery, for the health and pleasure of Mankinde* (1595); *The Ladies Cabinet opened where is found hidden several Experiments in Preserving and Conserving, Physicke and Surgery, Cookery and Huswifery* (1639); or *The Treasury of Hidden Secrets Commonly called, the Good-housewives Closets of provisions, for the health of her Household* (1659). These books made public women's oral traditions. If many were written by men, this may have been for the simple reason that many women could not write.

Cookery was to change dramatically in the course of the eighteenth century. Women were gradually proscribed from medical cookery as it changed in law from a trade to a profession (pharmacy). By the 1750s medical recipes had been dropped from cookbooks. In sharp contrast to the seventeenth-century cook, the cook of the late eighteenth century claimed not to "meddle in the physical ways."[51] Elizabeth Raffald swore that among her eight hundred recipes, none were offered as medicine: "neither have I meddled with physical receipts, leaving them to the physicians of superior judgement, whose proper province they are."[52] As medicine became increasingly distinct from cooking, women were discouraged from ministering to the sick in the absence of physicians. The medical doctor James M. Adair penned an attack on "Lady Doctors" announcing "it is time for the ladies to retire."[53]

By the 1770s one can see the outlines of modern professions and sexual divisions in health-related fields. On the one side, medicine, pharmacy, and botany became distinct sciences developed by men. Medicine lost its connection with basic nutrition (and Galenic humors); medicines were less often homemade and more often bought at a pharmacy. On the other side, domestic cooking and nutrition devolved into the nonmedical duties of the wife and mother.[54] At the same time, the professional preparation of food remained the preserve of the male chef. (Courts had always had male chefs to cook for royalty.)[55] After the 1770s domestic cookbooks specialized in dinner recipes or household management. Madam Johnson's *Every Young Woman's Companion in Useful and Universal Knowledge,* for example, contained no medical remedies; it focused instead on spelling, reading and writing, pickling and pastry, rules for carving, how to estimate family expenses (for families in the middle classes), and rules for maids.[56]

Legitimizing Exclusion

In the case of both midwifery and medical cookery, history was called upon to legitimate the newly expanded role of men in those fields. When Britain's leading man-midwife, William Smellie, tried to claim midwifery for men, one part of his strategy was to focus only on the achievements of men, ignoring entirely the fact that women had ever dominated the field.[57] Much the same happened with cookery. Though the inventions of both medicine and cooking had traditionally been attributed to women, in the seventeenth century there emerged a myth of male origin. Haller claimed that Apollo had invented both medicine and cookery; others said it was Charon the centaur; still others claimed that God himself had passed along this knowledge to Adam.[58] Women cookbook authors, in contrast, claimed a female heritage, tracing an exclusively female line back to the goddesses Ceres, Diana, and Pomona.[59] In the frontispiece to a seventeenth-century German cookbook (see Figure 13), Ceres is honored as the inventor of cooking. Christian

Figure 13. Frontispiece to a cookbook published in Nuremberg in 1691, depicting Ceres presiding as goddess over the kitchen. From her lap issues the "most useful food," bread. Diana lays wild game at Ceres' feet; Pomona offers her fruits, Neptune his fish; from the forest, Pan carries firewood. Yet, as the frontispiece teaches, these treasures are brought in vain if not combined with the art of "the experienced female cook" (shown with her pot). From *Der aus dem Parnasso ehmals entlaufenen vortrefflichen Köchin welche bey denen Göttinnen Ceres, Diana und Pomona viel Jahre gedienet* (Nuremberg, 1691). By permission of the Herzog-August Bibliothek, Wolfenbüttel.

women favored female forebears like Sara. (Some traced Eve's fall to the fact that she ate the apple uncooked.)[60] In her history of cookery, Eliza Smith denied that Esau was the first cook, for surely he learned how to cook from his mother Rebecca.

The ascendancy of the male expert had consequences far more serious than symbolic disputes over priority. The replacement of women midwives by male gynecologists changed the development of gynecological practices. Women lost control not only over their own health care, but over definitions of their own minds and bodies as well. Early in the seventeenth century physicians blocked midwives' attempts to develop their own professional corporations—bodies which would have provided an institutional base for the training of new members and centers for developing new techniques. At the same time doctors refused to admit midwives to their colleges, thereby guarding their monopoly on medical science. Because midwives were excluded from centers of learning, they did not have an opportunity to participate in the new developments in their field and consequently were portrayed as ignorant.[61]

But there was one further step in this political struggle between medical men on the one hand and midwives and medical cooks on the other. After the removal of women from the field of health care, this exclusion was made to seem natural. Physicians would—in the absence of women—develop a picture of female nature that suggested that women were inherently unscientific. As described in Chapter 7, the same period that saw the decline of the midwife also saw the first illustrations of the female (and distinctively feminine) skeleton as the medical community renewed its interest in exact descriptions of sexual difference.

5

Battles over Scholarly Style

If men can claim an Apollo as the author
of the sciences, women can claim a Minerva.

—Jacques Du Bosc, 1658

The frontispiece to Diderot and d'Alembert's *Encyclopédie* presents an elaborate allegory of feminine hegemony in science (see Figure 14).[1] Truth and Reason, personified as women, reign over the sciences—all equally regal, equally feminine. Although surprising to modern eyes, these grand figures are not exceptions; they wade in the mainstream of Western traditions. Indeed, woman is the *dominant* image of science throughout the seventeenth century and deep into the eighteenth.

The fact that science is represented as a woman is interesting in and of itself. Assessing what it means, however, is a difficult task, one that turns our attention from the struggles of real women to the more elusive subject of cultural meanings of masculinity and femininity. In studying conceptions of femininity we should distinguish carefully three elements: how gender is defined; how sex is understood; and how real men and women participated in science. Masculinity and femininity are not inherent characteristics that have a universal meaning above and beyond a historical context. These terms can mean very different things at different times and in different places, and they often refer as much to the manners of a particular class or people as to the characteristics of a particular sex.

Notions of gender have shaped in important ways the history of modern science. Consciously or unconsciously, the majority of men

and women imagined science as a woman during the early years of the scientific revolution. Only for a brief time did Baconians explicitly oppose this image and argue that science should properly be masculine. *Masculinity* served in this case as a term of approbation and referred only tangentially to men. Though the feminine image of science remained strong long into the eighteenth century, the feminine eventually came to represent a style of scholarship, a set of values, and a way of knowing to be excluded from the new scientific order.

Apart from its usual sense, *scientific style* can also mean the images used to project the demeanor and rules of conduct within scientific circles. I explore in this chapter the influence of gender in scientific culture and scholarly style—in particular, how struggles around issues of gender helped to forge the ethos of modern science. I shall highlight two moments in this struggle: the rise and fall of a feminine image of science and the battle over intellectual style played out in the salons of Paris.

When Science Was a Woman

In the age of Enlightenment, iconology—"discourse in images"—was supposed to serve as a universal language.[2] Yet when it came to

---

Figure 14. "Academy of Sciences, Arts, and Trades," the frontispiece to Diderot and d'Alembert's *Encyclopédie.* Under a temple of Ionic architecture, Truth is wrapped in a veil, bathed in a ray of light breaking through the clouds. To her right, Reason, wearing a crown, lifts the veil from Truth, while Philosophy snatches it away. At Truth's feet, Theology, on her knees, receives light from on high. Behind Philosophy, one finds Memory, and Ancient and Modern History. Immediately below these are Geometry, holding a scroll on which is drawn the Pythagorean theorem; Physics, with her right hand on an air pump; and Astronomy, with her crown of stars. Below them and to the right are Optics, with a microscope and a mirror; Botany, holding a cactus; Chemistry, with a retort and furnace; and Agriculture, in the bottom right-hand corner. At the upper left, to the left of Truth, we see Imagination preparing to adorn and crown Truth. Beneath Imagination are different genres of Poetry, along with Painting, Sculpture, and Music. Below the clouds are the arts and professions, which emanate from the sciences and most of which are pictured as men. By permission of Department of Special Collections and University Archives, Stanford University Libraries.

science there were at least two distinct allegories vying for attention. Baconians championed a masculine symbol—virile, ready to act and command. Others championed a feminine symbol—a feminine icon—of science as woman, discreetly mediating between the demands of male scientists and the secrets of female nature.[3]

In early modern science, the struggle between feminine and masculine allegories was played out within fixed parameters. Whether science itself was to be considered masculine or feminine, there never was serious debate about the gender of nature or the gender of the scientist. From ancient to modern times, nature—the object of scientific study—has been conceived as unquestionably female.[4] At the same time, it is abundantly clear that practitioners of science—scientists themselves—have overwhelmingly been men.

But what about science? What gender was it—as an activity and set of ideals—to have? In one tradition the answer was clear: science was a woman. This tradition, stretching back at least to Boethius's sixth-century portrayal of philosophy as a woman, was codified and explained in Cesare Ripa's *Iconologia*—the Renaissance bible of iconography.[5] Ripa portrayed each of the sciences as a woman. Scientia—knowledge or skill—was portrayed as a woman of serious demeanor wearing stately robes (see Figure 15). Physica—physical science—was a goddess with a terrestrial globe at her feet. Geometry was a woman holding a plumb line and compass. Astrology, too, was a woman, dressed in blue and wearing a crown of stars and wings signifying the elevation of her thoughts to the distant stars. With a compass in her right hand and the celestial sphere in her left, she studied the movement and symmetry of the skies.

Personification—the process of endowing inanimate objects or abstract notions with human attributes—flourished in European art and literature from the thirteenth to sometime in the late eighteenth century.[6] The early modern mind filled the universe with life, giving human attributes to vices, virtues, the arts, faculties of the soul, and also to beasts, flowers, jewels, and seasons of the year. Scientific illustrators also utilized these images. In astronomical illustrations the heavens are populated with bears, fish, and all manner of other figures. Anatomical illustrations show skeletons lean-

Figure 15. Cesare Ripa's portrayal of science as a woman in his 1618 *Iconologia.* The mirror in her hand symbolizes the study of appearances leading to knowledge of essences; her triangle invokes both the perfect number and the three parts of a proposition resulting in a proof.

ing on a shovel or gazing into an hourglass, contemplating the brevity of life.

To the extent that abstract principles or virtues were personified, they were also given a specific gender. For Ripa and his followers, the world was highly gendered. Portrayed as feminine were: reason (often portrayed as Minerva, armed with a sword and taming a lion), theory, peace, liberty, the rational soul, architecture, perspicacity, truth, wisdom, invention, economy, art, strength, logic, imagination, mechanics, statecraft, government, the academy, history, medicine, and metaphysics. Indeed, the vast majority of ab-

IL SAGGIATORE
Nel quale
Con bilancia eſquiſita e giuſta
ſi ponderano le coſe contenute
nella
LIBRA ASTRONOMICA E FILOSOFICA
DI LOTARIO SARSI SIGENSANO
Scritto in forma di lettera
All'Ill.mo et Reuer.mo Mons.r D.
VIRGINIO CESARINI
Acc.o Linceo M.o di Camera di N.S.
Dal Sig.r
GALILEO GALILEI
Acc.o Linceo Nobile Fiorentino
Filoſofo e Matematico Primario
del
Ser.mo Gran Duca di Toſcana.
IN ROMA MDCXXIII
Appreſſo Giacomo Maſcardi

Figure 16. Frontispiece to Galileo's *Il Saggiatore* (Rome, 1623), bordered by two statues. A feminine Natural Philosophy (left), radiant with the light of truth, holds in her left hand the celestial sphere, representing the sphere of perfect knowledge. In her right hand she holds a book—perhaps the book of nature, more likely the book of natural philosophy. Mathematics (right), as queen of the sciences, wears a crown; in her right hand she holds an armillary sphere, in her left hand she wields a compass. By permission of Special Collections and University Archives, Stanford University Libraries.

stract virtues were feminine; masculine virtues included intelligence, instruction, and natural instinct.

These gendered images of the Renaissance (specifically the feminine image of science) remained strong long after the scientific revolution. In the 1790s Charles-Nicolas Cochin, designer of the *Encyclopédie* frontispiece, published his own great *Iconologie*. Though

Figure 17. Frontispiece to Johannes Hevelius's *Firmanentum Sobiescianum,* depicting the patrons of seventeenth-century astronomy. Urania, the muse of astronomy, is flanked by her male courtiers—Tycho Brahe, Hipparchus, Ptolemy, Copernicus, among others—the greatest astronomers of the past and present. Approaching from Danzig (pictured below the clouds), Hevelius rests his right hand on the Sobieski shield, the symbol of his earthly patron, while paying tribute to Urania, his heavenly patron. Nearby are reminders of Hevelius's achievements: his sextant, his *Catalogus Fixarum,* and his celestial globe. From Johannes Hevelius, *Firmamentum Sobiescianum* (Danzig, 1687). By permission of Special Collections and University Archives, Stanford University Libraries.

certain accoutrements changed with the times (eighteenth-century illustrators often equipped Lady Science with the latest barometer, vacuum jar, telescope, or even dress and coiffure), science was consistently portrayed as a woman. Cochin portrayed his Lady Science standing atop a volume of the *Encyclopédie*—to his mind, the compendium of all human knowledge.[7] From the sixteenth through the end of the eighteenth century iconographers never faltered in their conviction that science was a woman.

How were the images of the iconographers put to use? Feminine representations of science appeared most prominently on the frontispieces of scientific texts. Galileo invoked the feminine icon for the title page of his *Il Saggiatore* (see Figure 16). Natural Philosophy, pictured on the left, is radiant with the light of truth; Mathematics with her crown, on the right, is queen of the sciences.[8]

Feminine images of science did not always stand alone, as on Galileo's title page; they were also built into elaborate allegories portraying aspects of the organization of early modern science. In his remarkable *Firmamentum Sobiescianum,* the astronomer Johannes Hevelius illustrated his conception of seventeenth-century patronage—both mythical and real (see Figure 17).[9] The frontispiece shows the muse Urania as a Renaissance princess holding court in the heavens, where she is attended by her courtiers—Brahe, Ptolemy, Copernicus—the greatest practitioners of her art. Hevelius bows as he approaches Urania, his heavenly patron; in his right hand he bears the Sobieski shield, a reminder of the Polish king, his earthly patron.

The frontispiece to the London edition of Nicolas Lémery's *Course of Chymistry* portrays real distinctions of class alongside mythical distinctions of gender. Here a bare-bosomed Chemistry contemplates her chart of elements while revealing to Lémery her secrets (see Figure 18). The fact that her breasts are bared affirms the truth of the secrets she is about to reveal.[10] Lémery, author of the work, figures in the frontispiece only in a portrait held in the right hand of his muse; his (male) assistant works in the background. The men (by virtue of representing real men) are easily identified by class and rank. The genteel Lémery sports a curled wig and ruffled cuff; he also holds a copy of his book. A workman of modest attire is shown doing the real work of experimental chemistry in a small laboratory in the background. The female "Chemistry," however, abstract and in some sense otherworldly, has no obvious class origins to reveal.

Especially on the Continent, the classical feminine image of the sciences remained strong late into the eighteenth century.[11] One of the most elaborate examples of these images is found in the work of a woman scientist. Though Emilie du Châtelet published her *Institutions de physique* anonymously, she apparently placed herself among the images prominent on the frontispiece (see Figure 19).[12] A female figure ascends the temple of truth, but, oddly, she is empty-handed and thereby stands outside the normal conventions of allegorical discourse. Who is the figure? Perhaps she is the "Minerva of France," as Voltaire liked to call du Châtelet. Or perhaps she is du Châtelet herself, clad in the dressing gowns in which she did much of her writing. If du Châtelet is indeed present in the frontispiece, she resides among the goddesses, *not* among the (male) scientists whose portraits frame the scene.

Du Châtelet's frontispiece demonstrates an important difference between works by male and female scientists. There is no obvious portrait of the author herself on this frontispiece. Her presence is very different from that of Galileo—self-confident and bold—in the frontispiece to his 1613 work on sunspots. Though many at this time argued that feminine icons represented real women (that Ceres, for example, was a real woman of ancient times and actually did invent agriculture), over the years feminine imagery grew increasingly abstract and distant. Unlike the men scientists of the period, such as Vesalius or Galileo, women scientists (with the exception of Margaret Cavendish) rarely presented self-portraits on the frontispieces of their works.[13]

Reading Allegories

What is the significance of the feminine icon? It would be wrong to think there is something essentially "feminine" about Hevelius's astronomy carried out under the auspices of Urania, or du Châtelet's physics done under the tutelage of Truth rendered as a woman. Yet the masculine and feminine symbols in the scientific culture of this time are not without significance. They exist in a symbolic or-

Figure 18. Title page and frontispiece to the fourth English edition of Nicolas Lémery's *A Course of Chymistry* (London, 1720). By permission of the Health Sciences Library, University of North Carolina at Chapel Hill.

A

COURSE

OF

CHYMISTRY:

CONTAINING

An eaſie Method of Preparing thoſe Chymical Medicines

Which are uſed in

PHYSICK.

WITH

Curious Remarks upon each Preparation, for the Benefit of ſuch as deſire to be inſtructed in the Knowledge of this ART.

By NICHOLAS LEMERY, M. D.
And Fellow of the Royal Academy of Sciences.

The Fourth Edition.

Tranſlated from the *Eleventh* Edition in the *French*, which has been reviſed, corrected, and much enlarged beyond any of the former, by the Author.

LONDON:

Printed for *A. Bell*, at the *Croſs-Keys* and *Bible* in *Cornhill*; *D. Midwinter*, at the *Three Crowns* in St. *Paul's Church-Yard*; *W. Taylor*, at the *Ship* in *Pater-Noſter-Row*; and *John Osborn*, at the *Oxford-Arms* in *Lombard-Street*, 1720.

Figure 19. Frontispiece to Emilie du Châtelet's *Institutions de physique* (Paris, 1740) showing the various sciences—Botany, Astronomy, Physics, Medicine, and Chemistry—seated on the ground. Mounting to the temple of naked Truth, through the clouds, is a female figure—perhaps the author herself. The portraits

INSTITUTIONS

DE

PHYSIQUE.

À PARIS,

Chez PRAULT fils, Quai de Conty, vis-à-vis la deſcente du Pont-Neuf, à la Charité.

M. DCC. XL.

Avec Approbation & Privilége du Roi.

of leading (male) scientists (most probably Descartes, Newton, and Copernicus) frame the scene. By permission of Special Collections and University Archives, Stanford University Libraries.

der where meaning goes well beyond the literal sense of the image. As Cochin suggests in his *Iconologie,* "it is under the veil of allegory that morality presents to men consoling truths and useful precepts."[14]

What, then, did these images mean to men and women of early modern Europe? One easy (perhaps too easy) explanation might be found in language. As Ripa pointed out in 1602, the image of a feminine science, nature, or truth simply suits "the way we speak."[15] In Latin, Italian, French, and German, abstract nouns generally carry feminine gender. Artists therefore may have simply reproduced in art the gender divisions in language. Voltaire—rather literal-minded in these matters—explicitly stated that gender in the language determined gender in the allegory.[16] Today one might even conjecture that the disfavor met by the feminine icon in England might be traced to the lack of gender in English nouns.

There is indeed a high correlation between gender as it exists in language and gender as it was captured in iconography. In a few cases, however, the match is not exact. In Jean Baudoin's *Iconologie,* for example, *valeur*—which is feminine in the French—is imaged as a man. Similarly *feu,* which is masculine in French and Latin, is depicted as a woman. In Cochin's *Iconologie,* too, a few examples of gender crossing can be found. *Toucher,* the sense of touch, is masculine in the written form but feminine in the language of icons. Then too, in some cases, related virtues are depicted as alternatively masculine or feminine. In Baudoin's work, *divine* knowledge is a woman, while *human* knowledge is a man. One should also note that Germans, who participated enthusiastically in traditional iconography, had a language often at odds with traditionally gendered images. The word *peace,* for example, which is feminine in both Latin and French and almost always personified as a woman, is masculine in German (*der Friede*).[17]

These, I should caution, are exceptions: by and large, language and images agree. Even so, the fact that gender is deeply embedded in most European languages merely pushes back one step the need for explanation: why, after all, is *scientia,* or *science,* or *Wissenschaft* feminine in the first place? The historical origins of grammatical gender remain for the most part unexplained.[18] The significance of these images runs deeper than the accidents of language. The fact that language dictates a feminine Scientia does not mean that a feminine image must be used. One might even imagine that, in an age of revolution, designers of scientific texts could have dispensed with these traditional images and not given them a prominent place on frontispieces.

Iconographers recorded with great richness the meanings of these images "consecrated through usage."[19] Baudoin, for example, decoded the meaning of the feminine icon—giving precise meaning to the wings on her head, the mirror in her hand, and her triangle. What the iconographers did not decode, however, was the gender of science.

To understand why science was a woman we must look beyond the iconographers. The most fruitful intellectual context for understanding the feminine icon is Christian neo-Platonism. In one sense, the feminine icon should be understood in its relationship to the male scientist: these frontispieces were drawn by men; the conceptual schemes in which they dwell were framed by men. It has been suggested to me more than once that these images are simply erotic in a Freudian sense, that in the celibate world of early modern science a nude Scientia titillated the male scientist. I would argue, however, that Scientia is not an icon of free-floating eros but part of the neo-Platonic worldview.

Neo-Platonists (along with alchemists and cabalists) held that creativity—both intellectual and material—resulted from a union of masculine and feminine principles. The neo-Platonists described creation as the union of opposing male and female elements and made the joining of those elements the basis of all creativity. Henry More, a neo-Platonist of the seventeenth century, stressed that man without woman is incomplete: the masculine principle, though capable of existence without the feminine, would always remain imperfect. For More, the masculine nature of man—his "ethereal, bold intellect"—needs to be balanced by feminine nature, what More described as the "kindely joy of the body."[20]

More saw the union of the sexes as a solution to the newly pervasive mind/body distinction. Thus in his scheme (more in line with Aristotle) the male represents intellect while the female represents body. For Renaissance neo-Platonists, as for Plato himself, however, it is the soul (or the indwelling rational principle) that is female. According to Plato's story of creation, the soul is the universal "ruler and mistress" of the world, permeating and enveloping the heavens, enduring throughout all time. For Plato the rational soul of the universe is feminine and gives form to the whole.[21]

For Renaissance neo-Platonists, each man's individual soul is feminine. This soul yearns to unite with the King of Glory, God the Father. According to Giovanni Pico della Mirandola, "she purifies herself, and dressing in the golden vesture of the many sciences as in a nuptial gown, receives him, not as a guest merely, but as a spouse."[22] For Renaissance neo-Platonists marriage was the basic metaphor for unity in the individual and the cosmos; order in the universe could not be preserved without the cooperation of male and female elements.[23] Pico conceived of the soul as the bride, the life force of the physical body, whether of man or of the world. The marriage of the soul to God brought harmony to the universe; the marriage of the philosopher to *Scientia* brought knowledge.

Christian elements also encouraged the perpetuation of a strong visual image of science. The church pictured the world as a place where vices and virtues struggled for the soul of the true believer. These abstractions were given a dramatic persona—often female—as in the appearance of Lady Philosophy to Boethius. In his *De consolatione philosophiae,* the sixth-century philosopher Boethius describes how Philosophy appeared to him in a dream while he was in prison. In his dream Boethius is torn between Philosophy—the love of wisdom—and Fortune, or worldly concerns. These two women—both figures in his mind—vie for his attentions. Nourishing him with her virtue, Philosophy, his soul's physician, finally wins; Boethius is consoled.[24] Boethius's image of philosophy as a woman is the predominant image of philosophy for more than ten centuries.

Scientia, then, is feminine in early modern culture because it is feminine in the language but also because the scientists—the framers of the scheme—are male: the feminine *Scientia* plays opposite the male scientist. In order to unite in creative union with the female, the male scientist images his science as his opposite. But more than that, the scientist imagines that a feminine science leads him to the secrets of nature or the rational soul. This is particularly evident in the sixteenth-century woodcut from Gregor Reisch's *Mar-*

garita philosophica showing Ptolemy guided by the muse of astronomy (see Figure 20).[25]

It is this general scheme I see played out on the frontispieces discussed above. If we look again at the frontispiece to Hevelius's or Lémery's work, we see Urania or Chemistry serving as guides to truth (see again Figures 17 and 18). In feminine allegories it is consistently the muse who conveys to the scientist the truths of nature. Female goddesses of astronomy or natural philosophy mediate between (female) nature and the (male) scientist. To seventeenth-century minds, muses, the daughters of Jupiter, embodied the "soules of the Sphears"; through the influence of the muse, the minds of mortal men are inspired with "sundry and divers delecta-

Figure 20. Gregor Reisch's *Margarita philosophica* (Basel, 1517), showing Ptolemy guided by the muse of astronomy. By permission of the Folger Shakespeare Library.

tions" of "gentler Sciences."[26] The scientist sees himself not as a free agent but as a supplicant. His access to the secrets of nature depends on the grace of the muse. As the eighteenth-century iconographer Cochin reports, Pythagoras sacrificed to the muses for having revealed to him his great theorem.[27] It was in this sense that Hevelius paid homage to his muse, Urania.

This scheme becomes particularly interesting when the scientist is a woman, not a man. Women scientists often identified themselves with the muses. The seventeenth-century astronomer Maria Cunitz juxtaposed the name of the muse *Urania* with her own name *Maria* (also the name of the Virgin) on the title page of her *Urania propitia*—each is highlighted in red. *Propitia* suggests that Cunitz, the woman scientist, is graciously received by the muse, who then reveals to Cunitz the movement of the planets.[28]

The notion of a feminine guide to truth emerged from ancient cosmologies, such as Platonism, where the purpose of natural philosophy was to unite all souls with that one mind which is above all minds. The feminine icon consistently personified ancient science and ancient conceptions of science. It is significant that soon after Charles Cochin drew his Lady Science, she disappeared. Published in the 1790s, Cochin's iconology departed radically from earlier iconographies by defining science as "knowledge acquired through study and founded on evidence." Lady Science no longer looks to the heavens but to the earth; her wings have been replaced by the owl of Minerva (which sits beside her) to show that "science can only be acquired through study."[29] Though Cochin dropped ancient conceptions of knowledge, he tried (in vain) to retain the feminine image, but that too was soon to disappear.

The Masculine Allegory

Not all frontispieces in works of import to modern science invoked images of a female Scientia. The feminine icon had a masculine rival. From its inception Baconian science was intended to be "masculine" science. In his ode "To the Royal Society," royalist Abraham Cowley wrote:

Philosophy, I say, and call it, He,
For whatsoe're the Painters Fancy be,
It a Male Virtu seems to me.[30]

Cowley's thoughts on philosophy were consistent with his views on how gender divided the world. About poetry he wrote, "there is a kind of variety of Sexes in Poetry, as well as in Mankind: that as the peculiar excellence of the Feminine Kind is smoothness and beauty; so strength is the chief praise of the Masculine."[31]

Cowley's call for "a masculine philosophy" echoed that of Francis Bacon in one of his early essays, "The Masculine Birth of Time."[32] Much has been made of this essay in recent years. I should note, however, that the call for an explicitly masculine science was short-lived and that what Baconians meant by the term *masculine* requires explanation. What *is* clear is that gender in this case did not map directly onto sex: Bacon's call for a masculine philosophy was at most a tangential attack upon women. For Bacon, calling something "masculine" was to praise it, while calling it "feminine"—or, worse, "effeminate"—was an insult. As Benjamin Farrington has shown, Bacon attacked ancient (and especially Aristotelian) philosophy as a female offspring—passive, weak, expectant. Bacon launched his shafts into Aristotle for his logic, Plato for his contemplation, and Galen for his skepticism. By contrast, Bacon's masculine philosophy was to be active, virile, and generative—an experimental science drawn from "the light of nature, not from the darkness of antiquity."[33]

Ironically, though Bacon attacked the ancients with some vigor, he chose as his weapon ancient notions of gender. In calling for a masculine philosophy, Bacon invoked well-worn Aristotelian categories by which masculinity signified hot and active spirit while femininity signified cold and sluggish matter. Rejecting a passive, speculative, and effeminate philosophy, Bacon called for an active philosophy, one which would act as a formative principle upon a feminine nature.

Masculinity in this period, however, had even broader connotations. Bacon's plea for a masculine philosophy was also an attempt to set English science apart from prevailing intellectual currents on the Continent. In his *History of the Royal Society of London,* Thomas

Sprat contrasted the "masculine arts" of England with the "feminine arts" of the Continent. To English eyes, French intellectual culture appeared effeminate, especially in light of the highly visible role played by French noblewomen in Parisian salons.

> As the *Feminine* Arts of *Pleasure,* and *Gallantry* have spread some of our Neighbouring Languages [e.g., French] to such a vast extent: so the *English Tongue* may also in time be more enlarg'd, by being the Instrument of conveying to the World, the *Masculine* Arts of *Knowledge*.[34]

From its beginning, the explicit goal of England's leading body of scientists—the Royal Society of London—was "to raise a Masculine Philosophy."[35] Masculine philosophy was to be distinctively English (not French), empirical (not speculative), and practical (not rhetorical). Consistent with the discourse of the day, each of these favored qualities was considered masculine.

If Bacon's call for a masculine philosophy was an attack of a modern against the ancients and an Englishman against the French, it was also an attack on the feminine icon. Advocates of the experimental method dispensed with "mistress Antiquity," with Minerva and the Muses as "barren virgins"; the natural philosopher was henceforth to take Vulcan as his mentor and engage in direct observation and experimentation.[36] The portrait in Andreas Vesalius's *De humani corporis fabrica,* for example, shows the author holding the dissected arm of a cadaver (see Figure 21).[37] Vesalius represented himself in this way to emphasize his radical break with past anatomical practices where the professor merely read from a text (often one of Galen's) while a demonstrator followed the description in dissecting the body. For our purposes, it is important to note that no muse stands between the philosopher and nature; instead, the philosopher himself cuts, dissects, and analyzes the object of his science. At the same time one should keep in mind that though the frontispiece as a whole presents a masculine allegory in line with Bacon's prescription, the allegory has no single symbol—there is no masculine icon. Here, as elsewhere, the man portrayed on this frontispiece represents a *real* person and not an abstract image of science, reason, or truth.

Interestingly, in England—where the Royal Society was intent

Figure 21. A portrait from Vesalius's *De humani corporis fabrica* (1543) shows the author at work. In the experimental tradition, the scientist cuts, analyzes and probes nature directly.

upon "raising a masculine philosophy"—the feminine icon did not flourish.[38] Isaac Newton's work, for example, did not utilize feminine images. His *Principia* (1687) has no frontispiece, his *Opticks* (1704) has virtually no decoration whatsoever. The English translation of his *Method of Fluxions* (1736) does sport a frontispiece (see Figure 22).[39] Here, the principle of velocity is illustrated with the example of men shooting on the hunt. The Greek motto below announces that the purpose of the work is to make general principles of nature easily understood. Accordingly, men in contemporary English garb demonstrate the principles of velocity by blasting birds from the sky. True to the spirit of Baconian science, this frontispiece celebrates the triumph of modern "action" over ancient "disputation." The ancients displaced by this new regime huddle in the corner; some record Newton's laws on a scroll, others argue amongst themselves.

Though feminine imagery did not flourish in England, the muses were not entirely absent from English science. In a great stroke of irony, the Royal Society—that virile champion of a masculine science—was at least on one occasion portrayed as a woman. In 1684 the *Saggi* (or transactions) of the Accademia del Cimento of Florence were translated for the Royal Society by Richard Waller of London (see Figure 23).[40] A beautifully feminine Royal Society sits expectantly on the edge of her chair, awaiting the *Saggi di Natura* handed her by the equally lovely Accademia del Cimento. A very feminine Divine Nature points to the volume, her nudity affirming that the truths of her divine body are contained within its pages. A rather grim and passive Aristotle looks on. The implication is that truth is to be found in experimental philosophy of the sort pursued in the young academies of science.

Not all scientists portrayed in the "active" masculinist tradition were men. The German entomologist Maria Sibylla Merian was shown at work on the frontispiece of the French translation of her *Histoire générale des insectes de Surinam,* published in the 1770s many years after her death.[41] This frontispiece provides a rare example of a heterodox iconographic tradition. In the foreground sits a serene muse of natural history (see Figure 24). The ideal of mothering emerging in this period (see Chapter 8), however, has trans-

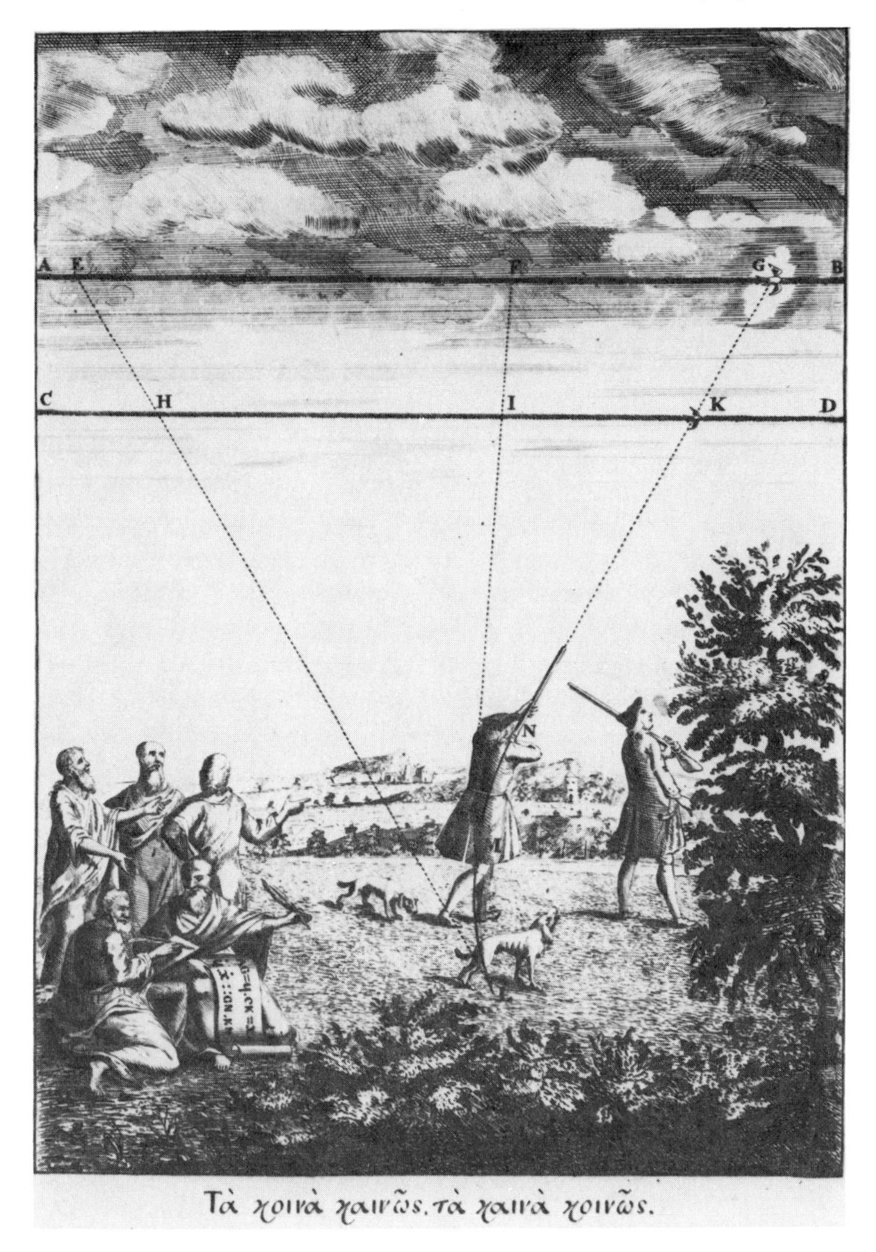

Figure 22. The frontispiece to Isaac Newton's *Method of Fluxions* (London, 1736) demonstrates the principle of velocity. By permission of Special Collections and University Archives, Stanford University Libraries.

Figure 23. The Royal Society of London portrayed as a woman. From *Essays of Natural Experiments, Made in the Academy del Cimento,* trans. Richard Waller (London, 1684). By permission of the Folger Shakespeare Library.

Figure 24. Maria Sibylla Merian at work in Surinam. The scene in the window shows Merian (portrayed in a good likeness) collecting insects in the tropics. From the frontispiece of her *Histoire générale des insectes de Surinam et de toute l'Europe* published in the 1770s, many years after her death. By permission of Dumbarton Oaks, Trustees for Harvard University.

formed the face of this muse. As much a mother as a muse, Natural History is surrounded by her *putti* children who—quarreling, collecting, and inspecting—gather butterflies. In the background, through a window, one sees a landscape from Surinam where Merian (portrayed in a good likeness) collects insects with her net, as she had done for many years. Dutch landowners, with whom she often quarreled, look on with some dismay; slaves carry supplies in the distance.

Did the Feminine Icon Represent Real Women?

I have argued that the feminine icon is best understood in the context of neo-Platonic thought. Interpreters from Christine de Pizan to Francis Bacon, however, thought of these images as remnants of a historic past. In their view, women had made significant contributions to the sciences and feminine icons recalled these distant, though real women. In her 1405 *Book of the City of Ladies,* for example, Christine de Pizan had celebrated women's contribution to "things mechanical." In the same way that men later praised Galileo for his invention of the telescope and military compass, de Pizan praised Minerva for her practical innovations—for the invention of numbers, for being the first to shear sheep of their wool, and for learning to extract oil from olives and press juice from fruit. According to de Pizan the Athenians, marveling at Minerva's great knowledge, made her their goddess of wisdom. Isis was a woman of such great learning that she, too, was made goddess by the Egyptians; Ceres was named a goddess for her invention of agriculture, the sowing of grain, and the making of bread.[42]

In the seventeenth and eighteenth centuries mythical muses or goddesses continued to serve as abstract representations of real historical women. In his *Generall History of Women* the Englishman Thomas Heywood noted that "the inventions of all good Arts and Disciplines have been ascribed to the Muses."[43] Heywood intended his description of these women to serve as examples for the women of his day. Just as men are incited to virtue by examples of great men, he wrote, what "properer object can there be of woman's emulation than the deeds of other famous women?" Or, as another Englishman asked, "if it were intended by Nature, that Man should

Monopolize all Learning to himself, why were the Muses Female, who . . . were the Mistresses of all the Sciences, and the Presidents of Music and Poetry?"[44] In Germany, too, Johann Zedler joined a recognition of women's ancient achievements with a call for their equal participation in intellectual culture. In his *Grosses vollständiges Universal-Lexicon aller Wissenschafften und Künste,* Zedler adorned his article on women with portraits of three muses. To the query whether women are capable of earning an academic degree, Zedler answered "why not?" and launched into a history of distinguished women in the sciences.[45]

The feminine image of science also had distinct social origins. If this image is about women of the seventeenth and eighteenth centuries, it is about elite women. Neo-Platonism flourished in both the Renaissance court and the seventeenth-century French salon, settings where elite women enjoyed prominence and prestige. The feminine icon was born and bred within elite culture and represented the role women played in that culture more, I would argue, that it did specific women of the past. Women's role in the court and salon was one of mediation. In Renaissance courts queens and duchesses served as patrons and ambassadors of culture. Women of the salons were active arbiters of public taste in the arts and sciences.[46] Neo-Platonism served these women well. First, it provided a philosophical justification for intellectual exchange between the sexes; neo-Platonists saw such exchanges as the fundamental source of human creativity and a way to insure balanced and tempered representations of the world. Second, it provided a model for exchange between the sexes which was something other than sexual. The Platonic union of male and female principles took place in the realm of spirit; Platonic love was spiritual and marriage was consummated in the mind. The feminine icon prospered within these aristocratic settings where courtiers and (male) scholars often of lesser rank were inclined to idealize a well-born and influential woman.

Whether or not the feminine icon represented real women, women identified with it and were identified with it. As illustrated in the frontispieces to the works of Maria Cunitz or Emilie du Châtelet, these women placed themselves (or were placed by the artist) among the muses and not among the historically real male scientists. These women scientists identified with an image of science which—by fusing real and mythical elements—linked them to a broader feminine heritage.

The Decline of Feminine Icons

At the same time that the feminine icon gave a luster to women's scientific heritage, an attack on that icon was often felt to be an attack on scientific women. Margaret Cavendish, for one, recognized that Bacon's call for a masculine science was an attempt to undermine women's already limited role in intellectual culture. In her 1666 *Observations upon Experimental Philosophy,* Cavendish wrote:

> For though the Muses, Graces and Sciences are all of the female gender, yet they were more esteemed in former ages, than they are now; nay, could it be done handsomely, they would turn them all from Females into Males; so great is grown the self-conseit of the Masculine, and the disregard of the Female Sex.[47]

The late eighteenth century saw the decline of the feminine icon in scientific culture. In his *Critique of Pure Reason* Kant banished Metaphysics—"Queen of all the Sciences . . . a matron outcast and forsaken, mourning like Hecuba"—from critical philosophy.[48] Kant coupled this attack on the feminine icon with an attack on learned women. The schooling of women, Kant believed, runs counter to natural law. Science has properly a "masculine mein"; he singled out two well-known French women for his invective:

> A woman who has a head full of Greek, like Madame Dacier, or one who engages in debate about the intricacies of mechanics, like the Marquise du Châtelet, might just as well have a beard; for that expresses in a more recognizable form the profundity for which she strives.[49]

Kant's attack on the feminine icon signaled a turning point in the historiography of science. The passing of the feminine icon also marked the passing of the classical appreciation of women's contributions to the sciences. Christian Harless in 1830 was the last to write a history of women in science that considered Isis, Minerva, and Hygeia important figures (real or mythical) in the history of science. Like de Pizan in the fifteenth century, Harless believed that myth was grounded in fact. Rejecting the romantic theories that myths were merely allegorical projections of human needs, Harless believed that ancient poets embroidered with poetic fantasy the accomplishments of persons distinguished for their strength, beauty, or understanding. Harless considered it "very possible" that Isis, Diana, and Hygeia were women who had been deified for their "outstanding talents, their knowledge of herbs, and their ability to treat illness."[50]

By the midnineteenth century, however, professional science found the ancient queens of the sciences nearly all dead and forgotten. Scientists no longer chose to represent their enterprise with a feminine face. Perhaps they thought femininity trivialized their pursuits. Perhaps the grace of the muse no longer fit the character of scientific work. But more than that, the social context of the feminine icon—the court and salon—had vanished along with aristocratic privilege. The French Revolution sounded the death knell of the privilege that aristocratic women had known in science. As science became more professional, ornamental images disappeared from its texts; the image of women became more exclusively tied to motherhood.

The rise of empirical science also contributed to the decline of the feminine icon. The new way of ideas—both the speculative tradition of Descartes and the empirical tradition of Locke—aroused a distrust of all rhetorical devices. Already in the 1690s Locke objected to the tendency of philosophers to personify the mind or soul.[51] In the new cosmologies of Bacon, Descartes, Locke, and Newton, the universe was to be explained in terms of regular and natural laws. Science ceased to have a persona—ideas were no longer to be clothed in allegory and emblems but were to be clear and distinct.

In this world, the feminine principle was excised from the imagined process of discovery. This can be seen in the contrast between the philosophical struggles of Boethius and Descartes. Of the figure of Philosophy in Boethius's *Consolation,* a twelfth-century writer remarked, "she appears as a woman because she softens the ferocities of souls."[52] In his dream Boethius struggles with Lady Philosophy until he finds resolution. Not so Descartes. Descartes—the modern philosopher—is profoundly alone. Personification is excised from his philosophy, as the feminine principle is banished from his world. His receptive mind is filled not with a comforting feminine presence but with the first possible unit of reason. Alone with his methodical doubt, Descartes can be certain only of one thing, *cogito ergo sum*.

Modern philosophy eventually did away with personification. Nonetheless the feminine icon evinced remarkable strength as a cultural symbol, remaining strong on frontispieces until its final gasp of glory in the *Encyclopédie*. Here, surprisingly, even the new sciences—botany, optics, and the new physics—were imaged as ancient goddesses. By the 1790s, however, the feminine icon had lost its former brilliance. Ripa's *Iconologia* itself, which had been printed and reprinted throughout Europe from 1593 to 1785, disappeared from sight. Ripa's treatise was attacked as a book of puzzles; allegory was pronounced dark, confused, multiplicitous, and enigmatic.[53] The feminine icon, then, went the way of all images in scientific literature. Over the course of the eighteenth century the illustrative tradition that had skeletons leaning on shovels or flourishing a cape disappeared as scientific illustration became increasingly abstract.

What happened to the image of science? Though feminine icons were to remain strong in the broader culture (as representatives of liberty, justice, and so on), they disappeared from frontispieces of scientific texts. By the beginning of the nineteenth century science seldom has a "face." The feminine icon is not replaced by a masculine icon, yet a new allegory begins to emerge. During the 1800s explicit images of science are replaced by implicit and popular images of the scientist as an efficient male, working in a modern laboratory, most often wearing a white lab coat (see Figure 25). Though this image is a literal one (being a photograph of a well-respected twentieth-century American scientist), it is also allegorical and points to a set of meanings that go beyond the literal image.[54] The scientist is now an isolated individual, profoundly alone. Not evident in the picture are the props and crews that keep this man at center stage—his colleagues, his technicians and graduate students, his secretaries and perhaps even his wife. Absent too are the patrons or politicians influencing his work. This self-sufficient individual is of serious demeanor, and he is active. The fact that he

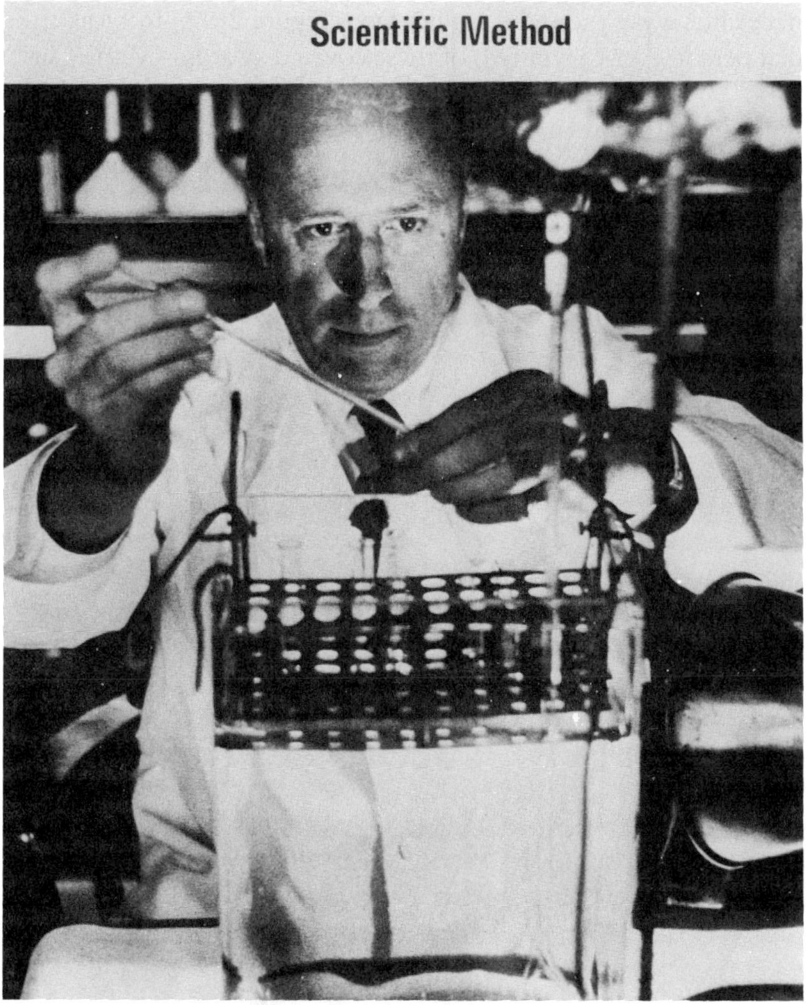

Scientific Method

Figure 25. Time-Life's volume on *The Scientist* portrays the "scientific method." From *The Scientist,* ed. Henry Margenau and David Bergamini (New York, 1964), p. 50.

is white and male is both descriptive and prescriptive; the image cultivates its own clientele.

And yet, some traditions never die. On the Nobel prize medals for chemistry and physics designed in 1902 we encounter familiar symbols and messages. On the front of the medal is a profile of the inventor and patron, Alfred Nobel. On the back we find a scene in which a female Natura holds a horn of plenty as Scientia (also female) lifts away the veil from her face (Figure 26).[55] Ironically, the first person ever to win two of these coveted awards—Marie Curie, in 1903 and 1911—was still unable to join on an equal basis the ranks of learned men. Woman could serve as the image of science, but women were not yet welcome in the fellowship of science.

Competing Scholarly Styles

The decline of the feminine icon paralleled broader changes in the style of science. Since the beginnings of modern science there have

Figure 26. The obverse of the Nobel medals for chemistry and physics, with both nature and science represented as women. Pictured here is Albert Einstein's medal, awarded in 1921. The inscription reads: "How good it is that man's life should be enriched by the arts he has invented." From Life Science Library, *The Scientist,* ed. Henry Margenau and David Bergamini. Photo by Phil Brodatz. Copyright 1971 Time-Life Books Inc. By permission of the Hebrew University of Jerusalem, Israel.

been a number of struggles over intellectual style. In the seventeenth century there was the struggle between the ancients and moderns over the desired character of scientific language: should language retain the allegorical richness of the ancients, or adopt the more flat-footed precision of the moderns? In the eighteenth century scientists tried to cleanse "nature, the earth, the human soul, and the sciences of all poetry."[56] The struggle, as Wolf Lepenies has described it, was one between science written in a literary or even poetic style versus science written in dry and technical terms, with many tables and few well-chosen words.[57]

One important axis in this larger struggle turned on the issue of gender. Many in the eighteenth century championed what they saw as a distinctively feminine style of scholarship. In many ways, the style denoted as feminine—a gallant politeness, a gently poetic grace—was simply the obverse and reverse of Bacon's virile and active masculine style. Women allegedly brought to scholarship "a more varied vocabulary, greater nobility in diction, and more facility in expression." These qualities, it was hoped, could undercut the reigning *pédantisme des classes* and foster a philosophy that was more pleasant and more humane.[58]

As in the case of the feminine icon, the idea that a distinct feminine style contributed something to scholarly discourse was linked to the question of women's proper place in intellectual life. With the decline of the universities in the seventeenth century, the question again arose whether women should participate in scholarship and science. Since ancient times, the very presence of women had been thought to disrupt serious intellectual endeavor. Ancient Hebrew traditions (at least in the interpretation given them by the *Encyclopédie*) held that by virtue of contact with women, men lost the power of prophecy. In Christian traditions of medieval Europe, monastic life—important to the life of the mind—was a celibate one. These traditions continued in universities. Professors at the universities of Oxford and Cambridge were not allowed to marry; until late into the nineteenth century celibacy was required of all faculty. In the eighteenth century, an English historian noted:

> By the learned and studious, it has often been objected to female company, that it so enervates and relaxes the mind, and gives it such a turn for trifling, levity, and dissipation, as renders it altogether unfit for that application which is necessary in order to

> become eminent in any of the sciences. In proof of this they allege, that the greatest philosophers seldom or never were men who enjoyed, or were fit for the company or conversation of women. Sir Isaac Newton hardly ever conversed with any of the sex . . . ([he] died a stranger to love). Bacon, Boyle, Des Cartes, and a variety of others, conspicuous for their learning and application, were but indifferent companions to the fair.[59]

The perceived dangers of women to the life of the mind were so great that a number of philosophers (among them Bacon, Locke, and Kant) never married. Francis Bacon clearly considered wife and children impediments to great enterprises; Pierre Bayle declared the marriage of a learned man a waste of national resources. Even Mary Wollstonecraft agreed that unmarried men and women without children proved the most productive scholars.[60]

Courtly culture and the salons with their distinctive style and free mix of men and women provided relief from the celibacy of intellectual life. When women were visible in intellectual life, it was perceived that the presence of women changed scholarship in important ways. This is what the nineteenth-century historian of philosophy Karl Joël meant when he wrote that the French Enlightenment was a time when "woman was philosophical and philosophy was womanly."[61] Joël did not, in fact, approve of this feminine influence. The excesses of the French Enlightenment represented, in his view, a mere interlude between the manly (*männliche*) philosophy of the English Enlightenment and the masculine epoch (*Manneszeitalter*) of German philosophy ushered in by Kant. For Joël, the severe and rigorous thought of Kant awakened at long last the masculine strength of German philosophy.

Many of those in favor of tempering masculine styles with a good dose of the feminine were also champions of the salon. Both Hume and Diderot (though not necessarily friends of women) considered women of the salons their best allies against the impotent philosophy of the scholastics—those "enemies of reason and beauty, people of dull heads and cold hearts." Hume stood ready to consign "into [women's] fair hands sovereign authority over the republic of letters," since women of sense and education, in his view, were the best judges of all polite writing.[62] In his essay on women Diderot argued that society with women promotes simplicity in scholarly discourse: "Women accustom us to discuss with charm and clearness the driest and thorniest subjects. We talk to them unceasingly: we wish them to listen; we are afraid of tiring or boring them. Hence we develop a particular method of explaining ourselves easily which passes from conversation into style."[63] Madame Lambert expressed this in even stronger terms: "men who separate themselves from women lose politeness, softness, and that fine delicacy which is acquired only in the presence of women."[64]

One important aspect of these struggles between masculine and feminine styles in scholarship concerned women's place in scholarly institutions. But, as in the case of gendered images of science, there was no essential connection between the sex of the participant and these styles labeled masculine or feminine. The elaborate pageantry of gallant society emerged not from qualities innate to women, but from the contours of aristocratic life. The "feminine" style described here, though couched in the language of gender, was, in fact, an artifact of urban elite culture. As Madame Lambert described it, salon life joined a politeness and delicacy of the grand world of rank to the energy of intellectual work. In the salon aristocratic women served as patrons to bourgeois men, showering the new rich with the *parfum de l'aristocratie*.[65]

That the feminine style was thought an attribute of aristocratic manners is demonstrated by the example of changing attitudes toward the natural historian, Georges Louis Leclerc, the self-made comte de Buffon. At midcentury (the height of salon culture in France) Buffon was hailed as a great stylist, and his best-seller *Histoire naturelle* was widely read as a "cosmological novel." Known as the "king" of the well-turned phrase, Buffon united the rigors of geometry and algebra with the softness of poetry and rhetoric. Style for the French count was an important part of science. In his eulogy to a new member of the Académie Française, Buffon referred to the age of the man as thirty rather than twenty-seven in order to preserve the cadence of the phrase.[66]

While the dictates of style might once have allowed lapses in such insignificant matters, increasingly by the end of the century Buffon was criticized for sacrificing scholarly exactitude to poetic flourish. We are inclined to think today that Buffon was simply inexact and that his declining reputation as a naturalist turned on the faults in his work. As Lepenies has shown, however, Buffon's decline had much to do with the impact of the French Revolution on science. It was during this period that the work of the Swedish botanist Linnaeus became preferred to Buffon's in France. During the ferment of revolution Linnaeus, the country parson's son, seemed more trustworthy than the noble comte de Buffon. Linnaeus (for all his clumsy Latin and rhymed verse) was said to have been "formed from nature"; his system of binomial nomenclature was hailed as so exact and perfectly ordered that its author was said to have been present at the Creation. Buffon, by contrast, was judged a naturalist "by order of the king." Revolutionaries judged his style pompous and aristocratic, while natural historians sought increasingly to dissociate themselves from the tainted world of literature. Interestingly, Buffon was attacked for being a *coquet,* language usually reserved for women.

Whether a distinctively feminine style existed, and to what extent it was created and claimed by women, are matters requiring more study. But why call this style "feminine"? Sociability was one of the qualities cultivated in Parisian salons and is perhaps rightly associated with *salonnières* as part of their goal of expanding traditional elites. But it is unclear what it meant to call this quality feminine, for men participated in salons as enthusiastically as did women.[67] To the extent that a poetic style was identified with women, however, other forces were at work. Poetic style was identified with the feminine at the same time that it was being cast out of scientific culture. Labeling poetry "feminine" heightened the sense that it was inappropriate to science and outdated. One need only recall that Bacon and Harvey portrayed the ancients as feminine (contemplative and passive) as part of their strategy to discredit ancient learning. By the middle of the nineteenth century, the elimination of poetry from science had been granted the status of a natural law. In Claude Bernard's view, poetry—the first (and most primitive) in three stages of scholarly style—was followed by philosophy, and finally science.[68] Even Goethe, himself a literary figure of some stature, attacked the idea that science could be presented as a novel; Madame d'Epinay (a devotee of Rousseau) feared Buffon's work more "poetic" than "true."

The feminine style was associated with aristocratic culture, but it was also associated with French culture. Voltaire argued that the styles of national languages were so distinct that one could distinguish the Italian, English, and French as easily by the style of their writing (even in Latin) as by their gait. Voltaire traced these differences to qualities of the language itself—the roughness of its consonants, or the softness of predominant vowels—but also to social factors, such as the presence or absence of liberty in government and religion, "a more or less free conversation between the two sexes," and the influence of early authors. Thus Voltaire saw the strength and energy of the English language coming from the nature of the government, which allows the English to speak in public. The idleness of Italian life "emasculates" the Italian tongue, producing (among other things) its soft and luxurious vowels. As for his own country, the freedom of society and the perspicuous turn of phrases qualified the French tongue for conversation.[69]

The visibility of elite French women and their reputation for learning as well as for sexual and cultural freedom encouraged an association between French life and feminine manners. In the seventeenth and eighteenth centuries, it was generally agreed among Englishmen that France was ruled by women. According to the architect Christopher Wren, women's taste dominated architecture, language, and fashion. Wren attributed the overly ornate style of Versailles to women's influence.[70] Englishmen were also suspicious of any society where both men and women joined in discussing politics or religion, topics thought unsuitable for women. In his history of women, the English physician William Alexander wondered how it could be that "French women do not even withdraw from the table after meals; nor do men discover that propensity to have them dismissed."[71]

At the same time that Madame Lambert and a number of women advocated retaining a gallant and feminine style, others insisted that such a thing was imposed upon them and indeed undesirable. In response to criticisms of her *Essay in Defence of the Female Sex,* the Englishwoman Judith Drake wrote that many criticized it for its masculine prose. Drake was among those who thought that with equal education for women, differences of style would fade to the point that critics would no more be able to discern "a Man's style from a Woman's, than they can tell whether this was written with a Goose-Quill or a Gander's."[72]

The Parisian salon, then, offered an alternative style of scholarly endeavor—one where men and women engaged freely in intellectual pursuit. The men and women of the salon called themselves *savants* to contrast their own way of knowing to that of the scholars of the schools. For the *salonnière,* the *savant* combined knowledge and refinement while the *pédant* pursued serious learning to the exclusion of social graces. A successful *savant* united "science with eloquence, the muses with the graces, and art with nature."[73] The salon offered an alternative to the style of the schools, which, in divorcing scholarship from gentility, devolved into the combative style advocated by Rousseau.

The Attack on the Salon: A Masculine Style?

In the 1750s Rousseau launched a vicious attack on the scholarly style issuing from the Parisian salon. Rousseau had served as secretary to Madame Dupin, and one of his tasks was to help compile a lexicon of well-known salon women. (The lexicon was written in the gallant style that Rousseau so detested.) Rousseau blamed the decay of French arts and letters on the strong influence of women in the salons: "Every woman in Paris," he complained, "gathers in her apartment a harem of men more womanish than she." Rousseau found women's influence on men unnatural, harmful to both their bodies and their minds. Under the influence of women men, too, "become effeminate." The decadence of the arts and letters in France could be traced to men's habit of "lowering their ideas to the range of women," for "everywhere that women dominate, their taste must also dominate; this is what determines the taste of our age."[74]

Rousseau identified a womanly style as one where gallantry and humor prevails—a style that stifles genius. In the presence of women, men are required to "clothe reason in gallantry," to polish their conversation and be satisfied by jokes or compliments. By themselves, separated from the influence of women, men devote themselves to "grave and serious" discourse. "If the turn of conversation becomes less polished," he argued, "reason takes on more weight."[75]

Rousseau did not label his prescriptions feminine or masculine, yet much of what he prescribed as the proper style of science fit cultural definitions of masculinity—both ancient and modern. Like Aristotle and Bacon before him, Rousseau thought that scholarship should display marks of vigor. To emphasize the dynamic he imagined to exist between true scholars, Rousseau deployed a military metaphor. Ideas, he held, cannot be cultivated in sedentary salons, but only on the field of battle. Men among themselves would not "humor" one another in dispute; rather each, feeling himself attacked by all the forces of his adversary, would feel obliged to use "all his own force to defend himself."[76] Only through this combative process did Rousseau believe that the mind gains precision and vigor. Rousseau seems to have advocated a style of philosophical exchange that Janice Moulton has recently critiqued as the "adversary method"—a method by which not the best but the best-defended argument succeeds.[77]

Rousseau's choice of a military metaphor was not unintentional. With it he hoped to reassert the privileges of a military class displaced by the courtier and his refined manners. As noted in Chapter 1, the courtly culture from which the salons had emerged had encouraged the greater participation of women by devaluing physical strength. Vigor of the body was not thought essential to vigor of the mind. Rousseau, however, reasserted the parallels between physical and mental strength, whereby the prerogatives of the former became prerogatives of the latter. The salons, this "indolent and soft life to which our dependence on women reduces us," he argued, stripped both men and letters of their strength and vigor.[78]

This emphasis on strength meshed well with notions of masculinity emerging from medical definitions of the male body. Privileges that strength had won for the feudal lord were reasserted as laws of nature. The greater stature of the male was contrasted to the delicate make of the female body (see Chapter 7). For Rousseau, participation in science required a certain strength that women simply lacked.[79]

While Rousseau sought to reinstate strength of mind as a prerequisite for engaging in scholarship, others denied that a culture privileging physical strength produced superior science. Advocates of women's equality, Condorcet and Buffon argued that science flourished not in the most powerful but rather in the most peaceful and sedentary societies—a style of life which, Condorcet emphasized, also encouraged greater equality between the sexes.[80] Buffon agreed with Rousseau that the male was physically larger and stronger than the female, yet he did not find this a mark of superiority. Instead, Buffon insisted that men had abused their advantage by exercising a "cruel and tyrannical dominion" over the weaker sex—a domination more suitable for savage than for civilized peoples. Only in nations "highly polished" had women obtained that equality of condition which is due to them; in such nations politeness of manners, the offspring of the softer sex, was considered superior to mere physical strength.[81]

Rousseau's prescriptions for a more vigorous style of scholarship also included an institutional dimension. Rousseau thought men should retire to "circles" or clubs on the English or Genevan model for intellectual activity. Women and girls, in contrast, should meet in one another's homes.[82] In order to cultivate a proper style in scholarship, Rousseau prescribed a greater social distance between the sexes.

I am not suggesting women have an innate poetic vision of the world or that we should return to an earlier, aristocratic time. Nor should the identification of the poetic with the feminine be seen as stable or unchanging (indeed in the sixteenth century public poetic traditions were seen as masculine; in the nineteenth century Romantic poetic imagination was considered manly and heroic). I am suggesting, however, that the "nature" of science is no more fixed or uncontested than is the "nature" of man or woman. By the late eighteenth century, scientists and philosophers were championing a science stripped of all metaphysics, poetry, and rhetorical ornament.[83] In Lavoisier's words, the language of a science should consist of three things: "the series of facts which are the objects of the science, the ideas which represent these facts, and the words by which these ideas are expressed."[84] Literature, which Claude Bernard called the "older sister of science," was to be distinct from science.[85] It was banished from science under the disgraceful title of the "feminine." The equation of the poetic and the feminine ratified the exclusion of women from science, but also set limits to the kind of language (male) scientists could use. Goethe's reputation as a poet was said to have ruined his reputation as a scientist. Albrecht von Haller was careful to keep his poetry separate from his science, and he disparaged his poetic urge as a sickness.

Though the subject of great controversy, salons were not to shape the future course of science. The Revolution made the *salonnière* and her education obsolete. Detached from the prerogatives of class, the appeal to a "feminine" style in scholarship lost much of its power. After 1790 and the demise of the great Parisian salons, Hume's "sovereigns of the empire of conversation" disappeared along with the notion that femininity has something to contribute to abstract knowledge.[86] In mourning their passing, the great *salonnière* Madame Lambert drew a parallel between salon women and the muses. "There were, in an earlier time," she wrote, "houses where [women] were allowed to talk and think, where the muses joined the society of the graces. The Hôtel de Rambouillet, greatly honored in the past century, has become the ridicule of ours." The passing of this epoch moved Lambert to ask: "cannot women say to men, what right have you to be the guardians of the sciences and fine arts?"[87]

6

Competing Cosmologies: Locating Sex and Gender in the Natural Order

Aristotle . . . pretends that women are but monsters. Who would not believe it, upon the authority of so renowned a person?

—François Poullain de la Barre, 1673

Modern science was both the root and the fruit of a series of revolutions. The historian Alexandre Koyré has described these changes as: the shift from the geocentric world of Greek and medieval astronomy to the heliocentric and (later) the centerless world of modern astronomy; the conversion of the European mind from *theoria* to *praxis;* the replacement of the teleological pattern of thinking and explanation by the mechanical world view.[1] Other scholars have focused on different aspects of these developments, but what is not usually recognized is that revolution occurred also in another quarter—in scientific understandings of biological sex and sexual temperament (what we today call gender).

The Renaissance inherited from the ancient world two dominant cosmologies accounting for women's place in nature and society: the Aristotelian-Galenic theory of humors (important especially to medical traditions) and the Judeo-Christian account of creation. In the ancient world, as in our own, men and women were carefully placed in the great chain of being—their positions were defined relative to plants, animals, and God. Much could be said about ancient views of sexuality; here I shall focus on the relation between biological sex and gender (or an ascribed sexual character). For the moderns, biology has all too often served as a starting point for explanations of that which distinguishes men from women—not just physique but character and social status. The ancients saw things quite differently. For them, sex organs were not considered the determining factor in sexual character. Gender (or sexual temperament, as they called it) was more fundamental than biological sex; gender shaped sex, rather than vice versa. Gender in the ancient world was a cosmological principle.

Ancient Cosmologies: Woman as Imperfect Man

In the ancient world, woman was viewed as a unique sexual and moral creature, distinct from and inferior to man. The inferiority of woman—odd as it may seem—depended on her lesser *heat*. Fire (or heat)—along with air, water, and earth—constituted the four elements in the terrestrial sphere. As summarized by a seventeenth-century interpreter, each of these elements possessed distinctive qualities: fire was hot and dry; air, moist and hot; water, cold and moist; and earth, cold and dry.[2] The human body was composed of four humors corresponding to the four elements of the cosmos: blood (like air) was hot and moist; phlegm (like water) was cold and moist; yellow bile (like fire) was hot and dry; and black bile or the melancholic humor (like earth) was cold and dry. These traits were influenced and changed by eight additional factors: age, sex, color, composure, time or season, region, diet, and occupation. Good fortune (like good health) depended on cultivating a fine balance between these diverse elements (heat must be tempered by cold, dryness by moisture). The elements also stood in a hierarchical relationship to one another: things hot and dry were superior to things cold and moist. Heat was the immortal substance of life (see Figure 27).

To the ancient mind, temperament (sexual or otherwise) was also defined by the propensity to be hot or cold, wet or dry. Everything in the universe had a temperament. Things hot and dry—the sun, for example—were considered masculine, while things cold and moist—like the moon, or western regions of the earth—were thought of as effeminate. Thus Osiris, identified with the sun, represented the male principle and was the king of heaven; while Isis, identified with the moon, was thought of as the female principle and queen of heaven. In this sense masculinity and femininity had

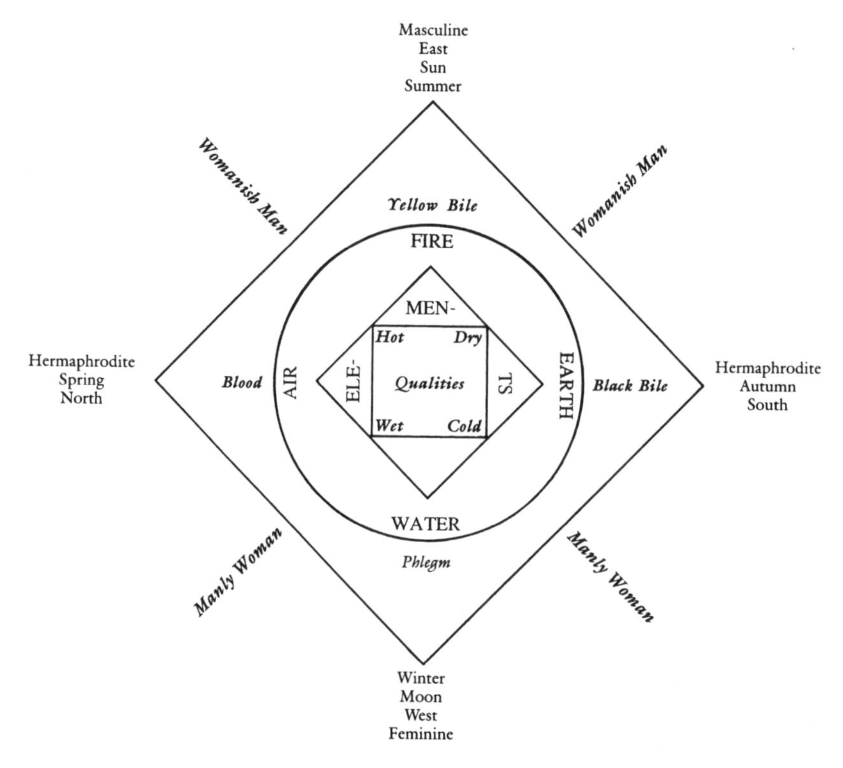

Figure 27. Diagram of the four humors. Sexual temperament (that which defined manliness or womanliness) derived not from the bodies of the male or female but from a specific mix of the qualities (the hotness or coldness, wetness or dryness) inhering in the four elements—air, fire, water, and earth. The perfect woman was wet and cold; the perfect man was hot and dry. The hermaphrodite being both wet and hot (or cold and dry) was ambiguous in its sex and temperament. To the extent that heat predominated, the hermaphrodite approached the manly axis and was thus considered a "womanish man"; to the extent that cold predominated, he or she was considered more womanly and thus a "manly woman."

nothing to do with the sexual nature of men or women but resulted from a specific mix of the four elements. Differences between the two sexes were reflections of a set of dualistic principles that penetrated the cosmos as well as the bodies of men and women. As Aristotle put it, maleness and femaleness were principles for which the organs of reproduction were mere instruments. "Taking the widest formulation of each of these two opposites," he wrote, "[we regard] the male *qua* active and causing movement, and the female *qua* passive and being set in movement." That is why Aristotle spoke of the nature of the earth as something female, calling it "mother," while the heaven and sun were given the title of "generation" and "father."[3]

Heat, which determined sexual character, also determined sex by shaping male and female genitalia into their characteristic forms. From the time of Galen until late into the sixteenth century, woman was thought to have a "spermatic vessel" (a polite word for penis) similar to man's.[4] Galen taught that male and female genitalia were not essentially different: "All parts that men have, women have too . . . the difference between them lies in only one thing . . . that in women the parts are within the body, whereas in men they are outside."[5] Even the womb was nothing special. As Galen pictured it, the neck of the womb was nothing other than the penis turned inward, and the bottom of the womb nothing but the scrotum inverted (see Figure 28). Women's organs were similar to men's in number and kind, but because women's organs were internal, women remained "imperfect and, as it were, mutilated." Again, this important difference between men and women depended on their varying degrees of heat. Women simply lacked the heat to propel their sex organs outward:

> Now just as mankind is the most perfect of all animals, so within mankind, the man is more perfect than the woman, and the reason for his perfection is his excess heat, for heat is Nature's primary instrument . . . the woman is less perfect than the man in respect to the generative parts. For the parts were formed within her when she was still a foetus, but could not because of the defect in heat emerge and project on the outside.[6]

As proof of the fact that women were men *manqué,* Galen, Pliny, and others recounted stories of women spontaneously transformed into men. Volateran, a cardinal, told of a woman in the time of Pope Alexander VI who, on the day of her marriage, "had suddenly a virile member grown out of her body." There was also a man at Auscis in Vasconia, at age sixty strong, grey, and hairy, who had been a woman until the age of fifteen. "At age fifteen, by accident of a fall, the Ligaments being broken, her privities came outward, and she changed her sex." Another woman, a citizen of Ebula, be-

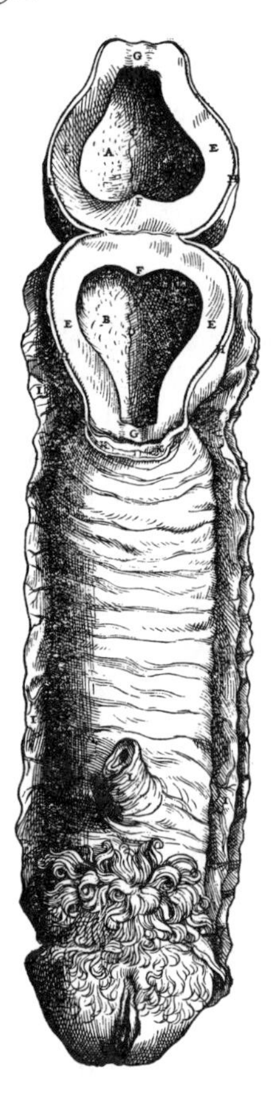

Figure 28. Female reproductive organs. Vesalius provided this visual rendering of Galen's conception in his *De humani corporis fabrica* (Basel, 1543, plate 60). By permission of the Boston Medical Library in the Francis A. Countway Library of Medicine.

came a man after twelve years of marriage.[7] The possibility of transformation was only one way, however: Galen had argued that, though a woman might become a man, a man could not become a woman. This, he explained, was because nature strives always for perfection.

Galen considered the most noble part of the human body to be neither the mind nor the heart, but the testicles, for their role in cooking the blood. A seventeenth-century writer attributed to him the view that the testicles are another fountain or well-spring of inbred heat—the "feu-place or fire-hearth"—whence the whole body receives an increase of heat. Women, too, had "testicles [ovaries] which boile the blood," but with less vigor. Should the testicles be absent or taken away, "in a word, all virility or manhoode vanisheth away."[8]

Varying degrees of heat were thought to distinguish not only the sexes, but also diverse peoples. The peoples of Africa and southern regions, for example, were thought to excel in intelligence and wit because the dry hot climate enlivened their temperament.

Renaissance and Early Modern Feminism

How did the philosophy of sex and gender differences change with the rise of modern science? Some historians have suggested that the new philosophy of Descartes and Locke made modern feminism possible. In this view, the new ideas promoted by the moderns in the seventeenth century served to wipe clean the slate of ancient prejudices in Aristotelian scholasticism, Roman law, and Christian ethics.[9] Prejudices against the intellectual capabilities of women were to be rethought on the basis of clear and distinct ideas. As we shall see, however, the fathers of modern science were curiously silent on the matter of gender differences. Furthermore, the urge to reevaluate women's moral and physical character can be traced to Renaissance humanism, which predated the new philosophy by several centuries. Though Joan Kelly has suggested that women did not reap the benefits of a Renaissance in the fifteenth century, this was a period of great debate about women and their intellectual character. These debates flourished in the feminine atmosphere cultivated at Italian courts where well-born ladies commanded respect (see Chapter 1).[10]

Renaissance humanism and the revival of Platonism chipped away at the oppressive Aristotelian dictates about the nature of women. Some humanists used these newly revived doctrines to subvert biblical authority. The neo-Platonic notion of a hierarchy of creation allowed Agrippa von Nettesheim, for example, to subvert biblical authority. Where scholastics argued that man was superior to woman because he was created first, Agrippa argued that woman is superior to man because she was created last.[11] Agrippa further justified his claim that woman is superior to man through five different kinds of arguments: *e nomine,* "Eve" means life and "Adam" means earth; *ex ordine,* Eve, created last, represents God's masterpiece of creation; *e materia,* Eve was created from human flesh, Adam from lifeless clay; *e loco,* Eve was created in paradise, Adam outside (and nobility follows from birthplace); *e conceptione,* a woman bore the son of God, man cannot.[12]

These were strong arguments, yet they were made to stand on standard authorities—that of Aristotle, the Bible, or Plato: feminists in this period rarely planted their feminism in new ground. Like other humanists, they employed traditional authorities and standard tools. Perhaps the most curious apologies for women were those employing the tools of that great enemy of women—scholasticism. Anna van Schurman (1607–1678), a renowned Dutch scholar, employed scholastic logic in her defense of women. Van Schurman's work serves as a prime example of a feminist attempting to make her argument understandable by locating it within philosophical traditions of the dominant culture. Her book, *The Learned Maid,* announced itself as an "exercise in logic upon the question whether a maid may be a scholar," and was drafted in proper scholastic form using major and minor premises.[13] Her argument began with the traditional definition of terms. She then proceeded to present fifteen theses defending the right of women to be educated, ending with a refutation of "the adversaries" of this right. In trying to make her voice heard, Schurman employed orthodox scholarly tools.

Other writers drew feminist conclusions by reversing the hierarchy of the sexes within ancient authorities. Where Aristotle had seen women as subordinate, feminists would claim that the female sex was equal to the male or even superior. The work of Marie le Jars de Gournay, the *fille d'alliance* of Montaigne, reveals the eclecticism inherent in this enterprise. In her *Egalité des hommes et des femmes,* written in Paris just before Descartes's *Discourse on Method,* Gournay ridiculed those who "trumpeted in the streets that women lack the dignity, the independence, the temperament, and the organs" to achieve the dignity of men. Yet, in her own argument for the equality of women, Gournay appealed to many of those same authorities. She, too, used the words of Plato, Aristotle, or St. Jerome, twisting them to her own purposes. In order to justify her claim that woman is the social equal of man, for example, she argued that sex differences have no significance beyond the purely physical. This she grounded in the teachings of the scholastics: the difference between the sexes, she wrote, is simply an accident required for the propagation of the species. Gournay backed up this properly phrased Aristotelian principle with the notion, drawn from scripture, of an androgynous creation: "Man was created male and female, the two were considered but one," thus male and female possess the same strength of soul. As for those who asked whether woman was created in the image of God, Gournay quipped, "it depends on how much value is attached to the beard." From these traditional authorities Gournay drew feminist conclusions, but like others of her time she did not challenge the world view upon which that authority was founded.[14]

With a bit of imagination, even the Aristotelian/Galenic humoral tradition could be turned to women's advantage. In his *L'Honneste femme* Jacques Du Bosc, for example, found women *more* capable than men of studying the arts and sciences. "It seems to me that . . . [women's] temperament, which the doctors judge more delicate than ours, is better disposed to the study of the arts and the sciences." Whereas the exercise of the body suited men, the exercise of mind suited women.[15] As at the Renaissance court, delicacy, and not strength, was seen as the primary virtue of the learned. Samuel Sorbière too, ever the courtier, wrote to Princess Elizabeth of Bohemia in 1660 assuring her that she—along with Queen Elizabeth of England, Mary Stuart, and Queen Christina of Sweden—served as a fine example of a woman who excels in knowledge. This was not to be a source of wonder, for (as Sorbière explained) "our doctors, who consider the brain the seat of reason and learning, find it as large in women as in men, and also claim that the softness of their constitution (whence perhaps the Latin name *mulier*) is much more suitable to actions of the mind than the dryness and hardness of ours."[16] This played into the hands of men (like Leibniz) who advocated learning for women precisely *because* women had leisure. One author in the *Guardian* argued that women's employments were of "a Domestick Nature, and not like those of the other Sex which are often inconsistent with Study and Contemplation."[17]

Proponents of feminism were not likely to rest their defense of women on one authority alone. "Approved Authors, evident reasons, Holy Writ, sanctions of both Civil and Canon Laws" all served as epistemological moorings. "Unquestioned Histories" were soon added to the flotilla. With the Renaissance we enter the heyday of lexicons listing female worthies in the arts and sciences.[18] Jean de La Forge drew on a number of trends favoring women in his history of learned women. Invoking neo-Platonism, he dedicated his *Le Cercle des femmes sçavantes* to a woman of noble standing (the Contesse de Fisque), admonishing her to join science with beauty.[19]

Let us look at one final example of the attempt to garner a new respect for women by placing them positively within traditional hierarchies. Marguerite Buffet, a "teacher for ladies" living in Paris at the end of the seventeenth century, began her defense of women with that most powerful authority—Christian theology. As others before her, Buffet held that souls have no sex. Since both men and women are created in the image of the divinity, she claimed that "beauty of the mind knows no differences between man or woman." Buffet found it necessary, however, to compare sex organs as well as souls. She countered Aristotle's claim that women are monsters of nature with the Galenic view that the "organs" (or "instruments," as she called them) are "similar and equal." If anything, she added, there is less difference between one sex and the other than there is among individuals of the same sex. Whatever differences might be found between the organs of the two sexes, Buffet held, are necessary for the preservation of the species and do not extend to the soul.[20]

Buffet also denied that female brains were "smaller and more narrow than those of the male," a condition thought to cause women's brains to retain acid and biting humors that "sting" female nerves and membranes. These accusations she judged to be based on prejudice and not on scientific fact. "Geometers and philosophers," Buffet wrote, have shown that buffaloes and cows—animals renowned for the grandeur of their heads—do not have the largest brains. Men, while glorying in their strength and large heads, have demonstrated not their superiority, but only that they have something in common with "stupid animals and large beasts."[21] Not satisfied with simply defending women's equality, Buffet attempted to turn the tide in favor of women, claiming that women have more lively minds than men and that women's beauty and virtue exceed that of men. Women can trace their greater beauty, Buffet taught, to the fact that female babies take longer to come to term, demonstrating that nature has taken more care in their making. Buffet did not let her opponents off easily. Plato and Aristotle, she wrote, were themselves very ugly and that is why they were so unkind to women.

The most important and powerful argument in the feminist arsenal, however, was a Christian one. After Agrippa's work in the midsixteenth century, it became common within feminist circles to assert that "the soul has no sex." Feminists stretched this idea, originally a Christian notion, far beyond the intentions of its author, Saint Augustine of Hippo. For quite different purposes, Augustine had asserted that both sexes, having been created in the image of God, possess a rational soul (though woman's rationality was of a lesser degree).[22] While woman might be inferior to man by nature, she was his equal by grace: in the afterlife souls had no sex—no legs, arms, or genitalia. Though some argued that all resurrected persons are male, Augustine envisioned paradise populated by asexual souls. Having shed their corporeality (and the sexuality required on earth for preserving the species), souls in heaven finally achieved the equality that at their creation was only potential.[23]

The vagueness of the term *soul* in this period allowed for an easy secularization of this claim, and feminists easily stretched the notion of soul to assert that the "mind has no sex." Some simply considered the soul itself as the seat of intelligence: "all things are thus equal between men and women with respect to the soul, which is the intelligent part and that which makes learned men and philosophers."[24] Others suggested that though equality might be granted women in the matter of salvation (their souls), it was not to be granted in matters of science (their minds). Richard Allestree, professor of divinity at Oxford, wrote in 1673, "but not to oppose a received opinion, let it be admitted, that in respect of their intellects they [women] are below men; yet sure in the sublimest part of Humanity, they are their equals: they have Souls of as Divine an Original . . . that spiritual Essence . . . owns no distinction of Sexes."[25]

This debate concerning the sex of the mind occurred as the brain was becoming established as the seat of reason. The work of Descartes and Locke sparked interest in philosophy of mind; Hobbes and Locke focused attention on reason as the cement of social contracts. At the same time, within medical circles interest was growing in the nature of the brain. In his summary of medical traditions written in 1615, Helkiah Crooke gave a new preeminence to the brain. Whereas Galen had deemed the testicles the most noble part of the body and Aristotle the heart (as the seat of the soul), Crooke yielded superiority to the brain (followed next by the heart and last by the liver). Crooke found the brain most noble and divine because it was the seat of all sense and voluntary motion, the habitation of wisdom, the shrine of memory, judgment and discourse, intelligence and understanding. In short, Crooke considered the head to be the "prince of the family," the "head of the tribe," to which all other parts are but attendants.[26]

Throughout the sixteenth and seventeenth centuries feminists attempted to locate women positively within ancient, and still powerful, cosmologies. But they did so without challenging those cosmologies and the social hierarchy they underwrote. In the end, feminists left themselves open to easy rebuttal. However brilliantly they stood Aristotle on his head, he was easily set upright again. A new philosophy, however, was in the making. How would that philosophy deal with the feminist claim that "the mind has no sex"?

Descartes and Locke: Is Neglect Benign?

Bacon and Descartes, Locke and Leibniz are often celebrated for having pioneered new scientific methods. Yet, curiously, these fathers of modern epistemology had little to say about the question of sex and sexual temperament. Though they wrote at a time when "the woman question" was much and hotly debated, neither Descartes nor Locke nor Leibniz focused attention on this question. They neither repeated the prejudices of Aristotle nor established their own positions on the nature and status of woman. The focus of their interest remained on the discovery and articulation of new methods (Bacon and Descartes) or on questioning the divine rights of kings (Locke) or establishing new techniques of calculation (Leibniz). Only Bacon and his fellows at the Royal Society called for an explicitly "masculine" philosophy (see Chapter 5).

Many of the principles of the new philosophy did allow (intentionally or not) for the participation of women in philosophy. By adopting a cool silence vis-à-vis the issue of feminism, however, Descartes, Locke, and Leibniz failed to reflect upon how gender relations might affect the foundations of modern philosophy. What might appear to be a "neutral stance" in effect left traditional male prerogative—both social and intellectual—unchallenged.

One might imagine that Descartes's emphasis on method should have provided opportunities that feminists could exploit. Descartes's attack on Artistotelian scholasticism was powerful and unremitting—indeed, it was better (in his view) for one never to have attended school at all than to have imbibed the myths propagated by these men. Those, he wrote, who "blindly follow Aristotle . . . are incapable of attaining knowledge of true philosophy." Those, by contrast, who have learned least about "all that which hitherto has been named philosophy" are the most capable of apprehending the truth.[27] Appealing to what he called an unsullied "natural intelligence," Descartes encouraged each person to learn to the limits of his natural ability.

This emphasis on a universally applicable method had the potential to broaden participation within intellectual culture. Reason, or the ability to manipulate words and signs, Descartes understood to be a universal instrument in mankind. "I have noticed," he wrote, "on examining the nature of many different minds, that there are almost none of them so dull or slow of understanding that they are incapable of high feeling, and even of attaining to all the profoundest sciences, were they trained in the right way."[28] Descartes believed that differences in human intellectual capabilities could be overcome by right method. "Those who walk slowly can," he wrote, "if they follow the right path, go much farther than those who run rapidly in the wrong direction." In contrast to the teachings of Aristotle, Descartes emphasized that reason is the same by nature in all mankind.[29]

Women appealed to natural intelligence in the Cartesian sense to justify their own philosophical endeavors. Though not a Cartesian, Margaret Cavendish, duchess of Newcastle, held that natural reason gave her authority to write about natural philosophy: "since I have not Scholastical Learning . . . I may Err in Words, yet I am Confident, I do not Err in Sense and Reason."[30] Descartes dined at

Cavendish's table, yet he never shared her interest in the question of women's intellectual abilities. He did, however, encourage a number of individual noblewomen to pursue the study of philosophy. In dedicating his *Principles* to Princess Elizabeth of Bohemia, Descartes praised her for her grasp of geometry and metaphysics. "The incomparable excellence of your intellect," he wrote to her, "is evident in the fact that in a very short time you have mastered the secrets of the sciences, and obtained a perfect knowledge of them all."[31] Though sincere in his praise, it should be noted that the women Descartes served—Queen Christina of Sweden, Princess Elizabeth, even the duchess of Newcastle—outranked him socially. Descartes's support in these cases was for particular individuals of high social station; he never made a general defense of the female sex.

In years to come, Descartes was often cited as a defender of women. In an Enlightenment comedy entitled *Le Club des dames, ou le Retour de Descartes,* the playwright Madame de Genlis called for Descartes, the "first person to defend women against the most barbarous opinions of men," to preside over a club for women where they were to cultivate their minds and reclaim their rights.[32] While Molière became notorious as an opponent of learned ladies, Descartes was celebrated as their defender.

Though Descartes never spoke directly to the woman question, he helped clear away much debris of older cosmologies. One important housecleaning took place around the crucial question of gender. Without comment, Descartes simply dropped all talk of sexual temperament. The only sexual differences Descartes acknowledged were those in reproductive organs. Masculinity and femininity for Descartes, unlike for Aristotle, were not cosmological opposites—the one hot and active, the other cold and sluggish.[33] Descartes (again, unlike Aristotle) did not conceive of women as having lesser reason; nor did he (unlike Rousseau) claim for women distinctive mental or moral faculties.

Much the same can be said for the work of John Locke. In his political philosophy, Locke followed traditional opinions that found it fit to subordinate a wife to the will of her husband.[34] In his epistemology and theory of education, however, Locke did not emphasize differences of sex. For him access to "learning and logick" was more a matter of class than of sex. Locke recognized that education was reserved for the leisured classes—of either sex. Where, he wrote, the hand is used to the plough and the spade, the head is seldom elevated to sublime notions or exercised in mysterious reasoning. "'Tis well if men of that rank (to say nothing of the other sex) can comprehend plain propositions, and a short reasoning about things familiar to their minds."[35]

In his 1693 *Thoughts on Education,* Locke suggested that girls of the upper classes be schooled like their brothers in the use of reason; Locke did not (as Rousseau would later) prescribe a radically different education for girls and for boys. Although his comments (originally addressed to a friend and distant relative, Edward Clarke) concerned the upbringing of a young gentleman, Locke stated that they could be used as a guide for raising children of either sex. Locke passed over the question of sex with the remark that "where the Difference of Sex requires different Treatment, 'twill be no hard Matter to distinguish." He did make certain allowances for the protection of a girl's beauty, yet he emphasized that the education of boys and girls should be the same, especially in the early years. In his letter to Mrs. Clarke about the education of her daughter, Locke wrote that he would rather have a wife with a healthy constitution—a stomach able to digest ordinary food, and a body that could endure upon occasion both wind and sun—than a paling, weak, sickly wretch that every breath of wind or least hardship puts in danger. For these reasons he recommended that girls have the same diet, lodging, and clothing as boys. Locke advised no more differences between boys and girls in moral education than in their physical education. Again he wrote to Mrs. Clarke, "since I acknowledge no difference of sex . . . relating . . . to truth, virtue and obedience, I think well to have no thing altered in it [an educational program for daughters] from what is [writ for the son]."[36]

These two or three references to the woman question suggest that Locke did not think that women of the upper classes were less capable of intellectual endeavor than men of those classes. Yet Locke did not treat this topic systematically. His comments on the education of girls were but a response to Mrs. Clarke's particular queries.

Like Descartes, Locke found disciples among women. In 1702, Catharine Cockburn wrote (at age twenty-two) a "Defense of Mr. Locke's Essay of Human Understanding." Thomas Birch, the eighteenth-century historian, has suggested that Cockburn's essay was one of the few pieces to support Locke's work at that time. This work, which won Locke's own approval, had as its purpose the "vindication of Locke's Christian Principles."[37]

What are we to make of Descartes's and Locke's relative silence on the subject of women, despite their innovative abandonment of age-old notions of femininity and masculinity (of the active masculine spirit and passive feminine matter)? Historian Paul Hoffmann has argued that Descartes's distinction between mind and body unseated the misogyny built into Aristotelian physiology.[38] The notion of a profound distinction between soul and body did not, of course, originate with Descartes. What Descartes had to offer was a grounding for this notion in a new epistemology and ontology. The idea of man as a machine undermined the Aristotelian dictum that because women are colder than men they have a lesser reason. If the mind operates independently of the body, as suggested by Cartesian epistemology, traditional allegations of female failings of body no longer imply feminine failings of spirit. All minds are created (at least potentially) equal.[39] Hoffmann has judged Descartes's contribution to feminism a positive one. Ruth Perry has gone further to suggest that the Cartesian method suited women's intellectual and also social condition. Meditative introspection required leisure and isolation, two things literate middle-class and aristocratic women had in abundance.[40]

Genevieve Lloyd, by contrast, has argued that the long-term effects of Cartesianism have been negative. Though Descartes intended to open the sciences (as he put it) "even to women," Lloyd argues that within the context of previously existing gender relations, the effect was quite different. Lloyd locates the seeds of a sexual division of intellectual labor in the Cartesian distinction between mind and body. The ideal of a pure, highly abstract reason separated from the demands of the body and also from the demands of ordinary life became a pillar of modern science. A philosophy that separated sharply the requirements of truth-seeking from the practical affairs of everyday life reinforced already existing distinctions between male and female lives.[41]

Though the philosophy of a radical mind/body distinction seemingly recognized the equality of women's minds by dissociating these from their (feeble) flesh, it failed to grapple with the male/female dualism that Lloyd has identified. The misogyny of the ancients or the silence of the moderns were not the only options open to seventeenth-century thinkers, however. A number of traditions—the neo-Platonic, cabala, and alchemical—dealt explicitly with the relationship between the masculine/feminine polarity and the mind/body duality (see also Chapter 5).[42] To take one example, Anne Finch, the Lady Conway, envisioned a unity of body and spirit resulting from a union of masculine and feminine principles. She held that spirit and body are one; the original man (what was later called the hermaphroditic Adam) was both masculine (in mind) and feminine (in body).[43] The unity of body and spirit Conway conceived as analogous to the union of man and woman in procreation: "for as the ordinary generation of men requires a conjunction and cooperation of male and female; so also all generations and productions . . . require an union . . . of those two principles." Conway also rejected the Hobbesian vision of a primitive war of each against all, postulating instead a fundamental unity within nature cemented by the kind of love or desire women and men feel toward their offspring.[44]

Anne Conway (and to some extent those associated with her way of thinking—Henry More and Franciscus van Helmont) found a solution to the mind/body problem in the union of the male intellect with the female body. At the same time Conway left intact the ancient subordination of the feminine to the masculine. Like others of her day, Conway accepted the familiar alignment of the *principium magis activum* with the male and the *principium magis passivum* with the female.

Poullain and an Anonymous Englishwoman

Though Descartes and Locke did not apply their principles to the question of women's participation in philosophy, others did. François Poullain de la Barre, who followed Descartes closely in method and doctrine, did what Descartes never dared: apply the principle of methodological doubt to the social domain. Poullain dared, in other words, to meddle in an area where Descartes had argued vehemently that one should not venture.[45]

According to the ex-Jesuit Poullain de la Barre (1647–1725), it was Descartes who had awakened him from his dogmatic slumbers. Poullain de la Barre reported that his training at the University of Paris had imbued him with a fervent antiwomanism: "When I was a scholastic, I considered [women] scholastically, that is to say, as monsters, as beings inferior to men, because Aristotle and some theologians whom I had read, considered them so."[46] Highly critical of scholasticism, Poullain claimed that a scholastic arrives at an opinion not by reason but by adherence to ancient custom. The use of Latin and ancient authority was in his opinion merely pretentious. Rejecting scholasticism, Poullain turned Descartes's new way of ideas to the inequality of the sexes—"of all prejudices, the most remarkable." Poullain thought that this prejudice, like all others, could be "absolutely renounced by clear and distinct knowledge."[47]

Using the tenets of Cartesianism, Poullain set out to demonstrate that there is no significant difference between the sexes. Central to his claim was that the mind—distinct from the body—has no sex. Sex indeed extends no further than to the organs of generation, "there being no other but that part which serves for the production of men." For Poullain, success in science required only reliable senses combined with right method. If sexual differences extend no further than reproductive organs, then everyone—both women and men—has equally reliable senses. Women have the same sense organs as men—their eyes see as clearly, their ears hear with the same degree of accuracy, their hands are as dexterous. And their heads are the same as men's. "The most exact anatomy has not discovered any difference in that part [the head] between men and women; the brain is the same in both, as are memory and imagination." On this basis Poullain argued that women were capable of doing anatomy. "They too have eyes and hands; may they not . . . perform dissections of a human body and consider the symmetry, and structure thereof?" For similar reasons women were capable of creative work in mathematics and logic, physics and engineering, metaphysics and astronomy, history and geography, medicine, theology and civil law; in short, there was "nothing too high for women."[48]

Poullain held that it was custom which subordinated women to men, making them "languish in idleness, softness, and ignorance, or otherwise grovel in low and base employments." That women had made no great advance in the sciences Poullain attributed not to their minds, but to the fact that they were employed in their "housewifery" and found "therein business enough."[49]

Poullain was an early advocate of higher education for women. Like Locke, who wrote his book for the education of a gentleman but allowed it to also serve for the education of a lady, Poullain wrote his book for ladies but suggested that it could also serve for gentlemen. Poullain set a rigorous curriculum: geometry for the exercise of the mind in forming a clear idea of truth; the *Logique de Port-Royal;* the method and meditations of Descartes; Géraud de Cordemoy's *Le Discernement du corps et de l'âme en six discours;* the *Physique* of Rohault; Descartes's *Traité de l'homme* with the *Remarques* by Louis de La Forge; also de La Forge's *Traité de l'esprit de l'homme;* the *Traité des passions,* the *Principes,* and *Lettres à la reine de Suède et à la princesse de Bohëme* by Descartes; the New Testament; Lesclache's philosophy; and an abridged edition of Gassendi's works.[50]

The anonymous English author of the *Defence of the Female Sex* (probably Judith Drake) also denied that there were differences of sex in the soul, basing her claim on the rationalism of Descartes and the empiricism of Locke:

> If we [women] be naturally defective, the Defect must be either in Soul or Body. In the Soul it can't be, if what I have hear'd some learned Men maintain, be true, that all Souls are equal, and alike, and that consequently there is no such distinction, as Male and Female Souls; that there are no innate *Idea's,* but that all Notions we have, are deriv'd from our External Senses, either immediately, or by Reflection.[51]

This author asserted that there is no difference in the quality of different minds (because the organs governing this part are not different), but merely in the quantity of experience guiding the mind.[52]

Both Poullain and the anonymous Englishwoman went beyond Descartes and Locke by applying the new way of ideas to the woman question. Furthermore, Poullain called immediately for philosophical principles to be put to work. For him, the new philosophy implied dramatic social change. If women's minds were, in fact, equal to men's, then why not open the world's professions to them?

Modern Anatomy and the Question of Sexual Difference

The idea that a wedge could be driven between mind and body held great potential for the liberation of women. By declaring that the mind has no sex, feminists denied that women had a special sexual character.[53] They hoped and assumed that modern anatomy could prove that in the brain—the seat of the soul—there were no sex differences. Feminists had reason to be optimistic about the new anatomy, for there were sympathetic stirrings in sixteenth- and seventeenth-century medical circles.[54] Two important reevaluations of female nature were under way: one pertaining to female sex organs and another concerning the question of the female's role in generation.

The most important reform in medical views of female nature was the change in attitude toward sex organs. Since ancient times the uterus had been much maligned. Plato thought it an animal with independent powers of movement; Democritus cited the uterus as the cause of a thousand sicknesses. Galen and even (for a time) Vesalius reported that horns bud from the sides of the womb (see Figure 29).[55] By the 1590s, however, anatomists had reversed this picture of women as "imperfect men" or monsters of nature. In 1615 Helkiah Crooke reported that many now considered Galen's views on the similarity of sex organs "very absurd." In her sexuality the female was now considered perfect. Crooke and others waxed eloquent on woman's unique womb, ordained by nature to "conceive and cherish the seed." The Parisian doctor L. Couvay, too, argued that women were to be esteemed because only they can produce children and thereby replenish the human race.[56]

A second reform was the new view of the female's role in generation.[57] By the seventeenth century, the Aristotelian view that the

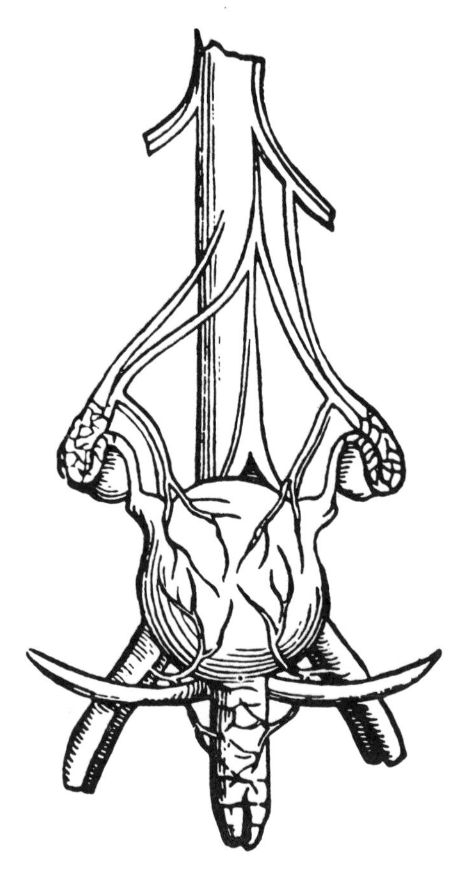

Figure 29. Female organs of generation with horns. From Vesalius, *Tabulae anatomicae sex* (1538), plate 87.

woman (through her menses) contributed only formless matter to generation, while the male (through his semen) contributed form or sentient soul, was called into question by one important school of thought among the "preformationists." Preformationists believed that a fully formed organism exists in miniature in the egg or sperm, in contrast with the epigenesists, who believed that a fetus gradually develops from a simple to a complex organism. In 1698 the London physician James Keill reported that there were two leading schools among preformationists: the ovists, who claimed that the female made the essential contribution (future generations were thought to preexist in the egg of the female and to be only "quickened" by the male); and the animalculists or spermatists, who claimed that the male made the essential contribution. A proponent of the latter view, such as Antoni van Leeuwenhoek, might argue that preformed animalcules—or miniature people—existed in male sperm, having been placed there directly by God at the time of creation in sufficient quantity to provide for the preservation of the species until the end of time.[58] Some physicians, however, were predisposed to claim—even in the absence of evidence—that the contributions of the male and female were equal. The Tory pamphleteer and medical doctor James Drake, feeling obliged to take a stand, returned to the notion that *both* partners contribute equally to generation: "however old and exploded the Opinion of a Plastick Power on both sides [male and female] be, I must . . . embrace it, even tho' I know not exactly wherein it lies."[59]

Buffon, the leading naturalist of the eighteenth century, also insisted that the male and female contribute equally to generation. Buffon explicitly rejected Aristotle's view that women have no "prolific fluid" and therefore supply only passive material for generation. More important, he rejected Leeuwenhoek's theory of "homunculi" because this theory implied that it was not the first woman, but rather the first man who contained all mankind in his own body. Buffon, an epigenesist, rallied experimental evidence to show that both men and women have seminal fluids that play an equal role in procreation, though this required him to revive the long-abandoned analogy between women's testicles (ovaries) and men's testicles.[60] Buffon thought that the resemblance that offspring bear to *both* parents required a theory of generation based on the mingling of seminal fluids from both sexes. He therefore asserted that "the seminal fluids of the male and female are equally active, and equally necessary for the purposes of propagation . . . these two fluids are perfectly analogous; they are composed of parts not only similar in their form, but in their action and movement."[61]

Thus, ancient prejudices were being reexamined—at least as far as reproduction was concerned. But what about nonreproductive parts of the body—the brain, the skeleton, or the chest? Were these still thought to be governed by an all-pervasive temperament, as Galen had believed? Though there was a revolution in views of sexual difference in reproductive organs, there was no equivalent revolution in views of secondary sexual differences in the seventeenth or early eighteenth century. Nor was there a revolution in views of sexual character (what we today call gender). Much like Descartes and Locke, anatomists of this period simply abandoned talk of cold and moist brains or melancholic humors, though their view of female nature still (implicitly, even at times explicitly) assumed the ancient theory of humors. A thoroughgoing revolution in views of sexuality was to come first in the eighteenth century (see Chapter 7).

If we look at a project central to the new anatomy in the sixteenth and seventeenth centuries—definitions and illustrations of the human body and skeleton—we see that pioneers like Andreas Vesalius did not openly subscribe to the theory of humors (as had the ancients); nor did he believe that sex differences penetrated the skeleton (as would the moderns). For Vesalius sex was only skin deep, limited to differences in the outline of the body and the organs of reproduction.

This can be seen in the *Epitome* of his great work on the fabric of the human body. Vesalius drew two manikins (paper dolls) designed to be cut out by medical students and "dressed" with their organs in an exercise intended to teach the position and relation of the viscera. One manikin represented a female form and displayed the system of nerves; the other represented a male figure and showed the muscles. Vesalius presented both the male and female manikins in order to demonstrate the position and nature of the organs of generation. Yet, when discussing parts of anatomy not having to do with reproductive organs, he did not differentiate the nonreproductive parts of the male from those of the female. Apart from the reproductive organs, Vesalius considered all other organs interchangeable between the two figures. In his instructions for constructing the manikins he made this explicit: "The sheet [of organs to be attached to the male manikin] differs in no way from that containing the figures to be joined to the last page [the drawing of the female manikin] except for the organs of generation."[62]

Nor did Vesalius find sex differences in the skeleton. In this same work, Vesalius drew male and female nudes and pointed out differences in the curves and lines of the two bodies and the two sets of reproductive organs (see Figure 30). To accompany his male and

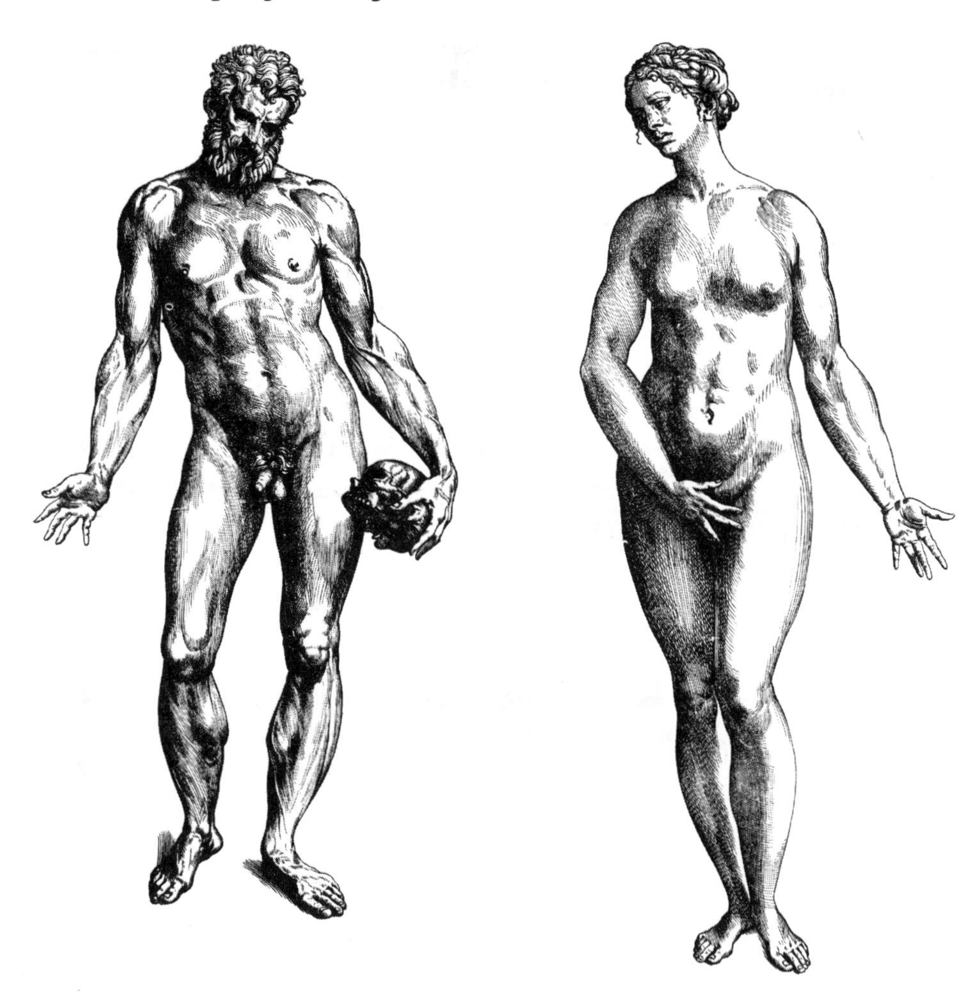

Figure 30. Male and female nudes used by Vesalius to illustrate differences in the shape of the two bodies and the two sets of reproductive organs. From his *Epitome* (Basel, 1543). By permission of the Boston Medical Library.

female nudes, Vesalius drew a single skeleton that he labeled a "human" skeleton (see Figure 31). Believing that one skeleton gives shape to both the male and the female body, Vesalius did not sexualize the bones of the "human" body. Though he made clear in textual notes that the skeleton was drawn from a seventeen- or eighteen-year-old male, Vesalius did not give a sex to his skeleton.

The indifference of early modern anatomists to the question of

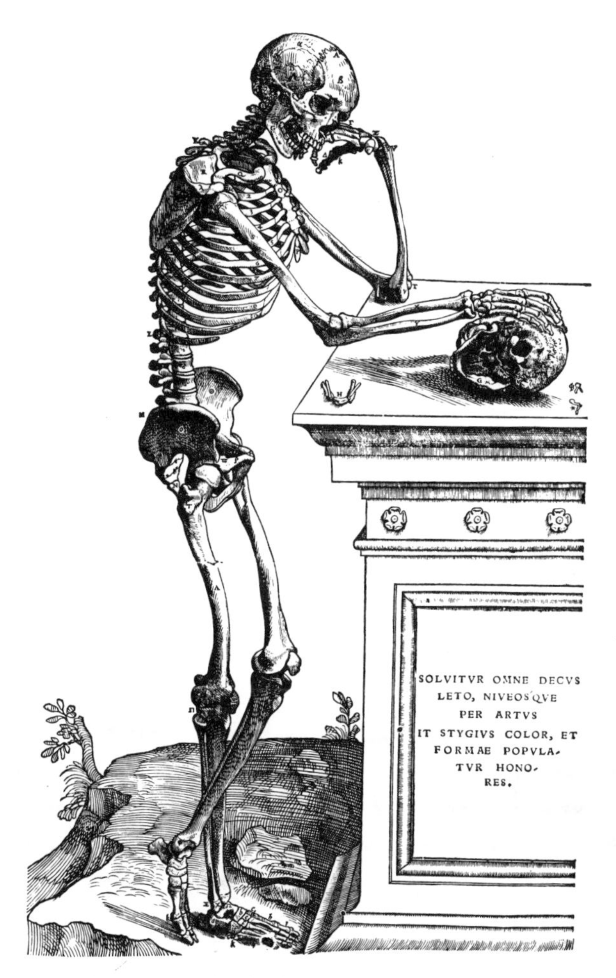

Figure 31. Bones of the human body, drawn to the same scale as the nudes. From Vesalius, *Epitome* (Basel, 1543). By permission of the Boston Medical Library.

sex differences did not derive from an ignorance of the female body. From as early as the fourteenth century women had, in fact, been dissected. The *Montpellier Codex* of 1363 includes an illustration showing the dissection of a female body.[63] The 1442 statutes of the University of Bologna reveal that the university received for dissection one male and one female body per year.[64] A statute enacted in France in 1560 required midwives to attend the dissection of female bodies so that they would know enough about female anatomy to be able to testify in abortion cases.[65] This trend did not change with the rise of modern anatomy. The frontispiece to Vesalius's 1543 *De humani corporis fabrica* depicts a public dissection in a theater teeming with men, dogs, a monkey, and one lone nun; on the table, under the knife, is a woman.[66] Vesalius based his drawings of female organs of reproduction in this work on dissections of at least nine female bodies. Vesalius did not procure these bodies without difficulties, however; at least one was stolen. Hearing that the mistress to a certain monk had died, Vesalius and his helpers snatched her body from the tomb.[67] This remained common practice for quite some time: William Cheselden, an English physician, reported in 1713 that he procured female bodies for dissection from "executed bodies and . . . a common whore that died suddenly."[68] Moreover, at least by the 1680s, skeletons of women were kept in major natural history collections. The Royal Society had one; the University of Leiden had two—"the Sceleton of a Woman of seventeen years old, who murdered her Son, and the Sceleton of a Woman called Catharine of Hamburgh strangled for Theft."[69]

The indifference of early modern anatomists to the question of sex differences (beyond reproductive organs) did not, however, lead them to "desexualize" the bodies they studied. On the contrary, until the nineteenth century the sex of bodies used for dissection was explicitly portrayed by including genitalia or breasts, or a wisp of hair falling over the shoulder (in the case of a woman) or a prominent beard (in the case of a man). In the late seventeenth century, the Dutch anatomist Godfried Bidloo produced a set of "true to life" plates unique for their explicit portrayal of the sex of the body dissected. At the same time, male and female bodies were used indiscriminately to illustrate various parts of the body. In William Cowper's 1697 publication of Bidloo's plates, a woman model appears in a series of plates describing the muscles in the upper half of the human body (see Figure 32).[70]

Godfried Bidloo and William Cowper, like Vesalius, focused on two major differences between men and women: external bodily form and reproductive organs. In 1697, in order to portray differences in symmetry and proportion between man and woman, Cowper reproduced two Bidloo figures that were drawn not from life

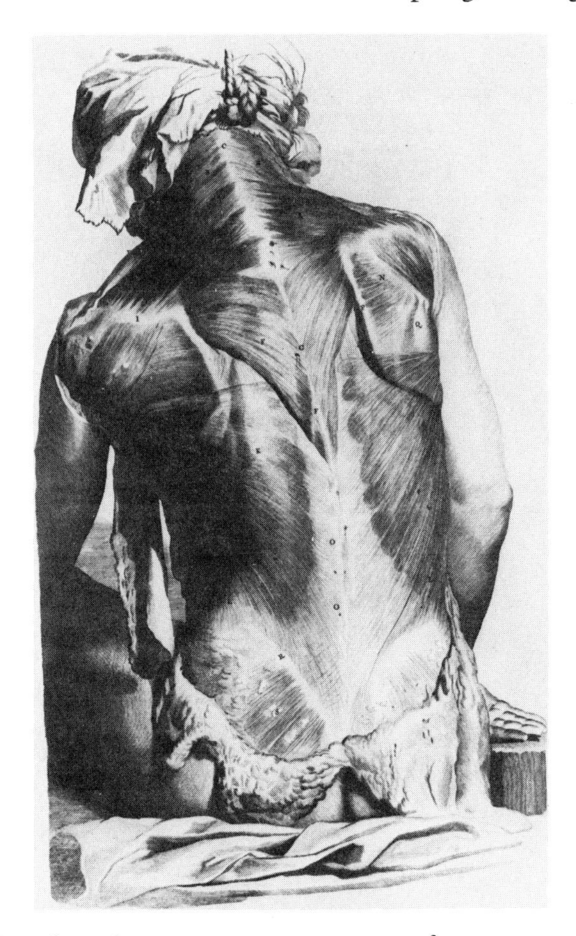

Figure 32. In early modern anatomy texts women, as often as men, were used as models to illustrate nonsexual parts of the body. From William Cowper, *The Anatomy of Humane Bodies* (1697; London 1737), plate 27. By permission of the Boston Medical Library.

but from classical statues; Bidloo claimed these figures exhibited "the most beautiful proportions of a man and woman as they were fixed by the ancients" (see Figure 33). The woman's distinctive parts (labeled A and B) are the breasts and genitalia, but Cowper did not attribute differences in the male and female outline to any structural differences "either in their whole frame, or in the intimate Structure of their Parts." Thus Cowper followed Bidloo in drawing abstractly "human" skeletons. Cowper's *Anatomy of Humane Bodies*

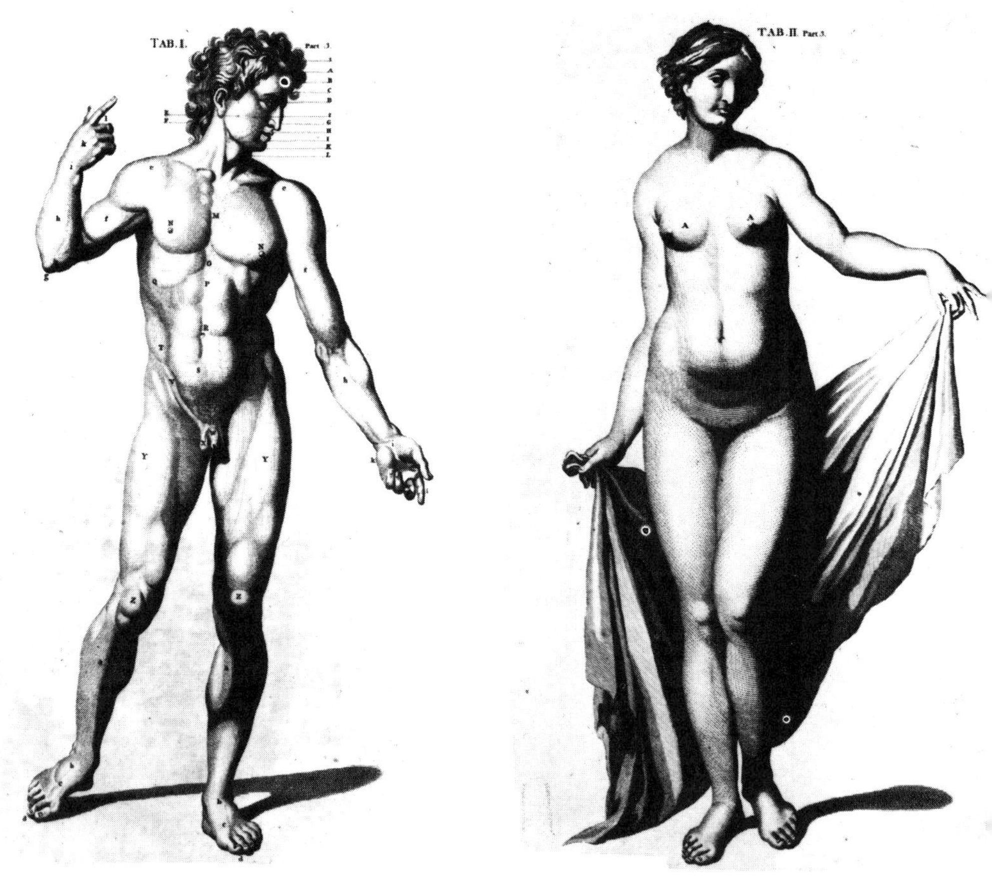

Figure 33. Cowper's male and female nudes drawn to classical proportions, showing differences in external form and reproductive organs. From Andrew Bell, *Anatomia Britannica: A System of Anatomy* (Edinburgh, 1798), plates 42 and 43. By permission of the Boston Medical Library.

begins with two skeletons displayed with various accoutrements—one with an hourglass, another with a drapery—but neither has an explicit gender.

What might at first blush seem an indifference toward the study of secondary sexual difference is, in fact, the product of an incomplete revolution. Though anatomists did not always subscribe explicitly to the ancient theory of humors, they continued for a time to fall back to the view that women are colder (even moister) than men; it took some time before new explanations took the place of the old. Cowper, for example, could ignore the question of sex in the bones of the body, because he had an altogether different explanation for differences in the shape of male and female bodies. According to Cowper, the distinctive form of the female body does not derive from the shape of her bones but from "the great quantity of Fat placed under the skins of women."[71] Fat, of course, accumulates in the absence of heat. William Harvey, too, thought women's distinctive shape could be traced to the layer of fat women wore "as a furred mantle." Harvey found fat virtuous for promoting health, yet the reason for its abundance in women could be traced to a cause traditionally considered a liability—an absence of heat. For Harvey, heat explained many bodily features. Heat expended in "excessive and unseasoned" coition makes men thin. Heat also accounts for the larger brain in men; greater heat increases the supply of blood, which in turn promotes the growth of the brain.[72]

If those in the vanguard of modern anatomy did not focus on secondary sexual character, others did. Helkiah Crooke, barber-surgeon and member of the London College of Physicians, collected and translated the best of medical knowledge (primarily the opinion of the ancients), along with the greatest controversies from the work of the Swiss Gaspard Bauhin and the French André Du Laurens in 1615. Among the "notable controversies" recorded by Crooke was that surrounding the question of the "differences of the sexes." It is important to note that in this section Crooke passed immediately from a discussion of male and female sex organs (here he followed the moderns) to the more important question (for him) of "temperament." In this matter Crooke followed the ancients completely. Weighed in the balance of "Philosophie and of Physicke," Crooke concluded that men are generally "hotter" than women. Many factors conspired to breed a colder woman. First, the female is generated from a colder seed originating in the man's "feminine" left testicle. The seed from the *feminium* testicle is not "exquisitely boiled," but "colder and thinner, having much whey in it." Second, female infants are generated in a colder place (the left side of the womb); the left side is colder because the liver generates heat on the right. The very bodies of women betray their weaker heat: "the habit of a woman is fatter, looser, and softer—fat is not generated but by a weaker heat."[73]

For Crooke and others of his time, heat—the key concept in the ancient view of sexual difference—also explained the perennial problem of ambiguity in sex. Though Crooke did not actually believe Galen's stories of women being changed into men, he did believe that hermaphrodites grew more manly by virtue of an increase in heat. "Hermaphrodites," he explained, "had the parts of both sexes, which because of the weakness of their heate in their nonage [immaturity] lay hid, and brake out afterward as their heate grew into strength."[74] Ambroise Paré, the French barber-surgeon, similarly noted that "the nature of the eunuch . . . may seem to have degenerated into a womanish nature, by deficiency of heat; their smooth body, soft and shrill voice does very much assimilate women."[75] By the same token, a "manly woman" might simply have heat in excess of that appropriate to a woman.

Anatomists, like philosophers of the seventeenth and early eighteenth centuries, left untouched the question of whether sexual character extends beyond the organs of reproduction. In the case of the anatomists, silence allowed ancient views to persist unchallenged even as other parts of the ancient world view were being overthrown. When pressed, most university-trained anatomists (like William Harvey) still traced women's character to their cold and moist nature; most barber-surgeons (like Helkiah Crooke) continued to state quite openly that women were colder than men, and thus defective in sexual temperament. Most anatomists in this period believed that the mind did indeed have a sex, but this was not a view they could reconcile with the ideas of post-Galenic medicine.

7

More Than Skin Deep: The Scientific Search for Sexual Difference

Sexual differences are not restricted merely to the organs of reproduction, but penetrate the entire organism. The entire life takes on a feminine or masculine character.

—Dr. J. J. Sachs, 1830

The revolution in scientific views of sexuality came in the late eighteenth century. Anatomists no longer satisfied with the embarrassingly outmoded theory of humors articulated a new vision for the origins and character of sexual differences, the relation of sex and gender, and the presence of sexuality in the body. From the 1750s through the 1790s, anatomists called for a finer delineation of sexual differences. Sexuality was no longer to be seen as residing exclusively in the sex organs. In 1775, for example, the French physician Pierre Roussel reproached his colleagues for considering women similar to men except in the matter of sexual organs: "the essence of sex," he explained, "is not confined to a single organ but extends, through more or less perceptible nuances, into every part."[1] The German anatomist Jakob Ackermann stated in 1788 that present definitions of sex differences were inadequate, failing as they did to describe the distinctively female body. In his two-hundred-page book detailing every imaginable difference between the bones, hair, mouths, eyes, voices, blood vessels, sweat, and brains of men and women, Ackermann called for anatomists to discover "the essential sexual difference" from which all others flow.[2]

Historians have come to recognize the middle of the eighteenth century as a time of profound transformation in views of sexuality and sexual temperament. Thomas Laqueur, for example, has argued that the old Galenic model of sexual difference, in which men and women were arrayed hierarchically according to their degree of metaphysical perfection (their vital heat), gave way to a new model of biological divergence. According to Laqueur, there emerged an anatomy and physiology of incommensurability in which the relation of men to women was not one of equality or inequality, but rather of difference.[3] But this is not the whole story.

What Laqueur describes is true of views of female genitalia.[4] One important element in the modern reinterpretation of sexual difference was the reevaluation of female sex organs, as we saw in Chapter 6. The emergence of the ideal of motherhood exerted a profound impact on medical views of the uterus. By the early years of the seventeenth century anatomists no longer thought of the uterus as an inadequate penis; instead the uterus was now a perfect instrument for carrying out that foremost task of women: providing their husbands with strong and healthy children.[5] Reviewing the history of medicine in 1829, Carl Klose rejected the comparison of male and female sex organs that had occupied natural scientists from Aristotle to Albrecht von Haller. Klose argued that the uterus, woman's most important sexual organ, had no analogue in man; the comparison with male organs was thus worthless.[6] Yet, in its uniqueness, the uterus still puzzled anatomists. As late as 1804 French physician Gabriel Jouard seemed unsure about how to classify the uterus. Was it a muscle? Part of the vascular system? Or perhaps one of the internal organs, like the liver or spleen? One thing was certain, however: the uterus was unique—*sui generis*—and comparable to no part in the male.[7]

The reevaluation of women's reproductive organs was simply one element in a much broader revolution. The revolution in views of sexuality of the eighteenth century did not limit sexuality to reproductive organs; sex would henceforth permeate the entirety of the human body. This *resexualization* of the body, along profoundly different lines from that of the ancient Galenic world, brought forth a host of new questions to the scientific community. Prominent among these was whether, *apart* from genitalia, there are significant differences between the sexes.[8] By the 1790s, European anatomists presented the male and female body as each having a distinct telos—physical and intellectual strength for the man, motherhood for the woman. Yet even in this age where males and females were considered essentially perfect in their difference, difference was arranged hierarchically. Despite the revolution in views of sex and gender differences, the age-old dominance of men over women remained in force (in spite of opposition to the fundamental premise of the revolution—that sex pervades the body).

The Female Skeleton Makes Her Debut

It was as part of this broader investigation into the nature of sexual differences that the first drawings of female skeletons appeared in Europe in the years between 1730 and 1790. The materialism of the age led anatomists to look first to the skeleton, as the hardest part of the body, to provide a "ground plan" for the body and give a "certain and natural" direction to the muscles and other parts of the body attached to it.[9] If sex differences could be found in the skeleton, then sexual identity would no longer depend on differing degrees of heat (as the ancients had taught), nor would it be a matter of sex organs appended to a neutral human body (as Vesalius had thought). Instead, sexuality would be seen as penetrating every muscle, vein, and organ attached to and molded by the skeleton.

In 1734 anatomist Bernard Albinus produced an illustration of the human skeleton that would serve as the model for anatomical illustration for more than three-quarters of a century (see Figure 34). The work was laborious, taking three months to complete. Albinus drew the skeleton from three different perspectives—front, side, and back—"not free hand as is customary, but from actual measure . . . brought down to scale, either from an indeterminate distance, as architects do . . . or from a distance of forty feet through diopters."[10] Having produced the most perfect possible drawing of the human skeleton (which, when overlaid with its muscles and reproductive parts, was clearly drawn from a male body), Albinus then lamented after the fashion of Genesis: "we lack a female skeleton."[11]

Albinus had good grounds for complaining that the study of female anatomy was inadequate before 1740. The standard studies of the human skeleton by Vesalius and Bidloo had been of the male.

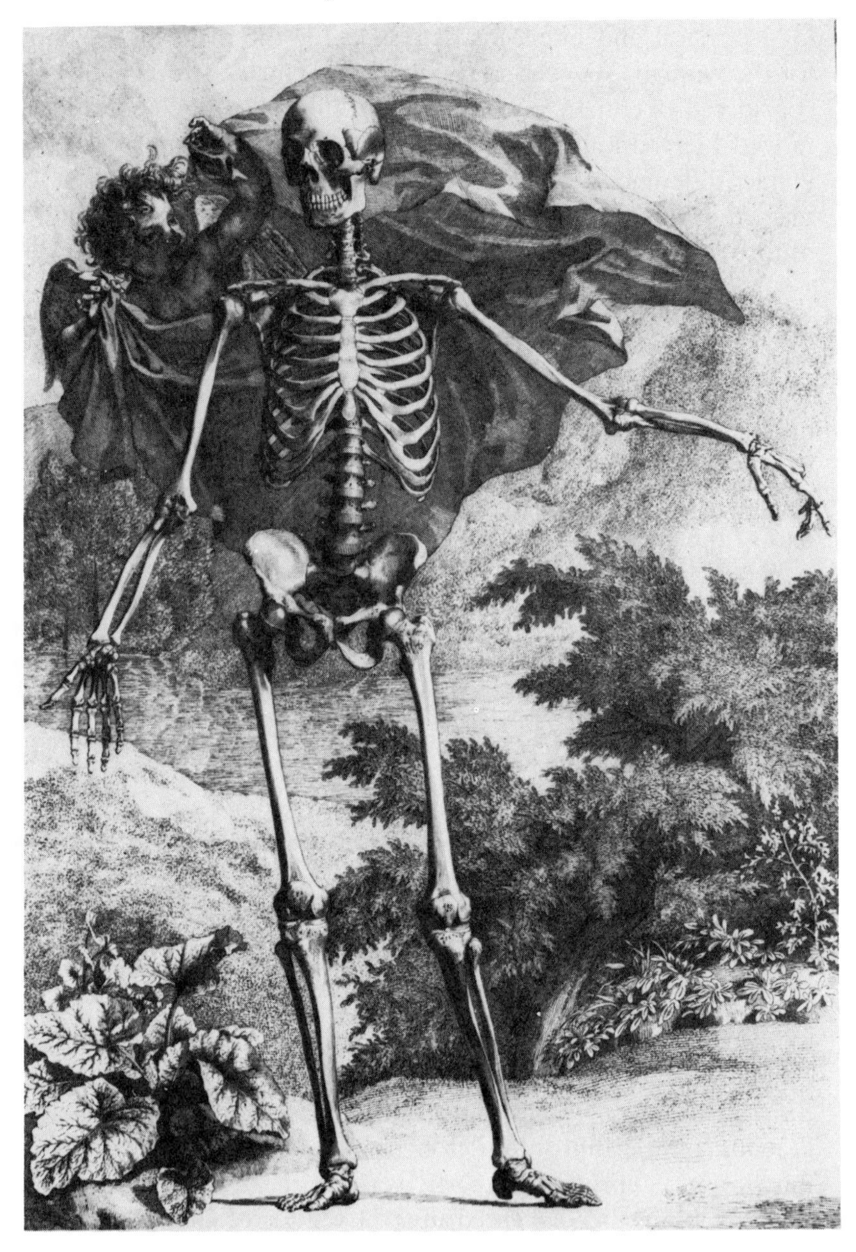

Figure 34. The definitive human skeleton of the eighteenth century. From Bernard Albinus, *Tabulae sceleti et musculorum corporis humani* (Leiden, 1747), plate 1. By permission of the Boston Medical Library.

Only one "crude" illustration of a female skeleton, published by Gaspard Bauhin in 1605, had appeared before the eighteenth century (see Figure 35).[12] Within fifty years of Albinus's plea, however, Europe was flooded with drawings of female skeletons, including those of William Cheselden (1733), Pierre Tarin (1753), Marie Thiroux d'Arconville (1753), and Samuel Thomas von Sommerring (1796).[13] Even though each of these drawings purported to represent *the* female skeleton, they varied greatly from one another.

In 1726 Alexander Monro, professor of anatomy in Edinburgh, appended to his text on *The Anatomy of the Humane Bones* one of the earliest descriptions of the bones of a female. To "finish the Description of the Bones . . . [so] that no part of this Subject might be left untouched," he wrote, "it [is] necessary to subjoin the distinguishing Marks of the Male and Female Sceleton." Though Monro was among the first to give attention to the female skeleton, he considered his study of the female of secondary import. Following a pattern well established since the time of Aristotle, Monro described the female as incomplete and deviant, using the male body as a standard of measure:

> The Bones of Women are frequently incomplete, and always of a Make in some Parts of the Body different from those of the robust Male, which agree to the Description already delivered, unless where the proper Specialities of the Female were particularly remarked, which could not be done in all Places where they occur, without perplexing the Order of this Treatise: Therefore I chose rather to sum them up here by Way of *Appendix*.[14]

Monro identified three causes shaping female bones. A weak constitution, he noted, makes the bones of women smaller in proportion to their length than those of men. A sedentary life makes their clavicles less crooked (their arms are hindered by their clothing and have been less forcibly pulled forward). And a frame proper for their procreative functions makes women's pelvic area larger and stronger to lodge and nourish their tender fetus.

Monro provided one of the earliest descriptions of the bones of the female body in his four-page appendix, but he did not supplement his description with illustrations, since he thought illustrations apt to mislead. The true anatomist, he held, must study anat-

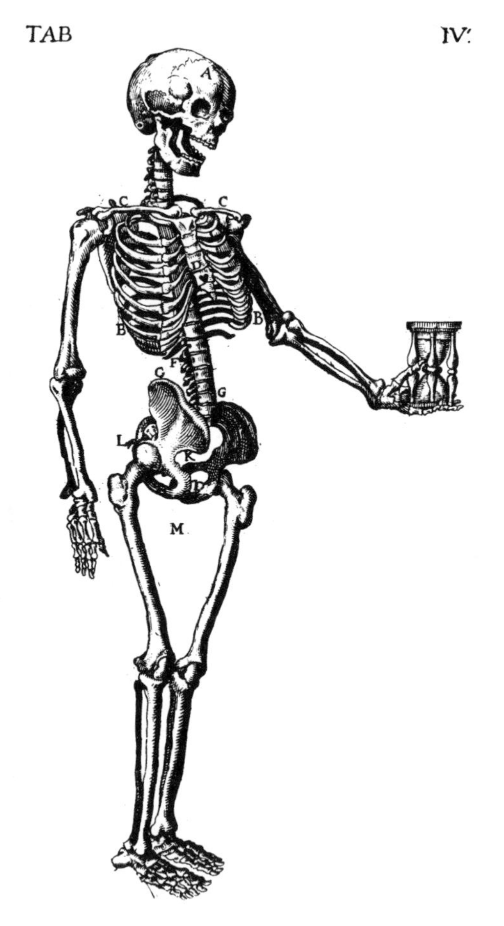

Figure 35. Skeleton of the "bones and gristles" of a woman. Note the heart inscribed on her breastbone. Gaspard Bauhin, *Theatrum anatomicum* (Frankfurt, 1605), plate 4. By permission of the Boston Medical Library.

omy from life. In fact, some of the very first drawings of a female skeleton came from two of his countrymen—James Drake in 1707 and William Cheselden in 1733. Neither of these illustrations, however, was accompanied by a descriptive text or any particular interest in the distinctive qualities of the female. The English anatomist James Drake, for example, included a female skeleton in his *Anthropologia Nova* of 1707. Drake did not, however, set his female skeleton alongside the male for the sake of comparison; rather, he used a male skeleton to show a view from the front and a female skeleton to show a view from the back. Apart from his note that the pelvis is larger in the female than in the male, the bones of both the male and the female are presented in such a way that each is supposed to represent the universally "human," abstracted from particularities of sex.[15]

William Cheselden's *Anatomy* replaced Drake's as a popular textbook as early as 1713. Yet, Cheselden did not draw his female skeleton until 1733. The female skeleton appeared in his elaborate (though commercially unsuccessful) *Osteographia* and was drawn more to the dictates of art than of anatomy. In this work, Cheselden drew his female skeleton in the "same proportion as the Medician Venus," then considered "the standard of all female beauty and softness"; he drew his male skeleton in the same proportion and attitude as the Apollo Belvedere.[16] Like Drake, Cheselden was only marginally interested in the distinctive qualities of female anatomy. Cheselden provided no comparative description of the male and female skeletons; nor did he give an explanation of his interest in the two skeletons. Even after 1733, no illustration of a female skeleton was incorporated into his publications designed for the use of medical students (his *Anatomy* or his *Engravings of the Human Bones*).[17] Nor was his drawing of the female skeleton thought to be particularly significant by his contemporaries. In his strident *Animadversions* on Cheselden's osteography, John Douglas found it curiously "superfluous" that Cheselden had bothered to include a frontal view of two skeletons (the male and the female).[18] This indifference to the bones of the female body may best be explained by the fact that Cheselden traced the lesser stature and strength of the female not to the skeleton but to a more general weakness resulting from menstruation.[19]

Text and image came together in the French rendering of a distinctively female skeleton published in 1759, capturing the imagination of medical doctors for more than half a century (see Figure 36).[20] Anatomist Marie Thiroux d'Arconville, who had studied at the Jardin du Roi, directed the drawings for this skeleton along with other illustrations for her French translation of Monro's *Anatomy*. This skeleton—one of the very few drawn by a woman anato-

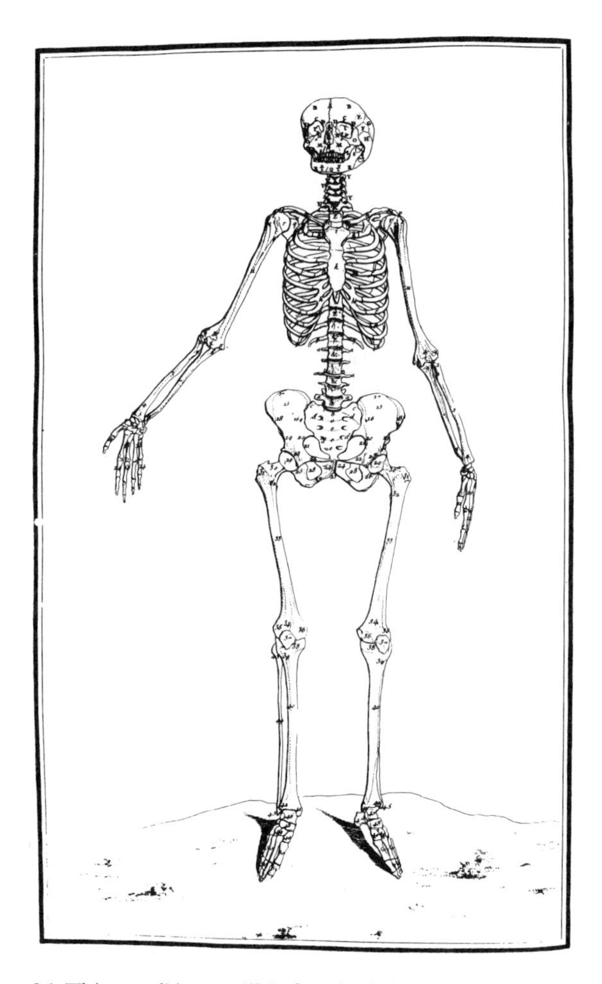

Figure 36. Thiroux d'Arconville's female skeleton, studied for its deviation from the male. From Jean-J. Sue, *Traité d'ostéologie* (Paris, 1759), plate 4. By permission of the Boston Medical Library.

mist—might also be called the most "sexist" portrayal of a female skeleton. (A woman of high social standing, Thiroux d'Arconville carefully guarded her anonymity and published her *Ostéologie* under the protection of Jean-J. Sue, member of the Académie Royale de Chirurgie. See Chapter 8.) In her portrayal of the female, Thiroux d'Arconville exaggerated—almost to the point of caricature—those parts of the body emerging as sites of political debate: the skull as a mark of intelligence and the pelvis as a measure of womanliness.[21] Thiroux d'Arconville depicted the female skull (incorrectly) as smaller in proportion to the body than the man's (see Figures 36 and 37). She also focused attention on the breadth of the pelvis by exaggerating the narrowness of the ribs. In her commentary to the plate, Thiroux d'Arconville emphasized that the chest of the female is narrower, the spine more curved, and the pelvis larger than in the male.[22]

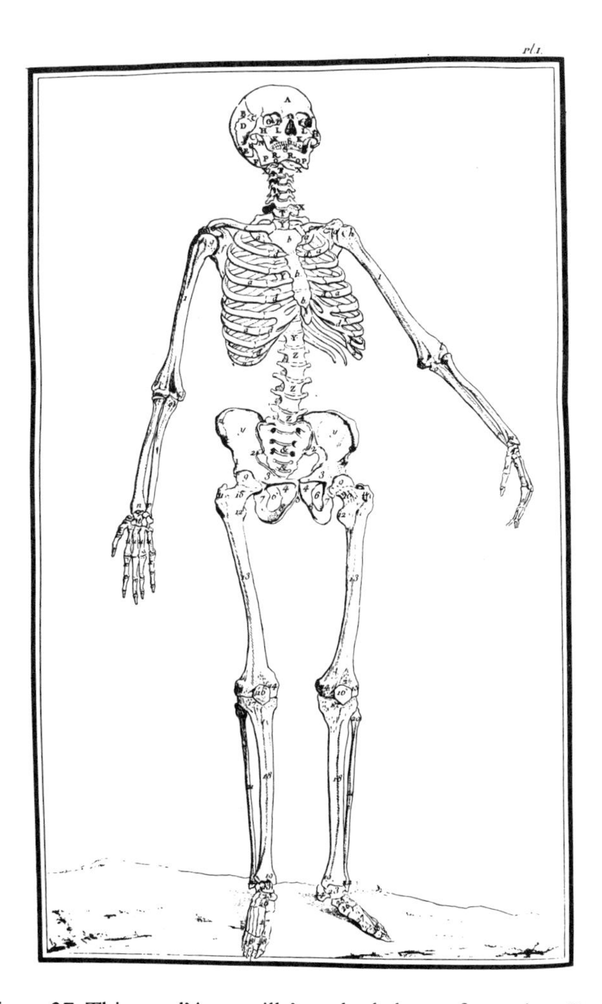

Figure 37. Thiroux d'Arconville's male skeleton, front view. From Jean-J. Sue, *Traité d'ostéologie* (Paris, 1759), plate 1.

Thiroux d'Arconville's skeleton is, in fact, remarkable for its proportions. The ribs are portrayed as extremely narrow and confining, making the pelvis appear excessively large. It would seem that Thiroux d'Arconville either intended to emphasize narrow ribs and wide hips as a mark of femininity, or she chose for her model a woman who had worn a corset throughout her life (see Figure 38). If this latter is the case, the illustration may not have been contorted to meet cultural expectations; rather, the cadaver itself had been disfigured over time. As early as 1741, the French anatomist J.-B. Winslow had warned that regular use of the corset deforms the ribs.[23]

In 1796, the German anatomist Samuel Thomas von Soemmerring produced a rival female skeleton (see Figure 39).[24] Although Thiroux d'Arconville's work was known in Germany, reviewers praised Soemmerring's female skeleton for "filling a gap which until now remained in all anatomy."[25] Directly answering Albinus's plea, Soemmerring spent years perfecting his female skeleton; when it was finished, he considered it to be of such "completeness and ex-

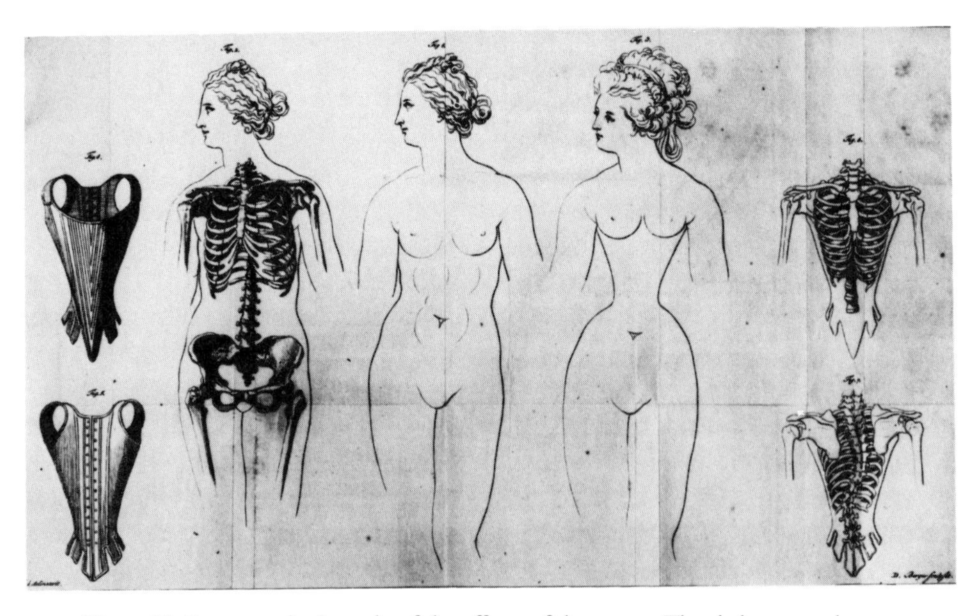

Figure 38. Soemmerring's study of the effects of the corset. The skeleton on the left shows healthy bones; the skeletons on the right show women's bones deformed by a life-long use of the corset. From his *Über die Wirkungen der Schnürbruste* (1785; Berlin, 1793). By permission of the Boston Medical Library.

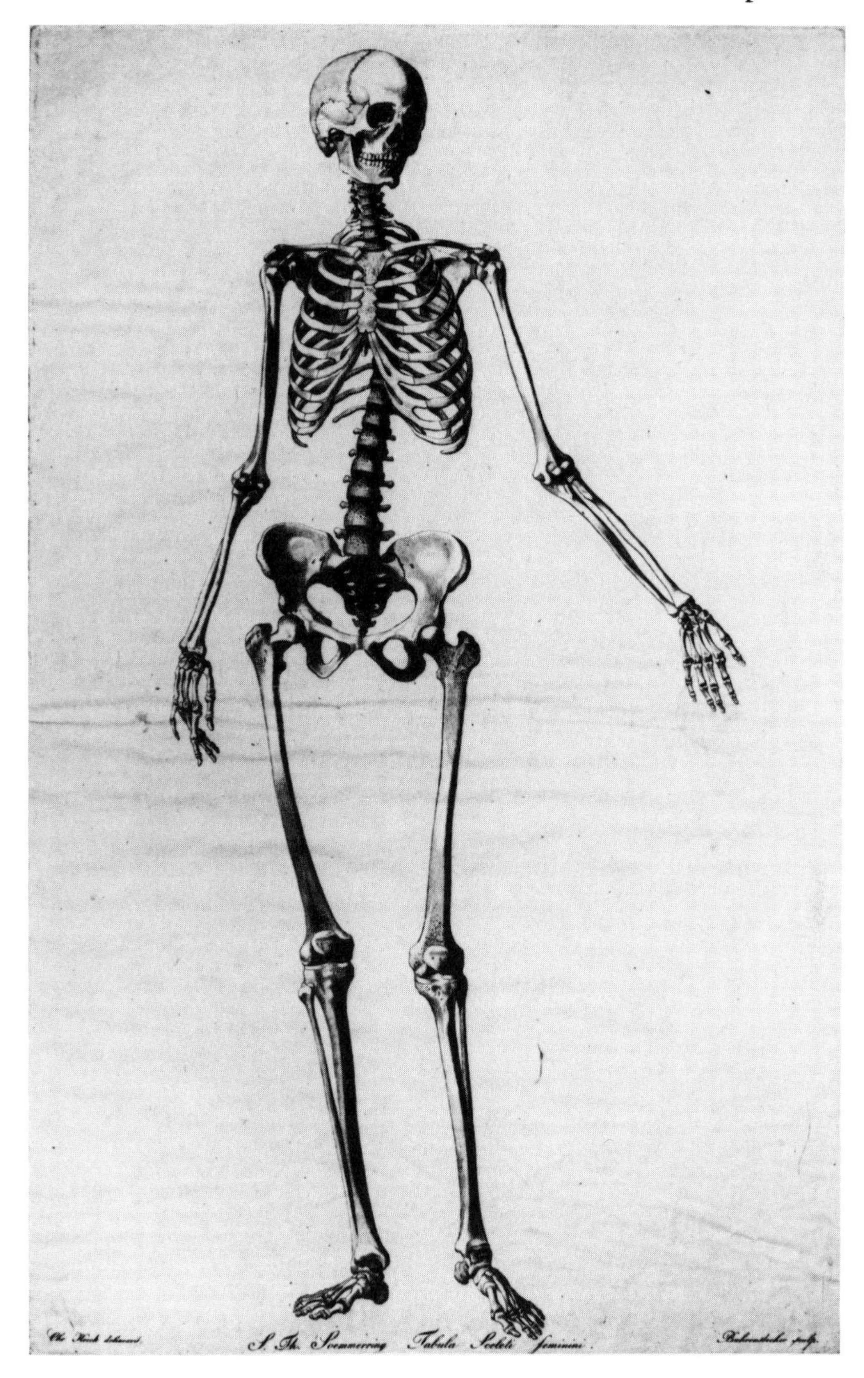

Figure 39. Soemmerring's distinctively female skeleton. From his *Tabula sceleti feminini* ([Utrecht], 1796).

actitude" that it made a perfect mate for the great Albinus male. As a model he selected the skeleton of a twenty-year-old woman from Mainz who had borne a child.[26] Not satisfied with the head of this particular woman, Soemmerring selected the skull of a Georgian woman from Johann Friedrich Blumenbach's famous collection. For proportions, posture, and contours of the body, he checked his drawing against the classical statues of the Venus de Medici and Venus of Dresden. Soemmerring intended his skeleton to represent not an individual woman but (as a nineteenth-century commentator put it) "the most beautiful woman as was imagined to exist in life, with all the carefully observed minutiae of the sexual character of the entire bony structure of woman."[27]

Although Thiroux d'Arconville and Soemmerring drew their female skeletons from nature and considered their work "exact," great debate erupted over the precise features of the female skeleton. In contrast to Thiroux d'Arconville, Soemmerring portrayed the skull of the female (correctly) as larger in proportion to the body than that of the male. Soemmerring drew the ribs smaller in proportion to the hips, but not remarkably so. As one of Soemmerring's students pointed out, the width of women's hips should not be overemphasized; they only appear larger than men's because their upper bodies are narrower, which by comparison makes the hips seem to protrude on both sides.

Despite (or perhaps because of) its exaggerations, the Thiroux d'Arconville skeleton became the favored drawing, especially in England. Soemmerring's skeleton, by contrast, was attacked for its "inaccuracies." Edinburgh physician John Barclay wrote, "although it be more graceful and elegant and suggested by men of eminence in modeling, sculpture, and painting, it contributes nothing to the comparison [between male and female skeletons] which is intended." Barclay criticized Soemmerring, in particular, for showing the incorrect proportion of the ribs to the hips; he defended this view with the argument that the female rib cage is much smaller than that shown by Soemmerring, because women's restricted life style requires that they breathe less vigorously.[28] Barclay concluded that Soemmerring was an artist, but no anatomist.

Crafting Ideals: "Homo perfectus" and "Femina perfecta"

What are we to make of this controversy? Did even the most exact illustrations of female and male skeletons represent the body accurately? One aspect of the scientific revolution in views of sexuality was the use of new methods; sexual differences were to be weighed and measured, described and represented exactly. Anatomists of the time tell us they spared no effort in achieving perfection in their illustrations. As Albinus recounted, he prepared his male skeleton carefully with water and vinegar so that it would not lose moisture and change appearance over the three months of drawings. At the same time that anatomists attempted to represent nature with painstaking precision, they also intended to represent the body in its most beautiful and universal form. Albinus quite consciously sought to capture the details not of a particular body but of *Homo perfectus*—a universal and ideal type. "I am of the opinion," he stated, "that what Nature, the arch workman, . . . has fashioned must be sifted with care and judgment, and that from the endless variety of Nature the best elements must be selected."[29]

Despite anatomists' intentions to represent the body precisely and in its most universal form, representations of the human body in the eighteenth century were laden with cultural values. Illustrations of male and female skeletons represented the bones of the male and the female body; but they also served to produce and reproduce contemporary ideals of masculinity and femininity. This occurred, in part, through the choice of models. Albinus tells us that he collected data from "one body after another" in an attempt to establish average dimensions of the male body. Though his drawing was to represent the true proportions of the male, Albinus was known to collect data only from skeletons pleasing to his eye. In drawing his great male, Albinus selected one "perfect" skeleton to serve as his model according to very specific criteria:

> As skeletons differ from one another, not only as to the age, sex, stature and perfection of the bones, but likewise in the marks of strength, beauty and make of the whole; I made choice of one that might discover signs both of strength and agility: the whole of it elegant, and at the same time not too delicate; so as neither to shew a juvenile or feminine roundness and slenderness, nor on the contrary an unpolished roughness and clumsiness.[30]

In the choice of his subjects, Albinus represented nature not (only) as it is, but (also) as it was most pleasing to the eighteenth-century eye.

Albinus drew his skeleton to conform to both contemporary ideals and to classic representations of the male form. As Hendrik Punt has pointed out, the legs of the skeleton are somewhat long in proportion to the rest of the body.[31] Punt has shown that Albinus selected a long-legged skeleton not necessarily because the average male has such legs, but in order to make his representation of the male body conform to Dürer's portrayal of Adam in his "Fall of Man" from 1504. Albinus also apparently took *himself* to be the perfect measure for his *Homo perfectus.* In an amazing act of self-affirmation, Albinus set the ideal height for his skeleton at 167 centimeters—exactly his own height.[32]

Though Albinus's fame rested on his reputation as a descriptive anatomist, at every step along the way he sacrificed objectivity to the ideal. Even in the drawing—where precise measurements of the subject were transferred exactly to paper—Albinus insisted that nature live up to his ideal: thus he eliminated anatomical details—fissures near small arteries and veins, for example—that would have destroyed the symmetry in the drawing.[33]

Much the same thing happened with drawings of the female skeleton. In the preface to Ackermann's book on sex differences, Joseph Wenzel (one of Soemmerring's students) argued that a sharp physiological delineation between the sexes was impossible, given the great variation among individual men and women. Wenzel stressed that individual variation was as important as group variation: "one can find male bodies with a feminine build, just as one can find female bodies with a masculine build."[34] In fact, he wrote, one can find skulls, brains, and breastbones of the "feminine type" in men. The physician Johann von Döllinger similarly claimed that certain parts of the male genitalia (such as the prostate) are feminine and parts of the female genitalia (such as the uterus) are masculine.[35]

Wenzel and Döllinger, however, were unusual in their emphasis on the ambiguities of sexual differences. By now most anatomists were playing down differences among males or females in order to heighten the contrast between the sexes. Though Wenzel had recognized the ambiguities inherent in sexual typing, he nonetheless established a standard of femininity that he felt transcended sexual ambiguity: "I have always observed that the female body which is the most beautiful and womanly in all its parts, is one in which the pelvis is the largest in relation to the rest of the body."[36] In the process of choosing models for their illustrations, anatomists reinforced the belief that the potential mother was the most womanly woman.

Soemmerring also made every possible effort to "approach nature as nearly as possible," yet, like Albinus, he advocated discriminating selectivity: when surveying the rich variety of nature, the physiologists were always to select the "most perfect and therefore most beautiful specimen" for their descriptions.[37] Ideals of beauty were important, for without them one was unable to detect those cases deviating from the perfect norm. Soemmerring revealed how he chose the "ideal" model for his illustration of the female skeleton:

> Above all I was anxious to obtain the body of a woman remarkable not only for her youth and aptitude for procreation, but also for the beauty and harmony of limbs, of the kind that the ancients used to ascribe to Venus.[38]

Soemmerring strived to create *Femina perfecta,* the perfect mate to Albinus's *Homo perfectus.*

In their illustrations of the female body, anatomists followed the example of painters who "draw a handsome face, and if there happens to be any blemish in it, they mend it in the picture."[39] Anatomists of the eighteenth century "mended" nature to fit emerging ideals of masculinity and femininity. By the nineteenth century, however, the bones of the human body took on additional overtones of masculinity and femininity. In 1829 John Barclay brought together the finest illustrations from the European tradition for the sake of comparison. As the finest example of a male skeleton Barclay chose the Albinus drawing. Then, looking to the animal kingdom, Barclay sought an animal skeleton, one which would highlight the distinctive features of the male skeleton. The animal he chose was the horse, remarkable for its strength and agility (see Figure 40). As the finest representation of the female skeleton Barclay chose the delicate Thiroux d'Arconville rendition. This he compared to an

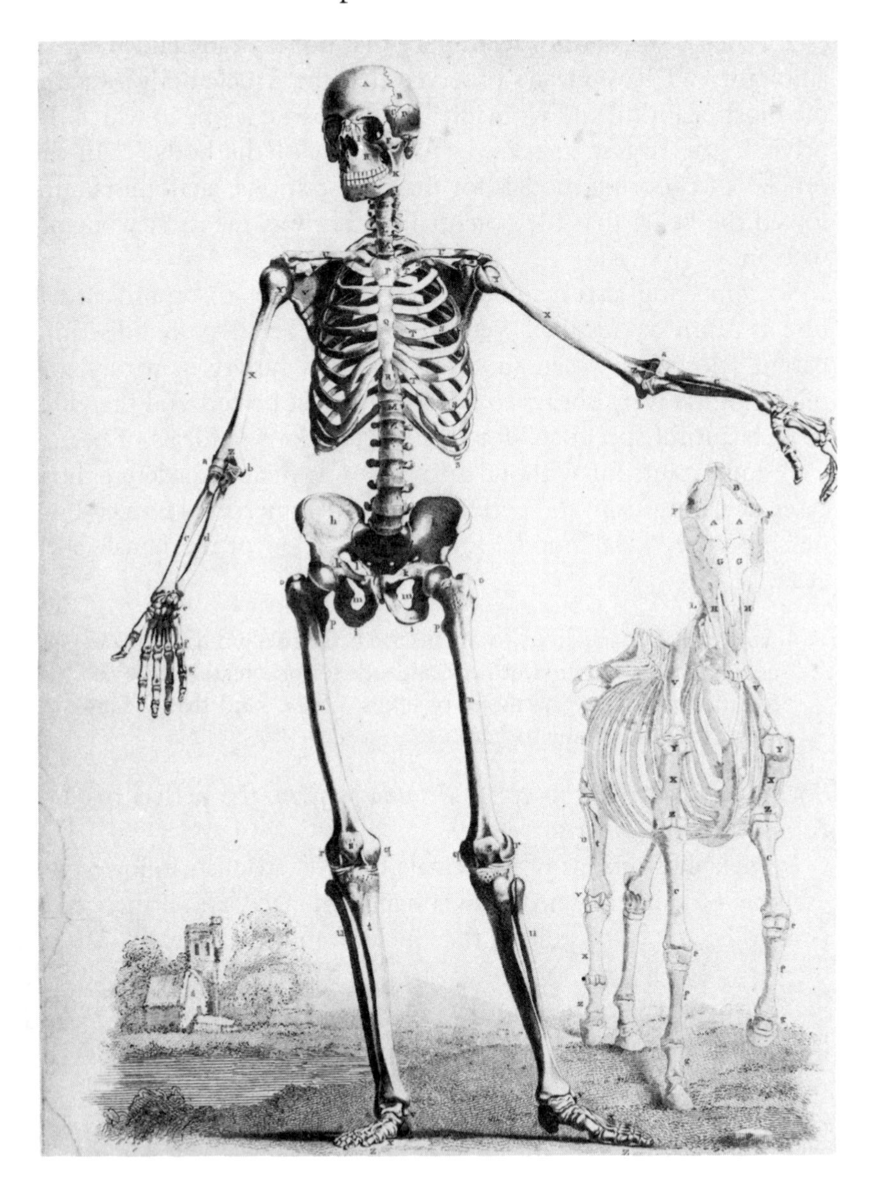

Figure 40. Masculinity reaffirmed. The male skeleton compared to the horse. From John Barclay, *The Anatomy of the Bones of the Human Body* (Edinburgh, 1829), plate 1. By permission of the Boston Medical Library.

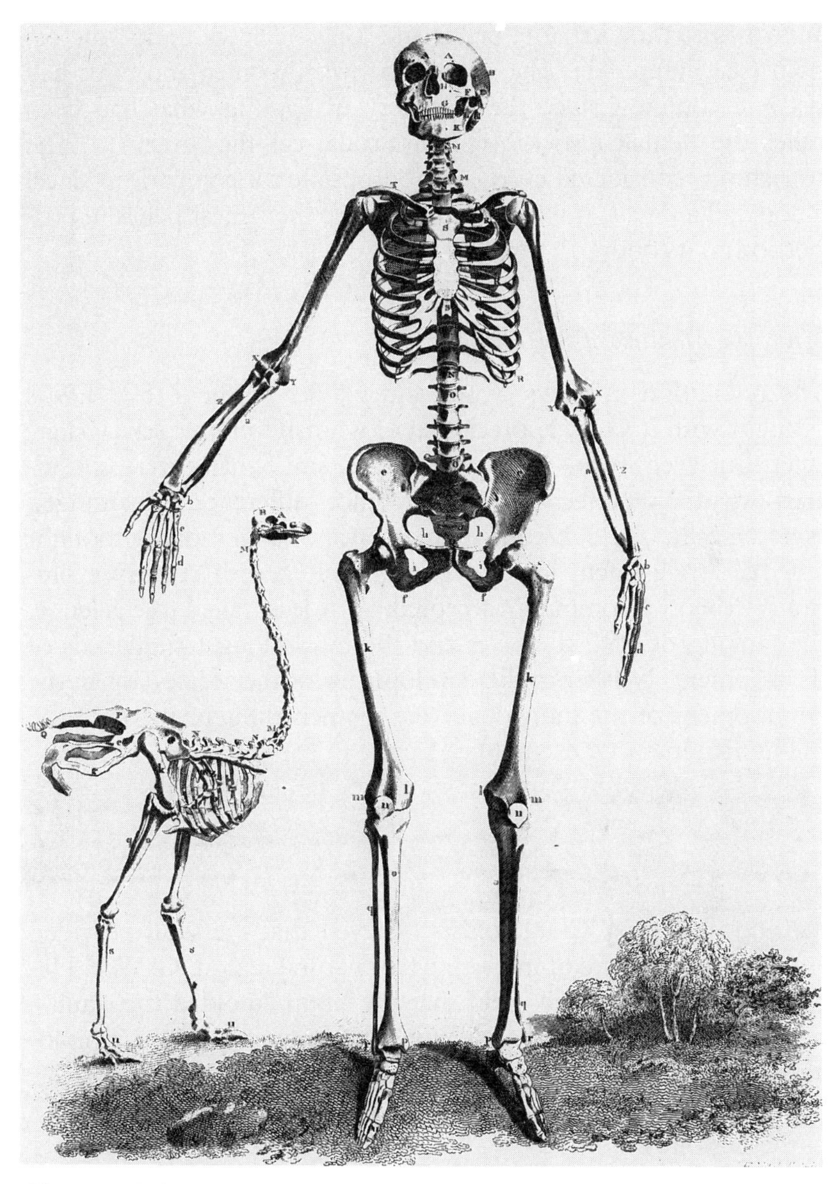

Figure 41. D'Arconville's female skeleton compared to an ostrich; both are remarkable for their large pelvis and long elegant neck. From Barclay, *The Anatomy of the Bones of the Human Body,* plate 4. By permission of the Boston Medical Library.

animal noted for its large pelvis and long, willowy neck—the ostrich (see Figure 41). The analogy between women and birds had become common since the discovery of eggs in what had been called the "female testicle" (what we today call the "ovary").[40] The ostrich was considered especially appropriate for comparison since, among all animals, it was thought to have the largest pelvis in proportion to its body.

Man, the Measure of All Things

The revolution in views of sex and gender of the 1750s–1790s brought with it a new appreciation of woman's unique sexual character. But even in this age, where males and females were considered essentially perfect in their difference, difference was arranged hierarchically. The subsequent emergence of evolutionary thought left this arrangement intact, for in the nineteenth century evolutionary theories commonly accepted, or at least failed to challenge, the ranking of both the sexes and the races along a single axis of development. Neither in the development of the species nor in the development of the individual were women thought to attain the full human maturity exemplified by the European male.[41]

Nineteenth-century anatomists claimed that women's development had been arrested at a lower stage of evolution, citing sexual differences again as evidence. As in the eighteenth century, attention remained focused on the skull and the pelvis as indices of human development. Craniologists believed that the skull provided an objective measure of intelligence or natural reason. G. W. F. Hegel, following F. J. Gall, held that the brain "molded the skull—here pressing it out, there widening or flattening it."[42] Craniologists analyzed the skulls of men and women, whites and blacks, hoping to measure more exactly the intellectual capacities of each of these groups.

Though the full flowering of the baroque age of craniology did not come until late in the nineteenth century, efforts to compare and contrast the male and female skull began in the eighteenth century.[43] When Soemmerring drew his female skeleton in the 1790s, he rejected Thiroux d'Arconville's portrait of the female skull. Where Thiroux d'Arconville had insisted that the female skull was smaller in proportion to the body than the male skull, Soemmerring pointed out that women's skulls are actually heavier than men's relative to total body weight (1:6 for women; 1:8 to 1:10 for men).[44] From this Ackermann (Soemmerring's student) concluded that women's brains are actually *larger* than men's, for the following reason:

> Women lead a sedentary life and consequently do not develop large bones, muscles, blood vessels and nerves as do men; since brain size increases as muscle size decreases, it is not surprising that women are more adept than men in intellectual pursuits.[45]

Like Renaissance scholars before him, Ackermann judged not the size and strength of the body but rather the delicacy of its make as the appropriate determinant of intelligence.

Soemmerring's findings and Ackermann's conclusions ran counter to perceptions that it was indeed men who were the more intelligent and creative of the species. In subsequent years, however, anatomists had to concede the truth of Soemmerring's depiction of the female skull as larger than the male skull in proportion to the rest of the body. Yet they did not conclude that women's large skulls were loaded with heavy and high-powered brains. Rather than a mark of intelligence, women's large skulls signaled their incomplete growth. The Edinburgh anatomist John Barclay in the 1820s used the proportionally larger size of the female skull as evidence that physiologically women resemble children, whose skulls are also large relative to their body size.

Barclay used this resemblance as an opportunity to realign the focus of comparative anatomy. In the eighteenth century, with the rise of companionate marriages (that is, marriages in which the partners are friends and lovers), anatomists had emphasized the comparison between the adult female and the adult male skeleton: the female skeleton never appeared except in the company of her (solidly built) mate. In the nineteenth century, however, anatomists played down the comparison between the male and female in order to draw attention to similarities between the skeletons of the adult female and the child. In 1829, Barclay presented for the first time a skeleton family (see Figure 42). Though anatomical drawings of children and fetuses had been published since the early eighteenth

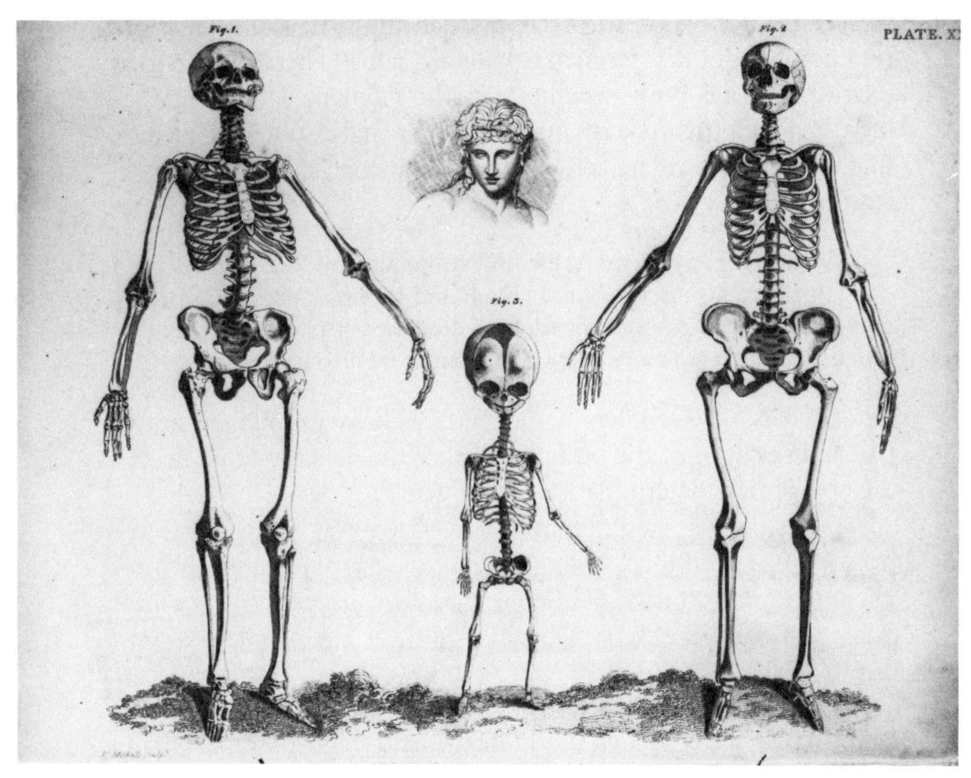

Figure 42. Barclay's skeleton family. Barclay rejected eighteenth-century comparisons of male and female skeletons and emphasized that the skeleton of the female more closely resembles that of the child. From Barclay, *The Anatomy of the Bones of the Human Body,* plate 32. By permission of the Boston Medical Library.

century, Barclay was the first to bring together the skeletons of man, woman, and child for the sake of comparison. As he noted in his commentary, Barclay introduced the child's skeleton (using the Thiroux d'Arconville plates) to show that many of those characteristics which Soemmerring had described as peculiar to the female, were actually "more obviously discernible in the fetal skeleton."[46] Barclay pointed out that in addition to equivalent skull sizes, both women and children have smaller bones compared to men; the rib cage, jaw shape, and feet size of women are also more similar to those of children than to those of men.

In choosing the Thiroux d'Arconville female for his comparison with the child, Barclay curiously chose a skeleton with a proportionately smaller skull. Barclay favored Thiroux d'Arconville's rendering, however, for its portrayal of the frontal suture (not shown in Soemmerring's depiction).

But skull size was not the only anatomical index of worth in the nineteenth century. Woman might be considered childlike in respect to her skull, yet in respect to her *pelvis* she set a standard of excellence. Barclay argued that it was "here alone that we perceive the strongly-marked and peculiar characters of the female skeleton."[47] Though inferior in skull size, in the realm of the pelvis woman was considered undeniably superior.

In the larger scheme of things, however, woman's superior pelvis size was not enough to set her equal to man. For in fact, the purpose of the superior pelvis was ultimately to provide passage for the superior skull. Craniologists were quick to point out that the European female pelvis must necessarily be large in order to accommodate in the birth canal the cranium of the European male. Thus male and female bodies were indeed complementary: the superior female pelvis complemented the superior male skull. And it was the man, after all, who was considered to hold in his larger cranium the seeds of civilization. The woman was simply designed to oblige him.

Focus on pelvis size served to naturalize woman's role as mother, but at a time when the image of the childlike woman was also becoming more and more popular in medical literature. In *Das Weib und das Kind* (1847) the German Dr. E. W. Posner described at great length the physical bases for the comparison between women and children. Women, like children, have shortened limbs and larger and rounder abdomens in relation to their chests; women's heads also tend toward "the childish type."

> The finer bone structure, the tender, less sharply developed facial features, the smaller nose, the larger, childishly round face, all clearly show this similarity . . . The nerves and blood vessels of women are also as delicate and fine as those of children . . . and the skin with its layer of rich fat is childish.[48]

Posner saw woman's "childish roundness" as a result of the fact that a woman stops growing at age fourteen (earlier than the male, at age eighteen) and thus never reaches full maturity.

While the comparison of woman to child was not in itself intended to be derogatory (children in the nineteenth century also represented innocence, freshness, and youth), the female portrayed by Posner never managed to reach the full human maturity exemplified by the European male body-type.

The comparison of women with children was by no means new in the nineteenth century. Xenocrates and Hermagoras in the ancient world had held that a woman can never be more than a child.[49] Galen thought that women and children suffered in a similar fashion from cold and moist humors, accounting for their lack of self-control. Aristotle grouped women, children, and slaves together in the three states of minority.[50] This attitude held strong in political literature of the seventeenth and early eighteenth century. Lord Chesterfield is remembered for his opinion that "women are but children of a larger growth."[51] By drawing parallels between women's and children's bodies, anatomists translated traditional views of women into the language of modern science.

One should recall that the image of a childlike woman painted by anatomists was not out of step with other aspects of European custom. At a time when middle-class wives were (on average) ten years younger than their husbands, it is not surprising that middle-class women should have appeared childish in comparison to their husbands.[52] Alexander Monro, who gave one of the first descriptions of the female skeleton, found this age difference quite proper. In an essay on "female conduct" written for his daughter, he stressed that "the most equal match surely is where the man is some years older than the woman he marries, that his greater experience may make him capable of maintaining the superiority he is entitled to."[53]

Women were not the only group singled out as childish. Characteristics viewed as sex-linked were also given surprising national connotations. In the seventeenth century the English viewed the French as unreformably feminine; in the nineteenth century (in the years following the Napoleonic wars) Germans accused the French of being immature and childish. The German educator Johann Ziegenbein found that "the French have all the deficiencies and advantages of children"; the French were also "womanly." As he wrote, all three groups—the French, women, and children—display a "lyric sensitivity," like to be flattered, are easily provoked, and love bright colors.[54]

The Analogy between Sex and Race

The study of sex differences was but one preoccupation of anatomists in the eighteenth century. Samuel Thomas von Soemmerring, who drew one of the first distinctively female skeletons, was not primarily interested in women's anatomy but in the anatomical basis of racial differences. In his major work on the comparative anatomy of the Negro and the European (*Über die körperliche Verschiedenheit des Negers vom Europäer*), Soemmerring defined racial differences in much the same way as he defined sexual difference: "if skin were the only difference, then the Negro might be considered a black European."[55] Difference, in other words, was more than skin deep. As in the case of the female skeleton, Soemmerring looked to that foundation of the body—the skeleton—for the essential differences from which all others derived. Racial differences, rooted in the skeleton, molded differences in muscles, nerves, and veins. Thus race, like sex, penetrated the entire life of the organism.

Anatomists attempted to rank the sexes and races in a single, hierarchical chain of being according to cranium and pelvis size (see Table 2). In respect to skull size, the European male represented the fully developed human type, outranking the African male, the European female, and the African female. In this racial and sexual hierarchy, the European male clearly held the superior position. The relative position of the African male and the European female, however, was not clear-cut; craniologists—in their efforts to highlight the comparison of each of these two groups to the white male—tended to emphasize their resemblance (and not difference) with words such as "men of the black races have a brain scarcely

Table 2. Sexual and racial hierarchy in the early nineteenth century

According to skull size	According to pelvis size
European male	European female
African male *or* European female	African female *or* European male
African female	African male

heavier than that of white women" or "the negro resembles the female in his love for children, his family, and his cabin—the black man is to the white man what woman is to man in general, a loving being and a being of pleasure."[56] Others found measurements indicating that the white woman's cranial capacity was slightly superior to that of the black man. But craniologists generally did not focus on how the black man (the dominant sex of an inferior race) should be ranked vis-à-vis the white woman (the inferior sex of the dominant race). Where the African female ranked, by contrast, was all too clear: in every culture and race, the absolute weight of the brain was always heavier in men than in women.[57] Moreover, craniologists emphasized that differences between the sexes, with regard to brain size, increase with the development of the race, so that "the male European excels much more the female than the Negro the Negress."[58] Thus craniologists believed (in direct contrast to *philosophes* such as Condorcet) that the inequality of the sexes increases with the progress of civilization.

Historians have focused attention on craniologists' measurement of skulls. In the eighteenth century, however, another measure—that of the pelvis—was thought equally important for understanding the physical and moral development (as it was then called) of the races. With pelvis size, sexual (though not racial) hierarchy was reversed. Here the European female represented the fully developed human type, outranking the European male.[59] The inconclusiveness that had plagued the relative position of the African male and European female in relation to cranial capacity was felt here in relation to the position of the African female and European male. Again, craniologists were more interested in the similarities between these two groups with regard to pelvis size than in their differences. What was clear, however, was that the African female stood well above the black male, whose pelvis was seen as so narrow that it was nearly apelike. Explaining these differences, a Cambridge University lecturer pointed to the fact that the pelvis in the black female was narrower in accordance with the slight inferiority of fetal head size among blacks.[60]

In the nineteenth century, sex and race increasingly came to define social worth. Anatomists (mostly male and European) studied sex and race, using the European male as the standard of excellence.

At the same time, anatomists believed their work free from bias, reflecting only the "cold-blooded" findings of science. According to Soemmerring, the anatomist did not have to take a *moral* stand in this matter because the body spoke for itself. Soemmerring considered sexual and racial differences not accidental, but "certain, definite, and distinctive." Interestingly, Soemmerring also assured his readers that personally he was not a racist; he considered blacks "no less human than the most beautiful Greeks" and expressed his firm opposition to slavery.[61] In what he considered an expression of neutrality on the question of racial superiority, Soemmerring wrote, "it would have made no difference if my results had shown that whites, rather than blacks, were nearer the ape." Soemmerring simply believed that nature, and not man, had created the inequalities between the sexes and races.

8

The Triumph of Complementarity

> The male body expresses positive strength, . . . sharpening male understanding and independence, and equipping men for life in government, in the arts and sciences. The female body expresses womanly softness and feeling . . . The roomy pelvis determines women for motherhood . . . The weak, soft limbs and delicate skin are witness of woman's narrower sphere of activity, of home-bodiness, and peaceful family life.
>
> —Dr. J. J. Sachs, 1830

Why did the comparative anatomy of men and women become a research project for the medical community in the late eighteenth century? Anatomists' interests in the female body were shaped, in part, by concerns for women's health. Jakob Ackermann, for example, who took a particular interest in the build of the female body, appended a chapter on women's health to his book on sex differences, arguing that doctors should consider differences in body-build that might influence the course of a sickness.[1] The comparative study of skeletons was thought particularly useful to the practice of medicine: as one physician argued, a bone or cavity of a particular size natural in a man might indicate disease in a woman.[2] This concern for women's health on the part of physicians came at a time, we should recall, when doctors were taking health care out of the hands of midwives because, it was argued, they lacked an academic understanding of female anatomy.

Developments internal to the medical community provide one explanation for anatomists' interests in the female body. The real force fueling the search for sex differences, however, was political. As we have seen, the *querelle des femmes* had long plagued European elites. This debate intensified and took on new meaning in the tumultuous years leading up to the French revolution.[3] Enlightenment thinkers faced a dilemma: how was the continued subordination of women to be reconciled with the axiom that all men are by nature equal? Grappling with this problem in Diderot's and d'Alembert's great *Encyclopédie,* the Huguenot medical doctor Louis de Jaucourt wrote, "it appears at first difficult to demonstrate that the authority of the husband comes from nature because that authority is contrary to the natural equality of all people."[4] How were Enlightenment thinkers to justify the inequality of women in the newly envisioned democratic order?

Crucial to these debates was the question of epistemological moorings. Throughout the sixteenth and seventeenth centuries women's subordinate status had been explained by a variety of authorities, including Aristotelian science and Christian teachings. By the 1750s, however, old authorities were succumbing to new. Though few agreed on the exact description of woman's nature, a new consensus was emerging on how the question should be discussed.

It is not immediately obvious that science should have been called upon to arbitrate an essentially political debate. In the eighteenth century, however, there was still great optimism that social issues—such as women's rights and abilities—could be resolved by science. The belief that science was impartial (*unparteyisch,* as Soemmerring and his colleagues called it) led to the hope that science could provide objective evidence in the debate over woman's intellectual and physical character. Perhaps the knife of the anatomist could find and define sexual difference once and for all. Perhaps sexual differences—even in the mind—could be weighed and measured.

"Nature" and its laws played a pivotal role in the rise of liberal political theory. Philosophers, such as John Locke and Immanuel Kant, attempted to set social convention on a natural basis by identifying the natural order underlying the well-ordered *polis.* Natural law (as distinct from the positive law of nations) was held to be immutable, given either by God or inherent in the material universe. Within this framework, an appeal to natural rights could be countered only by proof of natural inequalities. The marquis de Condorcet wrote in reference to the equality of women: if women are to be excluded from the *polis,* one must demonstrate a "natural difference" between men and women in order to legitimate that exclusion.[5] In other words, if social inequalities were to be justified within the framework of liberal thought, scientific evidence would have to show that human nature is not uniform, but differs according to age, race, and sex.

In his article on women in the *Encyclopédie,* Jaucourt employed this way of thinking to resolve his dilemma about women's position in the new democratic order. Marrying the laws of men to the perceived laws of nature, Jaucourt wrote with unabashed conviction that "the laws and customs of Europe unanimously and decisively give authority to the male because he is the one with the greatest strength of mind and body." After the 1750s the anatomy and physiology of sexual difference seemed to provide a kind of bedrock upon which to build relations between the sexes. The seemingly superior build of the male body (and mind) was cited more and more often in political documents to justify men's social dominance. The discovery of anatomical and physiological differences between men and women appeared to make male privilege (as stated in the Prussian code of 1750) "certain and universal."

The Domestic Imperative

The resolution of the "woman question" was clear and decisive. The revolution in European life and manners between the 1760s and the 1820s brought with it the triumph of the notion of sexual complementarity. Göttingen philosopher Christoph Meiners spoke for many when he wrote that those who claim women are capable of entering the professions must have reflected very little on the difference between the sexes: "what would happen to the happiness of families if women, formed to bear, suckle, and bring up children, and likewise to manage domestic concerns, were to leave their houses, their children, and their servants to take their seats with their husbands in the legislative assembly of the nation, the courts of law, public offices, and even to join the army and encounter the dangers of war?"[6] There was a clear sense that equality for women threatened their domestic duties—duties considered important to both middle-class life and national policy.

If the new rights of citizens were not to extend to women, liberal democratic theory had to be amended. The theory of *sexual comple-*

mentarity, a theory which taught that man and woman are not physical and moral equals but complementary opposites, fit neatly into dominant strands of liberal democratic thought, making inequalities seem natural while satisfying the needs of European society for a continued sexual division of labor by assigning women a unique place in society.[7] Henceforth, women were not to be viewed merely as *inferior to* men but as fundamentally *different from,* and thus *incomparable to,* men. The private, caring woman emerged as a foil to the public, rational man. As such, women were thought to have their own part to play in the new democracies—as mothers and nurturers.[8]

The theory of sexual complementarity represented an ideological resolution of problems inherited from the seventeenth century. From the time of Molière and Fénelon, there had been an attempt to rein in the privilege of aristocratic women of both court and salon. Like his successors in the eighteenth century, archbishop François Fénelon (1651–1715) had prescribed domesticity as an antidote to the power of aristocratic women. His ideal of domesticity, however, was backward looking—attempting as it did to reinstate pre-Valois restrictions limiting women even of the highest rank to a retired life, amidst their families. Domesticity, alongside royalism and clericalism, was fundamental to the seventeenth-century aristocracy's desire to preserve its political power.[9]

Many at this time expressed their fears that women were no longer fulfilling their duties as conscientious mothers. Satires of the learned lady, which began to appear in the 1660s along with the *virtuosa* herself, turned not on Aristotelian or Galenic views that women's brains are too moist to form solid judgments, but on the fear that the greater equality of women in intellectual culture would undermine established social hierarchies. Molière was much acclaimed for undercutting the *précieuses* with his ridicule.[10] He portrayed Cartesian women as running mad after philosophy and having no time for marriage or household duties. In *Les Femmes savantes* (1672), the best known of these satires, a husband (Chrysale) whose dinner has been neglected rails against his science-minded wife and women like her for ignoring their domestic obligations. "They want to write and become authors; no science is too deep for them . . . They know the motions of the moon, the pole star, Venus, Saturn, and Mars . . . and my food, which I need, is neglected."[11] Women's disregard for household duties upset the patriarchal household; women are also charged with encouraging their servants to rise above their proper station.[12] Chrysale thus complains that his daughter has taught her servants Greek and Latin: "My servants aspire to science in order to please you . . . One burns my roast while reading some story. Another dreams of some verse when I ask for a drink. In short, I see your example followed by them, and though I have servants, I am not served."[13]

Louis de Lesclache was another who suggested that the new experimental philosophy threatened women's traditional virtues. In his *Les Avantages que les femmes peuvent recevoir de la philosophie et principalement de la morale,* Lesclache criticized his wife for endangering the reputation of her house by mixing with unsavory people—astrologers, chemists, and palm readers—in unfashionable neighborhoods, and for no end but "to discover, by pouring the contents of one bottle into another, how much of the larger bottle could be held in the smaller." He also complained that she had spent an entire night on the roof observing the moon with her *grandes lunettes.* Her extravagance was so great that she imagined the moon to be inhabited, promising a rich reward to anyone who could invent a glass powerful enough to discover what might be the latest fashions of people on the moon. Lesclache concluded that women should not be taught natural philosophy, but rather moral philosophy, with lessons in prudence, temperance and justice; they should search only for that knowledge which "establishes order in their homes, serves their children, and increases their fear and love of God."[14]

Fears that the learned lady threatened to disrupt the status quo were justified: it was part of the political program of *salonnières* of the seventeenth and eighteenth century to eschew traditional forms of marriage and motherhood. With books to read and lectures to attend, noblewomen shifted the responsibilities of motherhood to wet nurses and governesses. From the thirteenth through the sixteenth centuries, it was common for an aristocratic woman to send her infant to the countryside within hours of birth to be suckled and reared by another woman, usually of the peasantry.[15]

Women's desires to engage like men in productive lives free of the cares of parenting came into conflict with growing beliefs that those nations were strongest which had the largest population. Though wet nursing had provided a solution to the problem of child rearing for (upper-class) mothers and fathers who preferred other pursuits, it also resulted in high infant mortality. Nurses often had too little milk for the infants in their care, and they often lived in poverty and were overworked. Fears began to grow that Europe was losing population; in France, philosophers and physiocrats believed (incorrectly) that the population was declining. The marquis de Mirabeau, a physiocrat, traced depopulation to the neglect of mothers for their children, alongside other factors such as the concentration of property in few hands, luxury, the financial system, and the decadence of agriculture. For mercantilists, too, the human being became a precious commodity for the state. Alphonse Didelot summarized these concerns in 1770: "a state is powerful only insofar as it is populated . . . Let the arms that manufacture and those that defend it be more numerous."[16] The preservation of family and maternal duties became important matters of state.

For state ministers, the simplest way to increase birth rates was to reduce infant mortality through improved training for surgeons, midwives, and—most especially—for mothers. An important element in this campaign was a series of health and conduct manuals written for women by medical doctors.[17] The increasing cultural authority of doctors in the eighteenth century shifted authorship of conduct manuals from churchmen (such as Thomas More or Fénelon) to medical doctors (such as Alexander Monro, who had also provided one of the earliest descriptions of the female skeleton).[18] These manuals emphasized the virtues of mothering, but also the need to establish a quiet and confined lifestyle appropriate to women's delicate physiology.

Through various means, then, a new value was attached to mothering. Noble and bourgeois women were admonished to abandon their salons and wet nurses and to take up the duties of breastfeeding their infants. Enthusiasm for breastfeeding within the learned community in France was so great that in 1763 the *Journal des Savants* listed all works on the subject.[19] This enthusiasm (along with other policies of the physiocrats) soon spread to other nations. In Prussia, the regime—with less finesse than its French neighbor—

simply passed a law requiring that "a healthy mother breastfeed her child."[20]

In this sense, attempts to domesticate women preceded Rousseau—often considered the father of efforts to establish an exclusively domestic sphere for women. By 1762, when *Emile* was published, the cult of domesticity was taking Europe by storm. Rousseau's work served as a bridge from seventeenth-century attempts to domesticate noblewomen to eighteenth-century ideals of motherhood, prescribed for (though not necessarily achieved or desired by) women of all classes.[21]

The Physicalists' Foundations of Complementarity

Rousseau abhorred the public influence of elite French women and the stirrings of feminism in prerevolutionary France. Social equality of the sexes, which he believed to follow from the "maxims of modern philosophy," led to an undesirable "confounding of the sexes"—equal duties, similar employment, and masculine women. He took on proponents of women's equality by arguing that woman is man's complement, not his equal. According to Rousseau, sexual differences prescribed vastly different roles for men and women in society. Rousseau's great success in framing a new brand of complementarity for the eighteenth century came from his clearly articulated solution to the two threats women posed to middle-class men: the threat of traditional power and privilege wielded by aristocratic women, and the new demands for equality coming from women of the third estate.

Rousseau's theory of sexual complementarity met with success because it developed new foundations for old arguments. The ancient theory of humors had correlated character (or temperament) with physical characteristics; a sedentary style of life was thought to cultivate a moist body, which, in turn, housed a soft mind. Modern science changed the causal structure of the argument but not its components. For Rousseau, physical asymmetries between the sexes explained the differences one observed in men and women's moral character and daily lives. Natural philosophy was to read in the book of nature "everything which suits the constitution of her [woman's] species and her sex in order to fulfill her place in the physical and moral order."[22]

Medical evidence played into Rousseau's hand. The first part of his argument centered on the physical complementarity of the sexes. Rousseau opened "Sophie, or the Woman," book 5 in *Emile,* not with a discussion of women's social condition but with a discussion of comparative anatomy, in the hope of discovering the "similarities and the differences between her sex and ours." Rousseau looked first to the medical sciences for a solution to the question of the proper relation between the sexes, picking up on a question central to anatomy at the time: does sex extend beyond the organs of generation? In answering this question, Rousseau (like Vesalius before him) concluded that in everything not connected with sex, "woman is a man." She has the same organs, the same needs, the same faculties. In everything connected with sex, however, woman and man are "in every respect related and in every respect different." That is, Rousseau considered woman a man except in her sexuality. But how far did sexuality penetrate woman's body? The dilemma for Rousseau—and one central to political theory of the time—was determining what in a woman's constitution is due to sex and what is not. Rousseau quickly concluded that sex permeates the entire life of the woman: "the female is female her entire life."[23]

Rousseau agreed with those in the medical community who emphasized difference between the sexes. For Rousseau, as for Thiroux d'Arconville and (later) Soemmerring, sex extended beyond genitalia and into the mind. Though he had no evidence of a connection between mind and body, Rousseau asserted that a perfect woman and a perfect man ought not to resemble each other in mind any more than in looks. Differences in mind and morals were connected to sex, Rousseau held, though "not by relations which we are in a position to perceive." Rousseau's argument was based on wishful thinking, and not on scientific evidence. Though the connection was at best tenuous, Rousseau insisted that the physical grounded the moral, and that woman's constitution determined her place in the physical and moral order.[24]

This leap from the physical to the moral was characteristic of those who, like Rousseau, wished to collapse distinctions between biological sex and socially constituted notions of gender. For Rousseau, there was no distinction between the female and the feminine. Biological sex differences molded intellectual and moral differences, which, in turn, suited men and women for different social spheres. Rousseau hoped to show that nature, and not men, had established the inequalities between the sexes.

Enlightenment enthusiasm for nature and its laws privileged the voice of medical doctors as those best able to understand human nature. Doctors overwhelmingly allied themselves with the complementarians; all evidence (as they saw it) pointed to a tripartite notion of physical, mental, and social complementarity. The *Encyclopédie* article of 1765 on "The Skeleton" devoted half its text to a comparison of the male and female skeleton, and concluded that differences between the male and female skull, spine, clavicle, sternum, coccyx, and pelvis proved that "the destiny of woman is to have children and to nourish them."[25]

Complementarians saw the body as the bedrock of organic life. In 1775 French medical doctor Pierre Roussel wrote an influential book where he emphasized that differences in the male and female skeleton shape the lives of men and women. Nature, he wrote, "has revealed through that special form given to the bones of woman that the differentiation of the sexes holds not only for a few superficial differences but is the result of perhaps as many differences as there are organs in the human body." Spirit or mind were among the organs Roussel listed. Roussel argued against writers who insisted that differences between men and women result from custom, education, or climate. To Roussel's mind, moral and intellectual qualities were as innate and as enduring as the bones of the body. It was, he believed, the task of medicine to provide a certain ground for ethics, for philosophy could not determine the moral powers of human beings without taking into account the influence of bodily organization.[26]

The medical community thus established a clear notion of causality: physical differences produce moral, and eventually social, differences. Soemmerring, the German anatomist who drew one of the first female skeletons, believed that gender differences were to be traced to nature, not to nurture. In his book on the comparative anatomy of the Negro and the European, Soemmerring also reported his observations on woman:

> A boy will always dominate a girl, without knowing that he dominates, and knowing even less that he dominates because of his solid, strong body. He will dominate even when he has received the same nourishment, love, and clothing as a girl. I have had the rare opportunity of seeing definite proof of this fact. From his earliest youth, Prince D . . . G was raised alongside his sister. Their training in all moral and physical matters was equivalent in every way. And yet, differences of masculinity and femininity in physical and moral character were always conspicuous. This is a fact of experience.[27]

There is an unacknowledged circularity to Soemmerring's thinking both here and in his work on female anatomy. Soemmerring feels quite confident that he is being objective by finding evidence of women's social station in their very bones. The evidence he uses for this conclusion, however, is a skeletal illustration into which he has inserted traces or elements of the very thesis he wants to prove. While complementarians favored medical evidence precisely because they believed it free from the imprint of social concerns, Soemmerring was unaware of or unconcerned by the deep cultural forces molding his own evidence.

Soemmerring and his ideas became influential. Relying on Soemmerring's work in an essay he read to the parents of a girls' school, Johann Ziegenbein insisted that already in the earliest stages of the embryo one finds sex differences: that boys will seize a stick while girls will take up a doll, or that men rule the affairs of state while women govern the affairs of the home, reflects nothing other than what is already in "the seed of the embryo."[28]

The appeal of the theory of sexual complementarity was its claim that physical differences reveal nature's design for social stability.[29] The Enlightenment project of building society according to nature's laws fanned the desire to see physical differences between the sexes as a blueprint for their social relations. Because relations between the sexes seemed grounded in nature and not in social convention, complementarians believed them to hold universally. Voltaire, the confidant of Emilie du Châtelet, believed that men everywhere dominate women. Because, as he wrote, men are superior to women in strength of body and mind "in all the world and in all races, from Lapland to the coast of Guinea, and from America to China," it is not surprising to find that all the world over men are the masters of women.[30]

The Political Foundations of Complementarity

Liberalism, which provided the theoretical foundation for the emerging democratic states of the eighteenth century, spoke about men's work in the state and society. Contractual relations, paid labor, and so forth, applied to men and their experience. Liberals did not consider women's work, even though that work—unpaid domestic and caring labor—was an integral part of the system as a whole. Complementarity theory, though not at the time recognized as a supplement to liberal theory, was one of the few theories that explicitly described the value of the private sphere and women's place in it. Complementarians articulated a vision of men and women not just as opposites but as interdependent parts of a physical and moral whole in which their complementary opposition (and not sameness or equality) was important to the smooth working of society.

The theory of sexual complementarity had several distinctive features. It was designed to keep women out of competition with men in the public sphere and, at the same time, to preserve the family within the state. Rousseau's work was pivotal, providing a much needed new rationale for separate spheres for men and women. It was a central ideological component of the complementarity theory that women, in their new role as free and empowered mothers, were equal in perfection (if not in civil liberties) to men. Unlike Aristotelians, complementarians did not rank men and women—masculinity and femininity—on a single chain of being where woman was viewed as man *manqué*. Rather, the relation between man and woman was considered incommensurate and complementary. Each, in fulfilling "nature's ends according to its own particular purpose," found its own distinctive perfection.[31] The wide appeal of complementarianism rested on its rhetorical guarantee of equality for women. As we shall see, however, setting men and women into separate spheres did not make them equal.

Complementarian literature was exported to Germany along with general unrest preceding the French revolution. The Prussian physiocrat Jakob Mauvillon, who wrote extensively on military strategy and economic policy, published his *Mann und Weib* toward the end of his life in 1791. For Mauvillon the central question concerning women was how they could best serve the state. He believed that woman's chief purpose was to bring forth a robust and handsome population. The successful propagation of the human species—nature's only purpose in creating two sexes—required that women remain subordinate within the state. Thus Mauvillon, like other complementarians, argued against formal equality for women, yet at the same time (and here we see what was unique to complementarians) he thought it essential that women willingly accept this subordination. Mauvillon hoped women would argue, as did Karoline von Woltmann some years later, that they *are* equal and free in their lives as wives and mothers.[32] He rejected the simple-minded subordination of women put forward in books like Ernst Brandes's *Über die Weiber* (Concerning women) where men asserted their authority, saying "I am master" (*Ich bin der Herr*).[33] Mauvillon felt sure that the state in which women felt free would produce the greatest population and become a powerful nation.

A second feature of the theory of sexual complementarity was the claim that the relationship between the sexes should not be contractual (as in liberal theory) but rather based on love, for man and woman are not equals but interdependent parts of one moral whole. Mauvillon recounted a distinctively heterosexual version of Plato's story of the origin of human love popular since the Renaissance: In the beginning mankind was a double being. Man and woman were but one creature with four arms, four legs, two sets of genitals, and two complete bodies. With their extraordinary strength they challenged the gods. As punishment, Jupiter split them into two. It is now the fate of each of these beings to search for their lost half; only by becoming united in love do they become whole once again.[34] All this was by way of saying that the social relationship between man and woman should be considered different from that between man and man. Women should not think in terms of equality because the same rules do not govern the relationship between man and woman as govern the relationship between man and man. Conceived as natural opposites, men and women unite harmoniously with one another. By assigning men and women separate yet complementary spheres of moral competence, complementarians felt that men and women would complete—rather than compete with—one another.

Important to this theory was the promise that woman's "equality" was guaranteed by her physical and mental difference. The Edinburgh physician John Gregory argued in his conduct manual, *A Father's Legacy to his Daughters,* that woman's equality depended on her retaining and cultivating her distinct character. Should a woman assume the freedoms of a man—that "hard and masculine spirit"—Gregory feared all would be lost. "Freedom," he held, "will not be gained by conversing with us with the same unreserved freedom as we do with one another . . . or by resembling us as nearly as they possibly can."[35] Rather the happiness of both sexes depended on the preservation and observation of their distinct characters. In her *Versuch einer Logic für Frauenzimmer* (Logic for ladies), Philippine von Knigge, too, held that learning was only to improve a woman's abilities in the home. It was the life ambition of the young Knigge, even as she set down her father's lectures on logic for the benefit of other young women, to be able to make good *Pfannkuchen.*[36] Complementarians thus identified the domestic sphere as the proper location of the positive qualities of femininity. In this view, with each sex set into its proper sphere, society avoided further vain disputes about the equality of the sexes.

By the 1790s, the theory of sexual complementarity had swept Europe and passed, in some cases, into national legislation. The revolutionary French government in a pivotal decision denied women political rights and the right to assemble, citing as justification newly established definitions of woman's nature. According to the National Convention, women did not possess "the moral and physical strength required for the exercise of . . . the rights [of citizenship]." According to the Convention, social stability results from, among other things, "the differences between man and woman." Each sex is called to the occupation that is fitting for it;

"nature, which has imposed these limits on man, commands imperiously and receives no law."[37] Theories of sexual complementarity proposed by Rousseau and others seemed plausible because they described divisions of labor between the sexes that were both ancient and real. As Condorcet pointed out, women seemed incapable of exercising the rights of the citizen only because they had never had the chance to exercise those rights.[38] Complementarians merely accepted what was customary or well established and called it natural.

Asymmetries in Medical Evidence

The theory of sexual complementarity did not arise unopposed. In the revolutionary fervor of the 1790s, treatises supporting the full equality of women in society appeared in England, France, and Germany. Mary Wollstonecraft's *Vindication of the Rights of Woman* called for women's access to learning and politics.[39] Condorcet argued in his *Sketch for a Historical Picture of the Progress of the Human Mind* that inequalities in rights between the sexes were based on nothing more than an abuse of physical strength.[40] Olympe de Gouges, the butcher's daughter, issued her "Declaration of the Rights of Woman and the Female Citizen" in 1791, calling for full rights and representation of women in the new French republic.[41] In Paris the Society of Revolutionary Republican Women was formed in 1793. Even in Germany, where liberalism did not have a strong foothold, the feminist cause was championed by a work entitled *On Improving the Status of Women,* published anonymously in 1792.[42]

Egalitarians generally shared with complementarians the Enlightenment faith in science; they shared also the belief that scientific fact was prior to and free from social imprint.[43] Though feminists grounded their call for equal rights in nature, they rarely appealed to medicine for evidence, and for good reason: there was little in the new anatomical texts that upheld a notion of social equality for women. Anatomists fell almost to the man inside the complementarist camp. Medical doctor Exupère-Joseph Bertin stood alone in opposing those who exaggerated differences in male and female skeletons. Bertin noted that the much debated difference in the male and female pelvis was not so great that it alone allowed one to distinguish a woman from a man. "It helps to make that distinction," he wrote in 1754, "but it is not enough."[44] After Bertin, few medical men made it their business to correct exaggerated claims of sexual differences.

The active collaboration between anatomists and complementarians was *not* built on incontrovertible evidence of nature. As we have seen in Chapter 7, drawings of female skeletons represented social ideals of femininity as much as they did the bones of the female body; in the worst case, corseted bodies had themselves been deformed by social expectations. Science provided uniform results supporting the complementarian cause because of the longstanding rules of admission to and exclusion from the scientific community. At a time when women were formally excluded from universities and informally excluded from scientific academies, the female body was consistently observed through the eyes of a socially homogeneous group. Elite European men (and, in the exceptional case of Madame Thiroux d'Arconville, women), sharing common interests and common backgrounds, saw the female body in similar light. Consensus, in this case, was achieved by controlling participation in medical circles to insulate against the presence of dissenting voices. Voices of dissent did not need to be silenced; they were not present.

Science was but one authority employed by egalitarians to make their case. As Rousseau had observed, the social equality of the sexes followed generally from the "maxims of modern philosophy." Feminism employed a variety of its tools—the critical use of reason, the principle of toleration, the appeal to natural rights and equality. In her "Declaration of the Rights of Woman and the Female Citizen," for example, Olympe de Gouges declared: "Woman is born free and lives equal to man in her rights." Condorcet, too, founded his call for equal rights in the credo that women, like men, are sentient beings, capable of acquiring moral ideas and of reasoning about those ideas; they thus possess all the necessary qualities of the citizen. Yet to the extent that scientific evidence was becoming a privileged mode of discourse, egalitarians worked at a disadvantage.

It was left to feminists—writing from outside the academy and without the sanction of science—to dispute the claims of those who argued for fundamental inequalities in male and female character. As early as 1744, Eliza Haywood criticized in her journal, *The Female Spectator,* anatomists' efforts to claim that women were physiologically incapable of deep thought. She argued that the supposed delicacy of the female brain did not necessarily render it less "strong" than the male brain. She doubted whether invention, memory, or judgment could really be found wanting in the female.[45] Haywood was not a natural philosopher, however, and thus qualified her authority to speak of such matters: "as I am not Anatomist enough to know whether there is really any such Difference or not between the Male and Female Brain, I will not pretend to reason on this Point." Like others of her time, Haywood bowed to claims that scientists have a privileged access to truth.

Lacking a large body of science that clearly supported their views, egalitarians could only deny that anatomists had found sex differences of political significance. One woman published a bitter "Letter to Women" in the French *Magasin encyclopédique* in 1796: "if nature has created two different sexes, it is merely the form and not the elements that has been altered . . . Leave the anatomist blind in his science . . . to calculate the force of a fiber . . . In all ages, men have sought to distance women from knowledge; but today this opinion has become more fashionable than ever."[46] Theodor von Hippel, a champion of women's rights and the mayor of Königsberg, similarly denied that medicine had uncovered significant differences between men and women beyond genitalia. Where Rousseau had insisted that differences of body implied differences of mind, von Hippel argued that differences other than those of an immediate sexual nature had eluded the anatomist's knife.[47] Von Hippel, however, was not a scientist. Egalitarians found it hard to find support for their cause: equality of men and women, at least in the eighteenth century, was simply not supported by science. Anatomists—in their desire to see the sexes as anatomically (and therefore socially) distinct—tended to discount the sexual ambiguities that some of their colleagues had identified in the shape of breast bones, pelvises, or even sexual organs. By emphasizing sexual difference, the medical community tended to reify sharp sexual distinctions upon which complementarians built their arguments for sharply differentiated social roles for men and women.

Masculinity, the Measure of Social Worth

In 1798 Immanuel Kant observed that a woman makes no secret of her wish that she were a man, but he also observed that "no man would ever want to be a woman."[48] Though complementarians and egalitarians offered different visions for establishing equality for women, both privileged masculinity. As in the anatomy of sexual and racial differences, where the European male body emerged as the standard of excellence against which women and blacks were measured, gender differences too were ranked hierarchically. Man—and more specifically, the cultural characteristics of masculinity—remained the measure of all things.

This is easy to see in the case of the complementarians. Though complementarians professed that women—in their new role as free and empowered mothers—were equal to men in their public employ, women still had no civil liberties. In absolute terms, the public sphere set policy for the private. Women's influence beyond the home was at best indirect. Complementarians feared what they identified as masculine women and they feared in particular that equal education would masculinize women. For the German historian of philosophy Christoph Meiners, learning since the ancients had borne a masculine imprint, though there were also many instances where women had been "manly" enough to teach at university. Ancient Greece and modern Italy, he wrote, "afford instances of females who, with masculine hand, broke the bolts and locks of their harems and with manly boldness placed themselves in the professor's chair."[49] This was a common notion: Voltaire's highest praise for Emilie du Châtelet was to call her a man. Madame Dacier's erudition was said to have changed her sex.

But it is important to realize that egalitarians, no less than complementarians, privileged masculinity.[50] In calling for equality, egalitarians stood firm in their conviction that it was *women* (and not men or society) who needed reform. For feminists of the prerevolutionary period, the desire to share the rights of men implied that women must learn the skills of men, not that men needed to learn skills traditionally cultivated by women. Though egalitarians called for the social equality of women, they rarely paid attention to the need to set the feminine equal to the masculine. They did not see it as problematic that, in the eighteenth century, the desire to participate in public life required women to assimilate to the dominant masculine culture. In order to fit women unobtrusively into public life, Jean Le Rond d'Alembert, editor with Diderot of the *Encyclopédie,* called for them to become "more solid and more male."[51]

Many revolutionaries (Wollstonecraft among them) believed that women's skills and aspirations were badly in need of reform, but also that women's difference was the source of their weakness. Sexual differences, bred from women's subordination, handicapped women in public employ. More than one revolutionary joined Mary Wollstonecraft in her plea for women to cast aside infirm femininity. In her influential *Vindication of the Rights of Woman,* Wollstonecraft roundly rebuked Rousseau for trying to make of women "a fanciful kind of *half* being"—delicate and demure. Wollstonecraft believed that for women to become equal, they must become masculine. "I presume that *rational* men will excuse me for endeavouring to persuade them [women] to become more masculine and respectable." Wollstonecraft did not, however, call for women to become indiscriminately masculine. She cautioned them against taking up "hunting, shooting, and gaming" and challenged them instead to become men's equals in virtue and knowledge. Mary Wollstonecraft encouraged women to refuse to cultivate the frailness expected of their sex and to develop instead a masculine strength of both mind and body.[52]

Wollstonecraft wrote her *Vindication of the Rights of Woman* in a self-consciously masculine style, as it had been defined in the battle over style (Chapter 5). Equating eloquence with femininity, Wollstonecraft vowed to avoid the "flowery diction" associated with the aristocratic women of her day. "Wishing rather to persuade by force of my arguments than dazzle by the elegance of my language," she wrote, "I shall be employed about things, not words!"[53] She, like Rousseau, wanted her adversaries to feel the full force of her argument.

Egalitarians like Condorcet, who did not call for the masculinization of women, thought women could retain their distinctive qualities while becoming equal within the state. Unlike the conservative complementarians, Condorcet saw no difference between men and women sufficient to exclude women from civic rights. To those who objected that pregnancy indisposed women from exercising their civic duties, Condorcet asked: what about those people who have gout the entire winter, or rheumatism? To those who said women have made no important scientific discoveries or have written no literary masterpieces, Condorcet answered that the same was true of the overwhelming majority of men. Condorcet thought women could be different yet equal within the state because for him extending the rights of citizenship to women did not imply a social revolution. He was interested exclusively in righting injustices of the past and in including women in the democratic process—and he downplayed (perhaps as a strategy) the social changes that the right to vote might entail. Voting, he reassured his opponents, would not disrupt the familiar rhythms of family life.[54]

At its most extreme the egalitarian argument required that women suppress not only their femininity but also their femaleness. There were those among feminists (men and women) who argued that women were indistinguishable from men in mind *and* body. Women were to enter the public realm as men—unencumbered by reproduction. Since childbearing was often cited as rendering women unreliable citizens and unworthy scholars, one anonymous writer attempted to show that childbearing was irrelevant in such matters. In her version of the prehistoric relations between the sexes (a popular subject of speculation at this time), this author reported that the women of America and Africa were not prevented by pregnancy from doing the same kinds of work as men. Without breaking stride, these women delivered in the middle of a wood or a field, washed themselves and their children in the next stream, and returned to their work "more freely than before."[55]

Advocates of equality for women put the burden of change almost exclusively upon women. Few argued that in order for women to attain equality, men should also learn the virtues and perfections of femininity—become more "chaste and modest."[56] Theodor von Hippel was one of those who did suggest that the equality of women required the reciprocal reform of both women *and* men. He proposed, for example, that men should share equally the burdens of child rearing: "is the raising of children only the duty of the mother," he asked, "or does this responsibility not also fall to the father?"[57] To those critics who cried, "women, you are mothers!" others replied, "do not forget that men are fathers!"[58]

Calls for women's equality envisioned an ideal of woman's relation to man and the state that differed vastly from the realities of eighteenth-century life. Egalitarians imagined an ideal world where women would participate in public life on the same footing as men. Yet the distance between the hopes of egalitarians and the reality of women's lives was so vast as to make these ideals unattainable. In the eighteenth century, no woman—rich or poor—was granted equality in either a formal (legal) or cultural sense; women were given neither the vote nor access to the liberal professions. In the face of these setbacks, egalitarians continued to call for women's equality—without, however, taking into account the fact that gender differences had their origins in the subordinate status of women in European society. They thereby underestimated the depth of social change required to bring about real equality. Equality was largely premised upon women abandoning their femininity and transforming themselves into men.

Purging the Feminine from Science

In recent years, a great deal of feminist literature has been devoted to the question: Is science masculine, and if so what can be done about it? There is, however, nothing inherently masculine about science; rather, science was part of the territory that fell to the masculine party in the struggles that divided social and intellectual labor between the sexes in European society. Because science inhabited the public realm where women (or femininity) dared not tread, science was seen as decidedly masculine.

The opposition between science and femininity formed a cornerstone of the doctrine of sexual complementarity. By the end of the eighteenth century conventional European wisdom taught that the sexes were not equals, but perfect complements of one another. Complementarity, developed with the enthusiastic participation of the scientific community, provided the fundamental justification for the continued exclusion of women from science. But it was not simply *women* who were to be excluded from science. Rather, a whole set of values, qualities, and characteristics subsumed under the term *femininity* was barred.

This opposition between science and femininity was not new to the eighteenth century.[59] Pythagoras articulated how gender fit into his understanding of the dualisms at work in the world:

female	vs.	male
unlimited	vs.	limited
even	vs.	odd
many	vs.	one
left	vs.	right
motion	vs.	rest
curved	vs.	straight
dark	vs.	light
bad	vs.	good
oblong	vs.	square

Aristotle also assigned gender to his series of dualisms:

female	vs.	male
passive	vs.	active
matter	vs.	form
imperfection	vs.	perfection
potential	vs.	actual

Although complementarians formulated relations between the sexes in terms compatible with Enlightenment thought, the location of gender in modern dualisms was remarkably similar to its place in ancient cosmologies. In this sense, the revolution in scientific understandings of gender differences was incomplete. As in the case of the childlike female skeleton, ancient prejudices against femininity were not overturned but merely translated into the language of modern science. Though feminine qualities changed in substance (from "oblong" to "feeling," for example), the superior masculine qualities remained in opposition to the less highly valued feminine qualities.

Modern thought has continued to emphasize a series of interrelated dualities: reason has been opposed to feeling, fact to value, culture to nature, science to belief, the public to the private, the masculine to the feminine. These normative developments emerged along with the growing distinctions between public and domestic life in European society. While science increasingly lost its amateur status and became a paid vocation, its ties to the public sphere strengthened. Social theorists taught that the public sphere of government and commerce, science and scholarship, was founded on the principles of reasoned impartiality—qualities increasingly associated with men and masculinity. At the same time, the rise of the sentimental family increasingly put the ideal mother in charge of child rearing and moral rectitude. The norms of femininity that developed in the eighteenth century portrayed femininity as a virtue in the sphere of motherhood and the home but as a handicap in the world of science. Within this framework, femininity came to represent a set of qualities antithetical to the methods of science. Natural philosophy, with its emphasis on rationality and objectivity, devalued precisely those qualities ascribed to women.

In setting the public sphere of the professions against the private sphere of the home, complementarians envisioned two distinct ways of living: each sphere had its own logic, ethic, and *modus operandi.* The purposes and activities of the public realm differed essentially from those of the home. As one complementarian put it, in the state, everything originates in abstraction, in concepts; while in the home everything originates in the physical needs of heart and soul.[60] A natural fit was also thought to exist between man and the public ethic, on the one hand, and between woman and the private ethic, on the other. As Hegel wrote, family piety, or the law of inner life, was the law of woman. This law, based in subjectivity and feeling, stood opposed to the universal character of the public law of the state. This opposition between family and public law Hegel defined as the "supreme opposition in ethics."[61] The complementarity theory was designed to remove men and women from competition in the public sphere, by removing women from participation in that sphere.

Only within this context can one understand the urgency of the notion that women simply are incapable of doing science. In defining why women could not do science, complementarians were not defining women so much as what was "unscientific." Women—as representatives of private life—became repositories for all that was not scientific. This was not the first time science had been set in a dualistic relationship with its perceived "other." During the seventeenth and eighteenth centuries scientists had made efforts to distinguish their enterprise from religion, politics, poetic expression, and so forth. In each case, particular limits were set to scientific knowledge by defining science as epistemologically distinct from its opposite. In the case of femininity, this process of defining what was and was not scientific was itself guaranteed by scientific definitions of sexual character. As Mary Wollstonecraft pointed out, the inability to do science had been defined by scientists as part of woman's innate sexual character.[62]

Rousseau, Antoine Thomas, Meiners, Wilhelm von Humboldt, and a growing number of anatomists and popular anthropologists, all held that creative work in the sciences lay beyond the natural capacities of women.[63] Each described this inability in similar terms: women, mired as they are in the immediate and practical, are incapable of discerning the abstract and universal. Complementarians taught with Rousseau that women lack the genius to engage in the search for abstract and speculative truths. Women might succeed in small works that require only quick wit, taste, or grace; they might even acquire erudition, talents, and other skills as a result of work. But their work is ultimately only cold and pretty, for women lack genius—that "celestial flame" which warms and sets fire to the soul.[64] For Rousseau, participation in science required a certain strength that women simply lack. For complementarians, defining what was not scientific—modesty, reserve, compassion, love, delicacy, and grace—sharpened notions of what science was.[65]

But if femininity endangered science, science was also seen as a blemish on the flower of true femininity. The anonymous reviewer of Amalia Holst's book on education for women rejected her notion that a learned woman could remain properly feminine. "Nature herself," the reviewer wrote, "appears to have drawn the boundary very exactly and correctly between feminine and masculine pursuits."[66] No man could bear to imagine his beloved with a bloody knife "burrowing into the bowels of a cadaver at an anatomical theater."

By embedding the theory of sexual difference in the theory of separate spheres, complementarians cemented the association of masculinity with science. To the extent that science was pursued in public institutions, it was to display a particular set of characteristics. In the eighteenth century these were also the prescribed characteristics of man. Quite another set of values was purged from science and preserved in the home.

Popular Science and the Decline of the Virtuosa

Even within popular science, limits were drawn to feminine competence. One measure of the changing conception of the proper gender of science was the use of pseudonyms. In the Middle Ages one might see women dressed in men's clothing attending universities or leading armies; women assumed male manners in order to enjoy male freedoms. For a time (from the Renaissance to the middle of the eighteenth century) men also used female pseudonyms in an attempt to encourage women to participate in intellectual endeavor. In *The Ladies' Diary,* the English almanac devoted to mathematics, one of the more prolific contributors of problems and solutions was "Ann Nichols," the feminine alias of William Wales. (No women are known to have written under male pseudonyms in this journal.)[67] The first journal in Germany to encourage women to sharpen their minds, *Die vernünftigen Tadlerinnen* (founded in 1725), was published by Iris, Phyllis, and Calliste—all pseudonyms for Johann Christoph Gottsched.[68]

At the turn of the nineteenth century, as the feminine came into greater conflict with the scientific, women hid once again behind masculine masks. Sophie Germain, the mathematician, did coursework at the newly opened Ecole Polytechnique in Paris (which, like all European universities, was closed to women at the time) under the pseudonym Antoine-August LeBlanc, the name of a student at the school. Under this name she also corresponded for six years with Karl Friedrich Gauss. When it was finally revealed to Gauss that she was, in fact, a woman, she explained that she had concealed her sex in order to escape the "ridicule attached to a *femme savante.*"[69] In choosing a masculine pseudonym Germain stood in the nineteenth-century tradition of George Sand (Amandine-Aurore-

Lucile Dupin, baronne de Dudevant) and George Eliot (Mary Ann Evans).

The growth of the cultural dictum that women are incapable of abstract thought coincided with a masculinization of *The Ladies' Diary,* which earlier in the century had encouraged women to sharpen their mathematical skills (see Chapter 2). Though the number of women contributing to the journal was never large, after 1720 the numbers of women answering mathematical puzzles fell off markedly; with the appearance of its companion work, *The Gentleman's Diary, or the Mathematical Depository* in 1741, *The Ladies' Diary* lost its monopoly on the popular mathematics market. The style of the *Diary* was also "masculinized" during this period, changing from verse to prose. At its inception in 1709 the editor had set down two ground rules: that the mathematical problems posed "be very *pleasant* and not too *hard*"; and that the problems be set out in verse in order to make mathematics more popular "among the Ladies."[70] After 1730 prose questions began squeezing out those set in verse, and by 1745 they predominated. In 1817, when a new edition of the *Diary* was issued, the early policy of both proposing and answering all questions in rhyme was judged harmful to the development of mathematical genius, and the editor insisted on changing all the "bad, and often hardly intelligible" verse into "plain but perspicuous prose."[71]

Other changes in popular science appeared in the last half of the eighteenth century. The early rapprochement between science and the wider public encouraged by the use of a more literary style began to break down. Dialogue form, which had been popular since Galileo's *Dialogue Concerning the Two Chief World Systems,* had attracted a wide audience by virtue of personalizing scientific views and distributing authority among participants (imaginary or real). Dialogues were widely used in science texts for women until the 1740s. In that decade, however, Noël-Antoine Pluche (like many writers of popular science to follow) abandoned dialogue for prose in his popular eight-volume *Spectacle de la nature.* Pluche's *Spectacle* began as a fiery exchange between a count, a chevalier, a prior, and a spunky countess-naturalist who (much like Maria Merian) investigated caterpillars, moths, and mussels in her kitchen and courtyard. By volume four, however, these characters had disappeared from Pluche's work, and the format of the work changed from dialogue to simple narrative. Discussions of natural or social philosophy were no longer presented as the personal views of a chevalier, a countess, or a prior, but rather as those of an unidentified, impersonal (though not impartial) narrator. This development had a dramatic effect on the reader. The reader's tendency to identify with any one of the protagonists—either male or female—vanished, as did the sense that the views presented were those of particular individuals. In Pluche's final volumes, science was presented not as an on-going process of discovery but as a set of simple and finished truths. Moreover, when Pluche's countess disappeared, so did his tolerant attitude toward women. By volume six of his *Spectacle* (published sixteen years after volume one), Pluche had reversed his earlier support for women's participation in science. He now argued that women should set aside the study of Latin, metaphysics, geometry, "the gloomy doctrine of vortices, and the mysterious dances of the planets," and focus instead on history, the gospels, and the Old Testament, from which they could quickly learn useful lessons in morality.[72]

A similar development took place in *La belle Wolfienne,* published from 1741 to 1753 by the French Huguenot and secretary of the Berlin Academy Jean Formey. The first three volumes of this six-volume work read like a philosophical novel. Espérance, a young lady from Berlin, outlines the principles of Christian Wolff's philosophy in her conversations with a male admirer during a stay at the country retreat of a certain Madame de B. By adopting the style of a philosophical novel, Formey fulfilled Wolff's own intention of adapting his system of philosophy to the mind of the *beau sexe* in the form of an exchange of letters between young people. By volume four, however, Formey had dropped the novel format. He could, he wrote, no longer sustain a style he found "tiring and inappropriate." It is difficult, however, to find a correlation between form and subject matter that explains why Formey felt he could not render his last three volumes in dialogue. Subjects he fit into courtly clothing in his first three volumes included "the immortality of the soul," "Chinese ethics," and (surprisingly) "logic." Subjects he later found improper for such a style were "metaphysics," "psychology," and "natural theology"—topics which by the nineteenth century would be viewed as somewhat feminine.[73]

As the century wore on, even science written specifically for women was stripped of poetic ornaments. In 1786 the leading astronomer of Paris, Jérôme de Lalande, published his *Astronomie des dames,* a book modeled on the highly successful astronomy for ladies written by Fontenelle a century earlier. Though Lalande shared Fontenelle's goal of presenting rudimentary knowledge of astronomy to the uninitiated, he was not "seduced," as he put it, by Fontenelle's style. "The pleasant allusions scattered throughout his conversations," Lalande wrote, "are not the taste of our century."[74] Lalande noted that the growing gap between the scientist and the public made it impossible to continue Fontenelle's project. Where Fontenelle had mixed astronomy with dialogue, episodes, allusions, or pleasantries—a style aimed at presenting astronomy in a form neither too "dry" for the general public nor too "banal" for the learned—Lalande wrote specifically for women in straightforward, simple prose. Lalande's work, however, failed—even with the ladies. As the editor of the *Bibliothèque des demoiselles* remarked, "Lalande's *Astronomie des dames,* though more instructive [than Fontenelle's], was less amusing and very little read."[75]

The style of popular science texts thus became more "masculine" in the eighteenth century. But the theory of sexual complementarity also taught that some fields of science were more appropriate for women than others. Indeed, as the century progressed, certain fields—such as botany—would be seen as feminine. In 1758, the same year as Rousseau's invective against "feminine influence" in letters, Pierre Boudier de Villemert began the task of sifting and sorting the various fields of science according to their perceived compatibility with feminine character and destiny. In *L'Ami des femmes* he stressed that women were to avoid all abstract learning and all thorny researches, noting that women of an earlier generation who excelled in theoretical work, such as Madame Dacier or Emilie du Châtelet, were rare figures, more to be admired than imitated. The studies Boudier de Villemert encouraged women to pursue were those which afforded great moral improvement: history and, surprisingly, physics. He limited women to a study of practical physics, from which he felt they could learn to appreciate the wonders of nature. From history, which Boudier de Villemert believed women competent to both read and write, women could learn interesting facts as well as lessons useful for life. Though Boudier de Villemert steered women away from abstract reasoning, he encouraged them to work in that field of uniquely feminine competence—imagination. This was a field where he felt women could compete effectively with men, even surpassing them without humiliating them.[76] For Boudier de Villemert, painting, music, and poetry—both their appreciation and execution—were fields particularly suitable for women.

Chemistry was another science thought particularly suited to women's talents and situation. The British author Maria Edgeworth identified the virtues of chemistry for her female readers: "chemistry is not a science of parade, it affords occupation and infinite variety; it demands no bodily strength, it can be pursued in retirement, it applies immediately to useful and domestic purposes."[77]

Thus women were encouraged to enjoy as amateurs those sciences which were domestically useful or which increased moral virtue. These were also often fields in which they were already a familiar presence; chemistry, like botany, was allied with pharmacy and medical cookery—fields well known to women.

Was Botany Feminine?

Of all the sciences recommended for women, botany became the feminine science *par excellence*. By the nineteenth century botany's reputation as "unmanly"—an ornamental branch suitable only for "ladies and effeminate youths"—was such that it was questioned whether able-bodied young men should pursue it at all. Hegel even compared the mind of woman to a plant because, in his view, both were essentially placid.[78] It is not surprising that botany was thought appropriate for women, for it (like pharmacy) was closely allied with (and indeed grew out of) herbal healing and gardening—fields in which women had long been active (see Chapter 4). But this was rarely the rationale given by contemporaries. Botany was thought an appropriate pastime for young middle-class women because it took them out into the air and taught them a certain intellectual discipline. Though after Linnaeus the study of plants seemed to require more of a focus on sexuality than might seem suitable for ladies, botany continued to be advocated (especially in England) as the science leading to the greatest appreciation of God and his universe.

Rousseau also found the study of botany an acceptable feminine pursuit. In response to queries from Madame de L. (Madeleine-Catherine Delessert), Rousseau wrote, "I think your idea of amusing your spirited daughter a little, and exercising her attention with such agreeable and varied objects as plants, is excellent; though I would not have dared to propose it myself." In explaining his willingness to assist her in teaching her daughter botany, Rousseau gave the familiar justification that the study of nature improves the mind. "I am convinced," he wrote, "that at all times of life, the study of nature lessens the taste for frivolous amusements, prevents the tumult of the passions, and provides the mind with profitable nourishment by filling it with an object worthy of contemplation."[79]

But what about the lurid world of plant sexuality? In 1735, when Linnaeus identified sexuality as the key to botanical classification, had he not remarked that the genital organs of plants are exposed to the view of all? Surely a science which studied the "shameful whoredom" of plants as they are caught in various acts of polyandry or polygamy was not an appropriate study for young women.[80] Despite these inconveniences, botany among all the sciences was considered least offensive to the delicate spirit. As Rousseau pointed out, the student of anatomy was faced with oozing blood and stinking cadavers, entomologists with vile insects, and geologists with dirt and filth while extracting minerals from the earth. And at the same time that Linnaeus emphasized the sexuality of plants, he made it palatable to young minds by assimilating plant life to European mores: through rich metaphors Linnaeus suggested that plants joined in lawful marriages whereby stamens and pistils met as brides and grooms on verdant nuptial beds.[81]

Perhaps the most important caveat issued to women was that their ambitions in botany should not transcend those of the amateur. Rousseau cautioned Delessert against her growing enthusiasm for botany: "You must not, my dear friend, give more importance to botany than it really deserves." The fact that women could engage in botany at all, Rousseau taught, is because it is neither complicated nor difficult and requires nothing but patience to begin. Botany, in his view, was "a study of pure curiosity."[82] Despite these reservations, Rousseau sent Delessert eight long letters on the structure and nomenclature of plants.

In 1785 Thomas Martyn, professor of botany at the University of Cambridge, translated Rousseau's letters on botany into English and added twenty-four of his own, written in the French philosopher's style and dedicated to the "Ladies of Great Britain." Martyn intended his letters "to be of use to such of my fair countrywomen and unlearned countrymen as wished to amuse themselves with natural history." He also cautioned against the pretensions of women amateurs in this field: "Nothing is more pedantic or ridiculous, when a woman (or one of those men who resemble women) are [*sic*] asking you the name of an herb or a flower in her garden, than to be under the necessity of answering by a long file of Latin words that have the appearance of a magical incantation." For the amateur, the study of nature was to be simple and pleasurable. The greatest pleasure of the study of botany came from the opportunity it afforded to enjoy the out of doors. "Botany," Martyn taught, "is not to be learned in the closet: you must go forth into the garden or the fields, and there become familiar with Nature herself."[83]

In England especially, botany was advocated as a natural branch of religion. With its emphasis on firsthand observation of the order and harmony of nature, the study of plants was acclaimed for inspiring an admiration for God. Educational writer Priscilla Wakefield emphasized that the structure of a feather or a flower could more easily impress the mind with ideas of infinite power and wisdom than the most profound discourses on such abstract subjects. Wakefield also stressed the benefits of botany for the health of the body and cheerfulness of disposition.[84]

Popular science for women did not decline toward the end of the eighteenth century. If anything, it increased.[85] The *Bibliothèque universelle des dames,* a series of 154 volumes first appearing in Paris in 1785, was designed to provide women's libraries with all necessary knowledge. Science occupied three of the series' ten classes.

Science was still considered a source of pleasure and virtue, appropriate for women's leisure hours. Antoine Mongez, for example, whose own wife was a well-known painter, contributed a volume on algebra intended to serve as an introduction to the mathematical work of Maria Agnesi. Even so, mathematics was to be reserved for leisure time.[86]

Though the *Bibliothèque des dames* emphasized that cultivation of the spirit enhances both youth and beauty, this Renaissance message—still a central one for women—was now out of step with mainstream science. As the gap between high and popular science widened, women, like all grand amateurs, were pushed more and more to the sidelines. With the professionalization of science, virtue and leisure lost their place as prerequisites for the practice of science.

The eighteenth century thus saw the triumph of the theory of sexual complementarity, a theory that articulated and justified the continued exclusion of women from science in terms acceptable to both liberal democratic theory and modern science. To many, the complementarian vision of gender relations seemed as convincing as it was comprehensive. The complementarians' special relation to the medical community had the virtue of making inequalities seem natural; women's continued disenfranchisement from civil rights and the liberal professions seemed the product of neither prejudice nor malice but of nature. At the same time, complementarians envisioned a role for women as mothers and nurturers in terms acceptable to and, at times, even championed by middle-class women. The theory of complementarity reformulated for a new era the qualities and virtues of masculinity and femininity in science, social pursuits, and power. These ideological constructions of gender—though invisible—served as very real barriers to women's continued progress in the sciences.

9

The Public Route Barred

> Learned women attract little attention as long as they limit their study to music and the arts, but when a woman dares to attend a university—when she qualifies for and receives a doctorate—she attracts a great deal of attention and the legality of such an undertaking must be investigated.
>
> —Johann Junker, 1754

In early modern Europe, noble networks and craft production gave women a definite—if limited—place in science. Though women were proscribed from universities and academies, they were often able to maneuver themselves, via less formal routes, into intellectual culture. With the breakdown of the old order (both the guild system and aristocratic privilege), however, women's place in science was to change dramatically.

Two developments—the privatization of the family and the professionalization of science—changed women's fortune in science. At the same time that the household was undergoing a process of privatization, science was being professionalized (a gradual process taking place over the span of two centuries). Astronomers, for example, ceased working in family attic-observatories as they had done in the days of the Winkelmann-Kirch family. With the increasing polarization of public and domestic spheres, the family moved into the private sphere of hearth and home while science migrated to the public sphere of industry and university.[1] Women wanting to pursue a career in science had two options. They could attempt to follow the course of public instruction and certification through the universities, like their male counterparts—an attempt that was to fail in the eighteenth century. Or they could continue to participate within the (now private) family sphere as "invisible assistants" to a scientific husband or brother—this became the normal pattern for women in science in the nineteenth century.[2]

Though much has been made of the opening of universities to women in the nineteenth century, the testing of academic waters began in the eighteenth century. In 1754 Dorothea Erxleben became the first woman ever to receive a medical degree in Germany. German universities were being revived at this time and new universities, including Halle (where Erxleben took her degree) and Göttingen, were founded to foster the diffusion of Enlightenment ideas within Germany.[3] In seeking and obtaining a medical degree at this time, Erxleben tested the new university at Halle, hoping through her example to open the doors of that institution to women.

Dorothea Schlözer was another who received public recognition of her intellectual achievement; in 1787, she became the first woman in Germany to be awarded a Ph.D. It was not unusual at this time for middle-class women to be accomplished in philosophy, music, or (in Schlözer's case) mineralogy. The next step, however, was more difficult; women needed to receive the public certification that a university degree offered in order to use that knowledge professionally. But women were not allowed to follow this route—at least not in the eighteenth century. Despite their efforts, Erxleben and Schlözer were unable to establish precedents for the regular admission of women to universities.

In this chapter the life of Marie Thiroux d'Arconville—the French anatomist who participated in drawing the first illustrations of the female skeleton—affords an example of how the breakup of noble networks brought about the decline of well-born women in science. Yet at the same time new options were opening to women, as is illustrated by the story of Dorothea Erxleben's and Dorothea Schlözer's quest for university education. University degrees awarded to women in the eighteenth century were unique to Italy and Germany; none were awarded women in England or France.[4] Yet nowhere in Europe did women gain access to universities as other than emphatically rare exceptions. Middle-class women who wished to participate in science—such as Caroline Herschel, whose story concludes this chapter—had to do so as invisible assistants to their fathers, husbands, or brothers in the increasingly private sphere of their own homes.

Marie Thiroux d'Arconville: A "Sexist" Anatomist

The decline of the aristocracy across Europe nearly put the noblewoman scientist out of business. The French anatomist Marie Thiroux d'Arconville (1720–1805), a woman of high social standing, was already something of an anachronism in her time. Though she taught that women should not meddle in medicine, she also produced one of the most noteworthy illustrations of the female skeleton in 1759.

When Thiroux d'Arconville was born in 1720, feminine influence in Parisian salons was near its peak; when she died in 1805, the revolution in views of sex and gender (to which she contributed) had come and gone, and feminine influence had become increasingly restricted to the now private sphere of the home. Learned women, even those working within the confines of their homes, were required to respect the prescribed limits of femininity. Thiroux d'Arconville's life followed the rules of conduct set down by Madame de Genlis, educator, playwright, and a follower of Rousseau, in 1811: if a woman does write books, she should avoid all publicity; she should show a great respect for religion and austere morals; she should not respond to critics of her work for fear that in the response she might transgress feminine delicacy, modesty, and softness.[5] Thiroux d'Arconville guarded herself and her family from public censure by publishing her works on science, history, and morality anonymously.

The details of Marie Thiroux d'Arconville's life come to us secondhand.[6] She was engaged at age fourteen to a *conseiller* to the parliament of Paris. Throughout her life she displayed a lively taste for learning but never let it interfere with her duties as wife or mother. Severely marked by smallpox at age twenty-three, she renounced society and took the dress and coiffeur of a seventy-year-old woman. Sequestering herself among her books, she occupied herself with history, moral essays, physics, medicine, chemistry, and natural history.[7]

Thiroux d'Arconville took courses in anatomy and probably also chemistry at the Jardin Royal des Plantes, where courses were free and open to the public. The royal gardens housed three major buildings—the cabinet of natural history, the chemical laboratory, and the anatomical amphitheater.[8] As many as eight or nine hundred people attended a course of thirty-eight lessons in chemistry beginning at 6:00 A.M. Four to five hundred people might see demonstrations of the entire anatomy of the human body over the course of three months (a course using at least twenty cadavers).[9] It is impossible to say how many women were among those present; there were no class rosters and no diplomas were given.

But, for the most part, Thiroux d'Arconville worked privately at home, where she had a laboratory and was able to read books and manuscripts brought to her from the Paris library. From her laboratory Thiroux d'Arconville hoped to serve the nation as a pioneer in a new field of study—putrefaction. One of her major contributions to this field was her study of the decomposition of organic materials, published in 1766. Concerned about the preservation of foodstuffs, she carried out experiments on some thirty-odd classes of substances to determine how decomposition could be controlled in each. Each day for more than five years she recorded how beef, for example, decomposes in air or water and how the speed of decomposition is affected by various mixtures of acids or mineral waters.[10]

Thiroux d'Arconville published her works anonymously, yet she also appended very personal prefaces, criticizing the errors of others working in the same field. The work of John Pringle, physician general of England, she found factually incorrect, and Hermann Boerhaave she denounced for believing that putrefaction does not extend beyond the vegetable kingdom. Her words were at times so sharp that one critic said that she wrote as if she "had a beard."[11]

In addition to her work on putrefaction, Thiroux d'Arconville published a translation of Peter Shaw's *Chemical Lectures*. The work that interests us here, however, is her illustrated translation of Alexander Monro's *Anatomy,* which contains her innovative drawing of the female skeleton. Because Thiroux d'Arconville published her works anonymously, they were often attributed to other people (one of her essays on morality, for example, was attributed to Diderot; her speech before the Berlin Academy was attributed to both her and Frederick the Great).[12] Much the same happened with her *Ostéologie,* the French translation (with her comments and illustrations) of Monro's *Anatomy.* This work, too, was generally (and wrongly) attributed to Jean-J. Sue (professor of anatomy at the royal college of surgery and royal censor for books on surgery) for the simple reason that his name appears on the title page. Even Alexander Monro seemed unaware that it was not Sue who translated his work into French.[13]

Thiroux d'Arconville agreed with Monro that illustrations and anatomical preparations—no matter how carefully they are done—rarely have the color, form, or consistency of nature. Nonetheless, she also believed that anatomical illustrations, when copied faithfully from the cadaver, could be very useful, though she added that it was always preferable to study nature rather than its representation. A radical empiricist, she spared neither time nor trouble perfecting her illustrations. "The figures were drawn under my eyes," she reports, "and there were many that I had redone several times in order to correct a slight fault." She believed that hers were the finest anatomical illustrations to date.[14]

Though her goal was to represent nature exactly, Thiroux d'Arconville's drawing of the female skeleton portrayed the female body as lithe and delicate—a rendering not out of step with Rousseau's vision of modern femininity. She became (unwittingly) party to the eighteenth-century revolution in views of sexuality. Thiroux d'Arconville, too, believed that women's activities should be curtailed. In her *Ostéologie,* Thiroux d'Arconville railed against women who wished to study medicine and anatomy—things, she believed, that fall beyond their sphere of competence. Women should, she taught, be satisfied with the power that their grace and beauty gives them and not extend their empire to include medicine.[15]

Her essay "Sur les femmes," where she fleshed out her views on women, is a rambling diatribe against women. Like Rousseau, she despised the women of her class and the women of the salon; she found them unredeemably frivolous, ignorant, and deceptive with their "small cares and childish conversations." Thiroux d'Arconville's view of women was uniformly bleak; unlike Rousseau and the complementarians, she did not prescribe a new, positive role for women. An aristocrat, she was not a champion of the new cult of domesticity.[16]

Though Thiroux d'Arconville often entertained Voltaire, whose lively mind she admired (though she was unable to accustom herself to his brand of humor), and received finance minister Anne-Robert Turgot, Madame de Kercado, botanist Bernard de Jussieu, chemists Pierre Macquer and Lavoisier, and others, she was not (nor did she intend to be) at the center of a salon.[17] She believed that intellectual women garner only ridicule; if their work is good, they are ignored; if it is bad, they are hissed at. As a consequence, she worked within the confines of her own home and veiled her science in a cloak of anonymity.

Dorothea Erxleben, Germany's First Woman M.D.

The story of Dorothea Erxleben, Thiroux d'Arconville's German contemporary, is quite different. A woman of the middle class, she practiced medicine publicly and eventually received university certification of her right to do so.

Dorothea Erxleben was born in 1715 in the small town of Quedlinburg (not far from Maria Winkelmann's Berlin). Like Winkelmann and Maria Merian, she was the daughter of a modest *Bürger* family.[18] Erxleben's father, Christian Polycarpus Leporin, was a medical doctor and her mother, Anna Sophia Meinecken, was the daughter of a pastor. At an early age Erxleben found she could read a book while doing different household chores and decided to let nothing stand in the way of her studies. As she wrote about these early years, "the more I learned . . . the more I came to believe that all young women who are sufficiently well-off should apply themselves as actively to studying as to household tasks. Since I judged that education benefits our sex and that women were united in their efforts to attain it, I never failed to learn all I could."[19]

As was typical for many women of her time, Erxleben's studies were mediated through her family. Her father, complaining about the talents of gifted women being wasted in the kitchen or at the sewing table, gave his daughter the same lessons as his son. Later, when her brother studied Latin with Tobias Eckhard (rector of the local Gymnasium), she again took the same lessons as her brother.

Since it was not proper for her to go to Eckhard's house for instruction, however, her lessons were carried home to her by her brother. At age sixteen Erxleben began learning her father's profession (medicine), again studying with her brother as he prepared for university exams. As Erxleben reflected in later years, she was supported in this early education by learned men who encouraged education for *Frauenspersonen* (women).[20]

When her brother entered the newly founded University of Halle in 1740, Erxleben petitioned King Frederick II for permission to accompany him and study for a degree. Frederick the Great never recorded his thoughts on women, yet under his reign the Berlin Academy of Sciences granted honorary membership to Catherine the Great of Russia and a handful of Prussian noblewomen. Erxleben's petition to Berlin met with better fortune than Maria Winkelmann's of thirty years earlier. On April 14, 1741, the Prussian state's Department for Intellectual Affairs warmly recommended that both brother and sister be admitted to the university.[21]

Erxleben's admission to the university met with a mixture of outrage and support.[22] Johann Rhetius immediately wrote a pamphlet claiming that women were forbidden by law to practice medicine and thus did not need a university degree.[23] Dorothea responded in 1742 with her book, *Inquiry into the Causes Preventing the Female Sex from Studying,* after her father found her notes and encouraged her to publish them. Her father also wrote a long introduction to the book, supporting university education for women as part of his plan to make German universities more democratic.

Dorothea Erxleben published her book out of respect for her father, yet she feared recrimination from all sides. Among men, she wrote, "some will feel as if I call them to war, or at least attempt to deprive them of their privilege"; among women, "many of my own sex will think I place myself above them." Most of her father's arguments on behalf of women and many of her own were grounded in biblical scholarship. Yet for Dorothea the theoretical question of women's intellectual abilities was overshadowed by their lack of educational opportunities. Most women, she wrote, have neither lessons nor books at home, others do not have the money for such things, and still others think it unseemly to study at schools with young men.[24]

No colleges or universities for women existed in Germany at this time, though some had been proposed. In 1747, several years before Erxleben received her medical degree, Göttingen professor Johann Michaelis submitted to Frederick the Great his petition to found a "university for the fair sex."[25] This was but one of several similar proposals: in 1707 Nicholas Gundling, professor at Halle, published his *Vorschlag einer Jungfern Academie* (Proposal for a School for Girls), and in 1744 Louise Gottsched (Johann Christoph's wife) suggested founding a *Frauenzimmer-Akademie* in Kant's hometown of Königsberg. None of these schemes ever came to fruition. Erxleben suggested that the next best solution was for women to be educated along with men at public universities, where a special part of the classroom could be set aside for women, as had been done for Anna van Schurman at the University of Utrecht. (Erxleben did not, however, endorse the suggestion often heard at this time that women should attend universities disguised as men).[26]

Advocates of women's education in Halle collected stories of women who had studied at university. In 1732 Erxleben's childhood Latin teacher wrote to her with enthusiasm about Laura Bassi, who had just received a doctorate in philosophy at the University of Bologna, and urged Dorothea to do the same. In her own work, Erxleben pointed to the many examples of learned women collected by the Frenchman Gilles Ménage in his history of women philosophers and by Christian Paullini in his book, *Learned German Ladies.* She also drew inspiration from the achievements of Elena Cornaro Piscopia, a woman who set an important precedent when she received a doctorate of philosophy at the University of Padua in 1678, more than sixty years earlier.[27]

A central question in the debates surrounding Erxleben's university study was the potential conflict between the demands of studying and her domestic duties. Her father, for example, cautioned that women should not use education "to throw off the yoke of men." A learned woman was to remain subordinate to the will of her husband and cover his ignorance with her learning. While Dorothea wholeheartedly agreed that women who make a profession of learning do not have time to run a household and care for children (she seemed to believe that women who wished to pursue an education should not marry), she disagreed with her father that education (particularly the study of chemistry and mathematics) made women arrogant and impious. "Are women responsible for atheism?" she asked.[28]

Dorothea Erxleben received permission to take her degree in 1740, but she was unable to attend university at this time. War broke out with Austria, and Dorothea's brother was required—against all efforts by both father and son—to go into military service. Dorothea did not go to university on her own; in a report to the king the minister overseeing her admission suggested that Dorothea did not "trust herself to go to university alone."[29] Instead, she married the local deacon Johann Erxleben, becoming at age twenty-six mother to his five children (she later had four more of her own). Six years after her marriage her father died, leaving the family burdened with debts. When her husband became ill and family responsibilities fell on her shoulders, she resumed her medical practice—still without a university degree.

But all was not well. As in the case of Maria Winkelmann, public exercise of an art brought recrimination.[30] After the death of one of Dorothea Erxleben's patients, three licensed doctors of Quedlinburg—Johann Herweg, Henricus Grasshoff, and Andreas Zeitz—demanded in 1753 that the authorities crack down on the medical "quackery" (*medicinischen Pfuscherey*) practiced by the likes of Frau Erxleben. In a letter to city authorities, they charged that medicine had been so completely ruined by quackery that respectable doctors could no longer make a living. Chief among the quacks cited—army surgeons, bathers, barbers, midwives, and others—was Dorothea Erxleben. The deacon Erxleben's wife, the doctors charged, treated patients with shameless audacity (*unverschämten Verwegenheit*), visiting patients publicly and allowing herself to be called "Frau Doctorin." The doctors complained that they were being deprived of their monopoly (given them by God and by law) and the honor attached to the practice of medicine.[31] In response the Prussian authorities enacted a law forbidding citizens to be treated by anyone but a licensed doctor. Those who did not comply were to be fined ten reichstalers.

The official in charge of the suit brought against Erxleben, Paul von Schellersheim, handed a copy of the doctors' letter to Dorothea with a notice that she had eight days to answer the charges.[32] In a sharp reply of sixteen pages, Erxleben objected that she had been confused with those who treat patients without proper medical knowledge. Her own qualifications, she urged, were in order. From an early age she had been taught the art of medicine by her father, a well-respected doctor. She had been admitted to university and had now, some years later, finished her doctoral dissertation, which she was ready to defend. "My adversaries," she charged, "are bold indeed to call my cures quackery; let the doctor who has never had a patient die cast the first stone." Of the charge that she allowed herself to be called "Frau Doctorin," she wrote: "I can hardly bring myself to comment on the ridiculous accusation that I allow myself to be called 'Frau Doctorin.' These gentlemen have never brought forward anyone who called me that, or heard someone call me that, without being severely reprimanded."[33] Erxleben did not deny that she often visited patients; nor did she go to them in secret. It was also true that she sometimes took money for her services. Yet countless examples could also be cited, she wrote, where "with God's help I cured people who gave me nothing but their best wishes. Would my gentlemen adversaries have me deny help to the poor?" Erxleben considered her enemies' accusations "gross insults to truth" and concluded her letter by offering to take a qualifying exam—but only on the condition that her accusers also take the exam.[34]

The doctors, of course, refused to take such an exam and claimed, "the dear lady considers herself a doctor, only by virtue of the fact that she can toss around some broken Latin and French. Such is her feminine understanding [*fämininischen Verstand*]." They called for public officials to try her for malpractice, and called her a witch for having treated a person she had not even seen. The doctors then turned their attack to the matter of her sex. It was not just that women were not intelligent enough to practice medicine; more was at stake. How could a woman such as Erxleben, constantly pregnant, practice medicine? they asked.[35]

On March 19, 1753, Schellersheim sent Dorothea Erxleben notice that if she wished to continue practicing medicine, she would have to take an examination at Halle within three months. Already in her last month of pregnancy, Dorothea found it impossible to meet the deadline. In January of the next year, she applied to Schellersheim for permission to sit for the examination. Schellersheim forwarded her request to Berlin, asking for a decision from the king. Once again Frederick gave his nod of approval and sent the request on to the medical faculty in Halle, where university rector Junker gave the matter careful consideration. Two questions in particular interested him. First, does the privilege to attend university extend to women? On the one hand, the imperial decree governing university admission used the masculine pronoun throughout and, for that reason, might seem to exclude women. Junker pointed out, however, that in Roman law the masculine pronoun was used in a way to *include* women: "one designates the sex to which the decree most often applies, but by affirming the one sex, the other is not excluded." Throughout the Erxleben case Junker emphasized that in respect to academic degrees, no legal distinction was to be made between the sexes. It would be, he concluded, an inexcusable injustice for men to exclude women from higher education.[36]

A second matter of import in Junker's decision was the status of medicine—was it a public profession, and therefore tantamount to holding public office? Junker acknowledged that Roman law clearly prohibited women from holding public office. The law stated that women should be excluded from all civil and public service; they should hold no magisterial office, nor sit as a voting member on a city council.[37] Herweg, Grasshoff, and Zeitz and their defenders held that this prohibited women from practicing of medicine; Junker ruled, however, that medicine was *not* a public office. Even if one conceded that women should be excluded from public service, he wrote, women should be allowed to attend university and practice medicine because these are not in the public domain. Junker also pointed to the fact that outside Germany, women had received university degrees and that the chancellor of Halle had already granted honorary degrees to women poets. At this time, the small town of Quedlinburg, where this entire controversy was unfolding, also boasted a woman lawyer, a Frau Dr. Siegelin.[38]

With Junker's approval, Dorothea Erxleben submitted her doctoral dissertation along with her request to sit for final exams. In her dissertation, *Concerning Swift and Pleasant but for that Reason often Unsure Treatment of Sicknesses,* Erxleben argued that doctors too often undertake unnecessary cures.[39] She claimed that some patients are too eager and some doctors too quick to intervene when, for example, the menses are irregular. She then discussed the proper use of purgatives, medications to promote urination or menstruation, and the proper use of opiates. Because she had many requests for this dissertation, especially from women, she translated it from Latin into German.

Erxleben's doctoral exam took place May 6, 1754. The rector Johann Junker reported that the *Frau Candidatin* answered in Latin all questions—both theoretical and practical—with such accuracy and modest eloquence that everyone present was satisfied. She passed the exam and, as Junker wrote, "proved herself a man [*sich männlich erwiesen*]."[40] The university received permission from the king to grant Erxleben a degree, and on June 12, 1754, Dorothea Erxleben became the first woman in Germany to receive a medical degree and the right to practice as a physician. There was a celebration in Halle that day. During her long and difficult struggle Erxleben showed no signs of "feminine weakness," but in her public speech on the day of victory she affected the modesty judged proper for women of her time: "My powers are limited, and I lack the art of well-turned phrases; even on this unusual occasion . . . I feel all of my weaknesses, not only those which affect all people, but especially those to which the weaker sex is accustomed." She did not hesitate, however, to enumerate her accomplishments—yet unknown to other women—which she acknowledged "without arrogance but also without fear."[41] With her newfound state support and university approval, Dorothea Erxleben practiced medicine without further incident until her death at age forty-seven, in 1762.

Erxleben's university degree was not an honorary award but a license to practice medicine. Her ordeal was symptomatic of the larger struggle concerning the position of women in European society. Erxleben and her family believed that a woman could lead a professional life and that university study was the proper route to that life. Her success depended on the goodwill of enlightened patriarchs—Frederick the Great, university rector Junker, and her own father. Yet the goodwill of these individuals was not enough to open universities to women on a regular basis. German women had to wait another hundred years until they were formally admitted to German universities—another woman did not graduate from the University of Halle's School of Medicine until 1901.[42]

Dorothea Schlözer, Germany's First Woman Ph.D.

Between 1754, when Dorothea Erxleben received her medical degree, and 1787, when Dorothea Schlözer received her degree in philosophy, much had changed. Rousseau's *Emile* had taken Europe by storm. New scientific definitions of female nature were settling in as a new orthodoxy. In Dessau, a small town near Halle, Johann Basedow had founded his first "Philanthropinum," where he put into practice Rousseau's ideas.[43]

Dissatisfied with notions of women's intellectual capabilities propagated by Rousseau and Basedow, August Schlözer, professor of history at the University of Göttingen, determined to raise his daughter as proof that women could master any scientific subject. From birth, Dorothea served as a pedagogical experiment designed to prove Rousseau and his cohorts wrong. As Schlözer wrote in 1787, "from the age of five, I called her [Dorothea] my anti-Basedow."[44]

This experiment took place, however, under the growing shadow of domestic femininity. Paramount to Schlözer's program was the attempt to demonstrate that a woman could be both learned and a perfect wife and mother. That Dorothea's education and university degree would lead to a desirable marriage, and not to an active professional life, never came into question. Throughout her years of training, Dorothea carefully considered what subjects she could or could not study and still find favor with a future husband. Unlike Dorothea Erxleben, whose degree served as the rite of passage to a professional career, Dorothea Schlözer's doctorate served merely as another badge of honor for an already illustrious academic family.

Dorothea Schlözer was born in 1770 in the university town of Göttingen, a contemporary of Karl Friedrich Gauss, the man who would lead the University of Göttingen to prominence in mathematics and the natural sciences. It was here that Dorothea Schlözer would receive her degree in philosophy—as with Erxleben, with the support of her father. A professor at the university, her father argued that knowledge belonged to "the essential destiny of the female sex" and was also an essential asset for girls from families who were comfortably well-to-do but not of the landed gentry.[45] While her mother taught her the necessary domestic arts—cooking, sewing, and the intricacies of managing a wine cellar, Schlözer's father took charge of her intellectual development. He believed that household activities, even when combined with feminine ornaments (drawing, singing, and dancing), left many empty hours which could be filled profitably with serious study, and he suggested that his daughter devote herself to mathematics and political history. From the age of six he trained Dorothea in languages (Plattdeutsch, French, English, Swedish, Italian, Dutch, Latin, and Greek), Euclidian geometry, algebra, trigonometry, optics, botany, and zoology, as well as religion and ancient and modern history.

Dorothea Schlözer's special field of study was mineralogy. Even as a young girl she traveled into the Harz mountains (sometimes on her own), where she studied mines for months at a time. When asked on a number of occasions why his daughter studied mines, Schlözer replied that though this would be of no use to her in the future, for the time being "it keeps her busy and, for a seven-year-old girl, that is the best protection against the temptations of the devil."[46]

The suggestion that Dorothea Schlözer be recognized for her academic achievement with a university degree came from Johann Michaelis, dean of the philosophical faculty at Göttingen and a man who, forty years earlier, had petitioned the Prussian king to set up a university for women. Dorothea described the proposal:

> I happened one day to visit my family friend Michaelis. He suggested that at the upcoming fifty-year jubilee of the founding of the University of Göttingen I should receive a degree from the university. Unsure whether Michaelis was serious or merely being polite, I returned home and reported the event to my father. Father, fearful that the degree would be merely an honorary one, suggested to Michaelis a full examination by the faculty. The date was set. On a Sunday afternoon in 1787, I was to be examined.[47]

The exam went according to schedule. Schlözer was excused (at her father's request) from a public disputation, and the exam was given in German—rather than Latin, as was customary—in the comfortable surroundings of Michaelis's house. A table was set with biscuits, the wine was poured. For two-and-one-half hours Schlözer was examined by a committee of professors in a variety of subjects: modern languages, mathematics, architecture, logic and metaphysics, classics, geography, and literature. The decision was unanimous; Dorothea would be awarded a degree.

Bizarre events accompanied this award. At the exam, Dorothea was dressed all in white and her hair decked with roses and pearls. The whole outfit, she wrote, was made to resemble that of "a bride, as my father wished it." Though she was to be awarded a degree in a public ceremony, Dorothea did not attend because her father thought it improper for her to take part in the public celebration. "Since I could not go into the church [where the degrees were awarded]," she wrote a friend, "I went to the library, where I could see and hear everything through a broken pane of glass."[48]

Shortly after her promotion, a local newspaper, reporting the event, satirized the "learned woman" while praising Schlözer, the "cultivated lady," for combining learning with feminine virtues:

> Usually one thinks of a learned woman as neurotic. And should she ever go beyond the study of literature into the higher sciences, one knows in advance that her clothing will be neglected and her hair will be done in antiquarian fashion, that she understands the culinary arts of the ancient Greeks but cannot cook a simple egg, that she forces her way into circles of men for whom she is nothing more than a book . . . For Mlle Schlözer, [however,] this is not at all the case! Mlle Schlözer sews, knits, and understands household economy perfectly well. She is healthy and loves to dance; she speaks freely with those of her own sex. One must gain her confidence before one comes to know the scholar in her.[49]

Four years after her degree, this bride of knowledge married a merchant and senator from Lübeck, Matthäus von Rodde. After her marriage Dorothea did little scholarly work. She published a series of letters on metal production in Clausthal and prepared the tables for her father's *Münz-, Geld-, und Bergwerksgeschichte des russischen Kaiserthums von Jahren 1700–1789*. She also published a cookbook of dinner recipes.[50] Despite her few publications, Dorothea Schlözer von Rodde was celebrated for her learning and was honored in 1801 at a public session of the Académie des Sciences in Paris.

Dorothea Schlözer's promotion was not debated publicly as Dorothea Erxleben's had been. Yet the reaction in Germany was severe. In the decade following Schlözer's much celebrated graduation, major works—including those by Ernst Brandes, Christian Pockels, and Christoph Meiners—were written to combat the intellectual independence of women. Meiners, one of her father's colleagues at the university, began a four-volume history of women one year after Dorothea's degree—a book which he hoped would help deliver Europe from the "calamity of pedantic women."[51] Though he did not mention Dorothea by name, Meiners made it clear that women had no business studying at university. Instead they should learn needlework (for which he praised Dorothea's mother) and domestic economy. If women of the upper and middle classes wished to pursue science, Meiners recommended that they should choose from those parts of natural philosophy and natural history that required neither profound knowledge of mathematics nor fatiguing study of systematic nomenclature.

Germany's first woman Ph.D., like the first woman physician, failed to set any trend in motion. Subsequent middle-class women, even those like Dorothea Schlözer who made it a point to combine perfect femininity with scientific training, did not succeed in finding a place in public institutions. After Schlözer, it was not until 1874 that another woman—the famous mathematician Sofia Kovalevskaia—was awarded a doctorate in philosophy (*in absentia*) from the University of Göttingen. Women were not admitted to Prussian universities on a regular basis until 1908.

Family Assistants: Caroline Herschel

Women were not to travel a public road in pursuit of science. With their exclusion from university reaffirmed, women had few options but to pursue science privately. In the nineteenth century, the normal pattern for women in science was that of the private assistant, usually a wife, sometimes a sister or niece, who devoted her life to a man as a loyal assistant and indefatigable aide. The wifely assistant is somewhat difficult to distinguish from the guild's unpaid artisan wife, for she is the legacy of that tradition. Yet changes both in the

Figure 43. Madame Lavoisier as the "loving assistant." Reprinted from Edouard Grimaux, *Lavoisier* (Paris, 1899).

structure of science and in the family served to distance female assistants from the world of science. A scientific wife became an increasingly private assistant, hidden from view within the domestic sanctuary (see Figure 43).

Caroline Herschel was one of those women who fit this mold.

Though better known than Maria Winkelmann or Maria Merian, Herschel did not display their independence of mind. Caroline wrote that she considered herself a "tool" for her brother, William; when he needed a soprano for his musical ventures she learned to sing, when he needed an astronomical assistant she learned to observe the heavens. Caroline Herschel did not even choose to become an astronomer. As she wrote of herself, "I found I was to be trained for an assistant-astronomer, and by way of encouragement a telescope adapted for 'sweeping' . . . was given to me."[52] For most of the rest of her life, she swept the skies for comets according to her brother's instructions.[53] It is impossible to say whether with more initiative she could have achieved greater independence in the astronomical world.

Caroline Herschel's obeisance (she once described herself as a "well-trained puppy-dog") resulted, in part, from the restrictions on women in intellectual culture. Without her family connections it is doubtful that she could have become an astronomer. She had few options. As a girl in Hanover she received an elementary education along with her brothers, learning to read and write at the garrison school. Although her father wished to give her the same lessons in music and philosophy that he gave William, her mother determined that Caroline's education would be rough and useful.[54] She was taught to knit and to sew in order to supply the family with socks and linen. When by the age of twenty-two she had not yet married, Caroline was delighted to join her brother William in England and become his housekeeper and general assistant, first as a music copyist and performer, then as an astronomical observer and recorder. William Herschel, astronomer to King George III, was a skilled telescope maker and constructed what were for that time the most powerful reflecting telescopes. Caroline learned all she knew about astronomy from William and, throughout the many years of their collaboration, remained his faithful assistant, rarely feeling it her place to set her own projects and tasks.

Caroline Herschel's work in astronomy was limited both by her position as assistant and by the instruments at her disposal. "The employment of writing down the observations, when my brother uses the 20-feet reflector," she wrote, "does not often allow me time to look at the heavens; but as he is now on a visit to Germany, I have taken the opportunity of his absence to sweep in the neighborhood of the sun, in search of comets."[55] Only with William away and using his telescope did Caroline discover her first comet. Her remark highlights a more general problem for women working in science in this period. For one thing, women seldom had access to the best scientific equipment, and their science suffered as a result. As Caroline once remarked, she did not discover her first comet with the "seven-foot" Newtonian telescope provided for her use but with her brother's more powerful device. Furthermore, working as assistants, women were kept busy recording observations and producing calculations and rarely had opportunities to undertake their own projects. Only in her brother's absence was Caroline able to follow her own inclinations.

If Caroline stood in William's shadow, she also shared in his glory. Within a period of about ten years (1786–1797) she discovered eight comets (having a claim to priority on five); in addition, she discovered three nebulae and published her *Catalogue of Stars* with the Royal Society.[56] She was commended by the Royal Society, and her discoveries were announced through the Society by letter to astronomers in Paris and Munich. After the discovery of her third comet, the French astronomer Jérôme de Lalande wrote her a letter of congratulations. She also won the attention of royalty. When William Prince of Orange stopped at the Herschel house at Slough to ask some questions about planets he was well aware of her astronomical expertise.[57] In 1787, King George III awarded her a pension of fifty pounds per year for her work as William's assistant (William received two hundred pounds as a royal astronomer).

Caroline Herschel became the first woman to publish scientific findings in the *Philosophical Transactions* of the Royal Society. Although she was never honored with membership in the Society (no woman was until 1945), she was awarded the Gold Medal of the Royal Astronomical Society and elected ("God knows what for," she writes in her memoirs) an honorary member in 1835.[58] As she was then eighty-five and living once again in Hanover, the election meant little.

Caroline Herschel's temperament was such that, under other circumstances, she might have acted no different. But she was hardly an exception. Women had few options in the nineteenth century.

With universities closed to them, they had to continue to rely on family members for ties to the scientific world. Though women may have served science well in their positions as invisible assistants, the growth of an increasingly public world of science distanced them from centers of scientific innovation. A wife like Maria Winkelmann-Kirch could no longer become assistant astronomer to a scientific academy through marriage. Such positions became reserved for those with public certification of their qualifications.

10

The Exclusion of Women and the Structure of Knowledge

> If a Man could thus divest the partiality attach'd to this self, and put on for a minute a state of neutrality . . . In a word, were the Men *Philosophers* in the strict sense of the term, they would be able to see that nature invincibly proves a perfect *equality* in our sex with their own.
>
> —Sophia, 1739

Since the Enlightenment, science has stirred hearts and minds with its promise of a "neutral" and privileged viewpoint, above and beyond the rough and tumble of political life. Many have hoped that science could serve as a neutral arbiter in social debates, including the "woman question." "Sophia," the anonymous author quoted above, hoped that reason could be neutral in this matter precisely because it had no sex: "We must be obliged to appeal to a more *impartial* judge, one incapable of siding with any party . . . This I apprehend to be *rectified reason,* as it is a pure intellectual faculty elevated above the consideration of any sex." Both sexes, she held, are equally disposed to the development of science: "as there is but one way of conveying food to the stomach, so there is but one method of supplying the mind with truth."[1]

Interestingly, impartiality was claimed by both sides of the debate. Jakob Mauvillon, for example, claimed that he had freed himself from all personal prejudices (as he put it, from *mein ganzes Ich)* in a book in which he asserted that women must remain subordinate to men.[2] Carl Pockels, too, insisted that his nine-volume work on sex differences (in which he asserted that learning masculinized women) was "impartial" and "purely empirical," following as it did the principles of reason. Pockels simply claimed that "one must take nature as she is."[3] On the other side of the question, Amalia Holst opened her 1802 *Higher Education for Women* with the charge that "men are always partial to their own sex when they judge ours, but they seldom allow us the same privilege."[4] Men, she stated, have written volumes about the female sex, and it was now time to hear from the female party. Like her predecessors, Holst intended to abstract herself as far as possible from favoritism toward either sex—though she also noted that in an attempt to right previous wrongs, she would try to sway the reader to her point of view.

Yet, inequalities in power made true impartiality impossible in several important ways.[5] First, science cannot be considered value-neutral while certain groups are systematically excluded from its institutions. Second, asymmetries in social power have given great authority to the voice of science. (In the modern intellectual world, "disinterest" carries with it the premium of "objectivity"; those, by contrast, without that cloak speak with a lesser voice.) Third, science cannot be considered neutral so long as systematic exclusions from its enterprise generate systematic neglect (or marginalization) of certain subject matters and problematics. We are aware of what these inequalities have meant for women, but these same inequalities have also had a profound effect on knowledge and its institutions. In what follows I discuss some of those effects.

Is Science Value-Neutral?

The notion that science should be "gender-neutral" arose alongside the revolution in views of sex and gender: both were integral parts of the revolution that brought modern science to life in the West. The claim that science was neutral rendered *invisible* the injustices in the system by sealing an already self-reinforcing system: Mauvillon's and Soemmerring's self-proclaimed neutrality was premised on the absence of dissenting points of view. Those who might have criticized the new scientific views were barred from the outset, and the findings of science (crafted in their absence) were used to justify their continued exclusion. The image of women developed in this context had the character of a self-fulfilling prophecy: women did not excel in science—but, then, they seldom had an opportunity to work in science. As we saw in the example of the growing interest in the female skeleton, an uncanny consensus developed within the medical community on the natural adeptness of women to bring forth children but not science. The paradoxical nature of this situation was captured by an anonymous "Lady": "Men have not only excluded the Women from partaking of the Sciences and Employ by long Prescription, but also pretend that this Exclusion is founded in their natural Inability. There is, however, nothing more chimerical."[6]

This paradox can be highlighted through the example of women's participation in medicine. The search for sex differences on the part of seventeenth- and eighteenth-century anatomists coincided with changes in the structure of medical care. Anatomists' interest in the distinctive character of women's body-build came at a time when medical science was taking women's health care out of the hands of midwives. Doctors stressed that practitioners of birthing techniques should be trained in the new principles of anatomy, but midwives were unable to receive training in anatomy because women were barred from universities and scientific academies. At the same time, the picture of female nature developing from the study of anatomy suggested that women were not suited for scientific pursuits. In later years the father of positivism, Auguste Comte, would find this process of men replacing women in step with "natural law."[7]

The collapse of midwifery also coincided with abrupt changes in gynecological and obstetrical practice. Where the midwife had emphasized the natural character of childbirth and saw her role as one of assisting the mother in birthing, the new man-midwife—trained as a surgeon—tended to set to work with his surgical instruments (such as the newly developed forceps). Because surgeons had traditionally been called in only in cases of emergency, few had ever seen a normal birth. Midwives had assisted the mother not only with the technical aspects of birthing but with other aspects of her daily regime, such as cooking and caring for the children, while the mother recuperated. Man-midwives, by contrast, attended the mother only during the hours of labor and eventually required women to give birth in hospitals—a practice that further undermined women's support systems. It was only in the 1960s and

1970s that the women's movement was able to begin to reverse these trends and return to women some control over their health care.

If the scientific argument against women played a role in removing women from the medical profession in the eighteenth century, it was also a key element in keeping them out of medical schools a century later. In 1872, medical professor Theodor von Bischoff argued against the admission of women to medical school, using what he termed the "impartial and certain" methods of science to prove that the "pure and unadulterated feminine nature" was not a scientific one. Bischoff's central argument rested on anatomical definitions of sexual differences. Equality, he stated, can only be achieved where initial conditions are equal. He then spun off the (by then) familiar list of sex differences—in bones, muscles, eye sockets, and skulls—relying, in some instances, on the work of Soemmerring and the supporting passages from Kant's *Anthropologie*.[8]

Though anatomists proclaimed their neutrality, the evidence they used was not itself free from the imprint of social concerns; research carried out on the female body told as much about cultural notions of the "feminine" as it did about female bones, muscles, or skulls. Though flawed, this evidence served as the basis for the continued exclusion of women from science. At the same time, the elimination of dissenting voices insulated the scientific profession against immediate correction of these misreadings of female nature.

The Privileged Voice of Science

The ascendence of the theory of sexual complementarity in the eighteenth century is one example of the power of modern science in the political arena. Another is the use of science in a debate between Auguste Comte and J. S. Mill in the middle decades of the nineteenth century. The debate centered on the question raised by Jaucourt in the *Encyclopédie* nearly a century before: how to reconcile the continued subordination of women with the Enlightenment axiom that all men are by nature equal. In his epic *Cours de philosophie positive,* Comte had claimed that the "sound philosophy of biology" could offer a resolution to the much acclaimed equality of the sexes (he was especially fond of the phrenologist Gall's theories).[9] Comte believed that one could expect the gradual liberation of subordinated men because there is "no organic difference between the dominant and the dominated," but he expected the subordination of women to continue forever "because it rests directly on a natural inferiority."[10]

Shortly after Comte's book appeared, Mill challenged Comte on the question of sexual differences, asserting that it was the theory of natural equality that had served to advance the liberation of women in the previous half century.[11] Mill questioned the value Comte placed on abstract thought and turned the question from women's ability to do philosophy to the question of the values dominating philosophy, arguing that philosophy should be supplemented by "the experience of women" because their point of view has been neglected. Moreover, Mill challenged women to write like women, "with their sentiments and their feminine experience," and not like the majority of women writers who write only "for men." Mill claimed that philosophy is "often abstract" and devoid of a consideration of issues of real interest; women, he held, might serve to ground philosophy in more concrete, practical matters.[12]

Mill's appeal for women to write in their own voices was to no avail. Feminists, barred as they were from universities, necessarily spoke with a lesser voice. The nineteenth-century feminist Hedwig Dohm was not reticent in her critique of science—yet, because she wrote from outside the academy, her work was commonly viewed as merely polemical.[13] Jenny d'Héricourt, a woman well trained in physiology, too was marked when she raised her voice in opposition to the prevailing wisdom.[14] In 1875 Austrian health minister Friedrich Ravoth articulated a common assumption of the day: that feminism was self-serving, while science followed universal laws of nature. In a speech to the Assembly of German Scientists and Doctors, Ravoth proclaimed that "competent and qualified scientific research can and must face the dialectical, or rather, sophistical talk about the so-called woman question with a categorical imperative, and uphold unchangeable laws."[15]

Feminists today continue to suffer from the assumption that their scholarship is marred by their allegiance to an explicitly political agenda. Their work is often discounted by those in positions of authority who are unaware of (or refuse to recognize) the ways in which gender structures knowledge and power. Thus while the academy continues to operate under the flag of an imperfect neutrality, feminists continue to fall victim to the very power relations they are trying to uncover and alter.

Building the Canon: The Case of Kant

Historical memory is highly selective: books that are ignored are lost, their message forgotten. Women of the seventeenth and eighteenth centuries failed to win a place for themselves in the academy. Even the memories of those struggles were eventually also lost, as the words of those who opposed orthodox views of women's nature were seldom preserved in libraries or taught in university lectures. They did not, in other words, pass into the Western canon. In the preface to Erxleben's 1742 defense of women's right to higher education, the medical doctor Christian Leporin noted that Anna von Schurman had published a book on the education of women in the previous century, but that "despite all my efforts, it was not to be had."[16] With Schurman's work misplaced by tradition, Dorothea Erxleben—Leporin's daughter—never had the opportunity to sharpen her young mind on Schurman's mature thought. Erxleben could not have known that the same fate would befall her own work. Amalia Holst some fifty years later noted that Erxleben's *Inquiry into the Causes Preventing the Female Sex from Studying* was "no longer available." Holst could not procure a copy, nor could Erxleben's stepson—a professor.[17] Feminists such as Theodor von Hippel and Mary Wollstonecraft, known in their day for having contributed to vital social questions, also fell victim to neglect. In 1806, Weissenborn lamented that Hippel's book published only a decade earlier was nearly forgotten, and Wollstonecraft's almost completely ignored.[18]

The problem continues to plague us today. In university curricula where attention is focused on "great works" (for example, Plato and Aristotle through Descartes, Hobbes, Marx, and Kafka), there is rarely mention of what these scholars had to say about women.[19] Moreover, expressly feminist tracts such as Mary Wollstonecraft's have until very recently been omitted. Even women writers who were major figures in their day—Christine de Pizan or Madeleine de Scudéry—are only now being introduced into university curricula. Issues of gender, like women themselves, have been held at bay—treated unsystematically outside the academy and little understood from within.

Even when the authors of acknowledged great works did address issues of gender, their efforts were often sidelined or ignored.[20] Take the case of Immanuel Kant. Kant wrote his philosophy in the 1760s through the 1790s at the peak of Enlightenment debates about women. Though often overlooked by experts in Kantian philosophy, Kant did in fact have quite a lot to say about women—though much of it consisted of rather unstudied pronouncements on women's character. Without reflecting on his position in the debate, Kant held the complementarian position that creative work in the sciences lies beyond the natural capabilities of woman. In his *Observations on the Feeling of the Beautiful and Sublime,* Kant claimed that deep meditation and long-sustained reflection are not suited to a person in whom unconstrained charms should show nothing other than a beautiful nature:

> The fair sex can leave Descartes's vortices to whirl forever without troubling itself about them, even though the well-mannered Fontenelle wished to secure ladies a place amongst the planets. The attraction of their charms loses none of its force, even if they know nothing of what Algarotti has written for their benefit about Newton's theory of gravitational attraction.[21]

Kant associated woman's "beautiful understanding" not with science, but with feeling: "her philosophy is not to reason, but to sense."

It is essential to look beyond Kant's words to the place of the woman question in his philosophical works. Kant was surprisingly "uncritical" in his treatment of women's place in intellectual life. The works in which he discusses women are usually considered "precritical." Interestingly, the words "lady" or "woman" (*Frauenzimmer, Weib,* and so forth) are mentioned only five times in all of his critical philosophy, compared with more than three hundred times in his precritical and anthropological writings (these latter began as a series of lectures given over the course of thirty years and were first published in 1798, only six years before his death).[22] Kant scholars seem to have generally overlooked his relegation of half of humanity to his precritical work.[23]

Kant uncritically accepted Rousseau's dictum that woman is made for man. He did not, however, point out that this dictum stood at odds with the categorical imperative grounding his own moral philosophy, according to which persons should be treated as ends, never as means. Kant's naiveté in this regard cannot be dismissed with the excuse that everyone treated the woman question in this fashion. Several years before Kant published his *Anthropologie,* the Prussian minister Mauvillon denounced Rousseau's dictum as offensive. Mauvillon held that all reasoning beings have the right to determine their own ends—to exist for themselves alone and not as the chattel of another (explicitly referring to woman's relationship to man).[24]

Nor can Kant be excused on the grounds that he was unfamiliar with these issues. Theodor von Hippel published his lengthy *On Improving the Status of Women* in 1792, six years before Kant published his *Anthropologie.* It is unlikely that Kant was unfamiliar with this work, since Hippel was among Kant's circle of friends. (Kant expressed his annoyance with Hippel for having published some of Kant's ideas before he himself got them into print.)[25] Hippel took exception to Kant's proclamation that women who wished to engage in science "should have a beard"; Hippel pointed out that it would be wrong to think that American Indians were a lower class of human beings simply because they did not sport that European "badge of honor."[26]

Lesser lights of this period were alternatively outraged by or apologetic for Kant's myopia. Weissenborn in 1806 pointed out that the German philosopher's views on women contradicted the axioms of his own system.[27] An anonymous "Henriette" described Kant's words as unjust: man, she wrote, is made for woman as much as woman is made for man; both are equally human.[28] Amalia Holst, by contrast, provided one possible explanation for Kant's shortsightedness: Kant, she pointed out, simply did not have a wife.[29] The words of these scholars, though often more thoughtful and scholarly than Kant's, were generally ignored.

The Scientific Guarantee of Difference

I began this book citing Poullain de la Barre's 1673 slogan "the mind has no sex." Before and since Poullain's time, the question of sex in the mind has been the topic of intense political debate—a debate whose history has, until recently, been neglected: when the sociologist Ferdinand Tönnies reported his discovery of highly gendered social systems in his 1887 *Gemeinschaft und Gesellschaft,* he claimed (partly because he was unaware of their ancestry) that gender relations followed from the laws of nature; when more recently Carol Gilligan found that women speak "in a different voice," the seeming novelty of her work captured the imagination of hundreds of thousands of readers.[30] If we are to understand this debate, it is important to examine the specific political forces surrounding arguments for or against gender differences as they have arisen in different contexts.

Historically, views of women's "special nature" have clustered around three fundamental positions: essentialism, liberal feminism, and cultural feminism. *Essentialists* from Aristotle to the modern sociobiologist believe that women simply cannot do science as well as men—that something in the physical or psychological nature of the female prohibits women from creative intellectual work.[31] It is primarily the essentialist position that has been guaranteed by science, yet, as I have argued, scientists have not made their claims from a disinterested position. The doctrine of sexual complementarity developed in the eighteenth century as part of the ideological apparatus associated with the professionalization of science and the rise of the ideal of motherhood. Complementarians saw women as repositories of the virtues of an earlier age with important work to do: in a scientific age women were to be religious; in a secular age they were to be the keepers of morals; in a contractual society they were to provide the bonds of love. Complementarians conceived femininity as a necessary ballast to masculinity: each gender was incomplete in itself, but together they constituted a workable whole. Science guaranteed this system by offering evidence that sex and also gender differences inhere in male and female bodies.

Essentialists have also commonly held that science is and should be *manly.* This belief was forcefully stated at the end of the nineteenth century, especially in France and Germany. For Karl Joël it was not just the nature of woman's body but also the distinct character of her mind that made "woman and philosophy strangers, even enemies."[32] The notorious sexologist of Vienna, Otto Weininger, maintained in 1903 that all the great women of history (Queen Christina of Sweden, Catherine the Great of Russia, mathematician Sofia Kovalevskaia, artist Rosa Bonheur) had been either homosexual or bisexual. Why else would intellectual women (George Sand and George Eliot, for example) take masculine names?[33]

Liberal feminists (today sometimes also called scientific feminists) since at least the seventeenth century have confronted essentialists by claiming that nurture, not nature, explains women's poor showing in the sciences. Liberals hold that the many supposed differences in the minds and bodies of men and women are the product of efforts to keep women in subordinate roles. Since the time of Poullain de la Barre, liberal feminists have tried to fight science with science, claiming that because anatomists have found no significant difference between men's and women's brains or sense organs, women are as capable as men of contributing to science.

Robert Merton exemplifies a related point of view in his attempt to come to grips with the fact of anti-Semitism in German science in the 1930s. Merton has reduced the normative structure of modern science to four distinctive ideals—universalism, communism, disinterestedness, and organized skepticism. Universalism underwrites the international, impersonal, even anonymous character of science, and finds expression in the demand that careers be open to talent. In explaining the conflict between the fact of anti-Semitism and the ideal of universalism, Merton has defined the problem as one of liberal science unable to realize its goals in an illiberal (that is, racist) society. When "caste-standards" of society at large are smuggled into science, social inferiors may be closed out of the scientific process and elaborate ideologies may be called forth in an attempt to show that social inferiors are inherently incapable of scientific work.[34] Many of the goals described by Merton are attractive. One problem with this view, however, is that the goals and ideals of scientific practice often follow those of society more generally. In the eighteenth century, scientists and social theorists helped to crystallize gender roles by constructing scientific definitions of male and female nature that bolstered emerging norms of masculinity and femininity. Yet science and philosophy did not do so from a privileged vantage point above and untouched by social struggle, for the future of science itself was under dispute.

At least since the Enlightenment, the argument for equality has served as the pillar of modern feminism, yet there is a second problem with the liberals' neglect of gender differences. In their desire to secure rights for women, liberal egalitarians (as described in Chapter 8) have tended to deny that gender differences exist, claiming instead that, for all practical purposes, women think and act in the same way as do men. Liberals tend to see *sameness* as the only ground for equality, and, as we have seen, this all too often *requires* that women be like men—culturally or even biologically (as when expectations are that working women need not take off time to have children). Thus a simple call for equality ignores the complexities of gender in modern life.

A third distinct approach to difference—what we today call *cultural* or *romantic* feminism—emerged in the nineteenth century. Elise Oelsner, for example, used notions of femininity similar to those of her conservative countrymen Joël and Weininger, but whereas her *confrères* viewed femininity as a liability, she saw it as an asset. Oelsner believed that traditional feminine traits could be marshaled in efforts to overcome social ills, suggesting, for example, that the "superior [*idealeren*] nature of women" could reform science by directing knowledge away from the pursuit of power and toward greater equality, freedom, and fraternity [*Brüderlichkeit*] for all mankind. Unlike the essentialists, Elise Oelsner did not believe that feminine qualities were the privileged qualities of women alone. For her, "the eternal feminine" had animated Jesus, Plato, and Schiller, men whose lives displayed feminine values: "a world-conquering goodness, ready self-sacrifice, warmheartedness and devotion."[35]

It is true that many cultural feminists (especially some of the French feminists fashionable today) are essentialists who, like Rousseau, believe that sexual character is fixed in the character of the species.[36] It is also true that other cultural feminists (especially those allied with the so-called standpoint theory of socialist feminism) have tried to set themselves apart from essentialists and conservative complementarians by emphasizing that gender differences have arisen historically as part of efforts to bolster traditional divisions in labor and power between the sexes.[37] Cultural feminism in this sense represents an important corrective to liberal feminists who encourage women to succeed by assimilating to the dominant culture; values defined as "feminine" have thus served as an assay with which to test for gender distortions in scientific knowledge. Cultural feminists today do what J. S. Mill suggested more than a century ago—consider how the institutions, goals, and research priorities of science might be restructured by taking into account the experience of women because that is a perspective that has been neglected.

Yet, one should be careful not to embrace feminine values as (necessarily) superior to masculine values. To attempt to build an epistemology on traditional femininity simply reverses patterns of domination without challenging that domination itself; it must be recognized that the much-touted feminine often represents little more than the obverse and reverse of the culturally dominant masculine. It should also be kept in mind that gender has not historically mapped directly onto biological sex nor onto women as historical actors. Baconians attacked ancient (and especially Aristotelian) philosophy for being weak and "feminine"; Englishmen often called their French rivals (in this case, male and aristocratic) feminine. Thus, the kind of epistemologies feminists see arising from women's labor—"caring" (Noddings), "holism" (Rose), or "maternal thinking" (Ruddick)—may in fact be part of a larger set of values that have been attributed to a broader group of outsiders who, like women, have been (until recently) barred from the practices of modern science.[38]

Lionizing "femininity" also runs the risk of playing into the hands of conservatives (as the recent Sears case has shown) who use rigid notions of gender differences to justify inequalities. It is important to keep in mind that the supposed virtues of femininity—caring, grace, cooperation, nurturing—have been molded by women's subordinate status and stand at odds with the demands of possessive individualism as it has emerged in the West. Though the virtues of femininity should be considered, the road from the kind of unreconstituted femininity that has been used to encourage women to internalize their own oppression to a new kind of femininity that might allow women to be both womanly (in whatever way they see fit) and equal will be a long one.

Recent feminism has been caught on the horns of the "difference dilemma"—the problem that "both ignoring and focusing on difference risk recreating it."[39] On the one hand, gender differences must continue to be analyzed because women as a group have been, and continue to be, discriminated against. Moreover, patterns of discrimination are evident not only in the absence of women from science but in the systematic exclusion of particular problematics, voices, values, and experiences from intellectual pursuits. And yet, on the other hand, emphasizing differences between men and women (whether these differences are conceived as innate or historical) runs the risk of perpetuating such divisions. Excessive emphasis on male and female differences can also ignore other forms of human diversity—for example, those of class, race, sexual orientation, religion, region, among others. As we have seen, women of science in the seventeenth and eighteenth centuries faced certain common problems because of their sex, yet the shape of their lives also depended on their social position (aristocracy vs. artisanry), place of residence (Paris vs. small-town Germany), marital status, and individual character—factors that created differences in how they responded to discrimination, what kind of science they did, and how they did it.

We cannot give up a careful analysis of gender differences at least until they cease to plague us—inequalities between men and women (economic, political, ideological, and cultural) are still significant. I have emphasized the opposition between science and femininity because "femininity" represents a consistent set of values expelled from modern science. Science and femininity share an intimate history, shaped as they both have been by similar social, political, and economic forces. By burying gender in science, European culture lost part of its past. It is time to unearth that history; it is time to transform both science and society so that power and privilege no longer follow gender lines.

Notes

Introduction

1. Pierre Roussel, *Système physique et moral de la femme, ou Tableau philosophique de la constitution, de l'état organique, du tempérament, des moeurs, et des fonctions propres au sexe* (Paris, 1775), p. 2.
2. [William Whewell], "On the Connexion of the Physical Sciences, by Mrs. Somerville," *Quarterly Review* 51 (March 1834): 65. I thank Robert Merton for calling this passage to my attention. See also Elizabeth Patterson, *Mary Somerville and the Cultivation of Science, 1815–1840* (The Hague, 1983), p. 138.
3. Christian Harless, *Die Verdienste der Frauen um Naturwissenschaft, Gesundheits- und Heilkunde* (Göttingen, 1830), p. ix.
4. Christine de Pizan, *The Book of the City of Ladies* (1405), slightly modified from the translation by Earl Jeffrey Richards (New York, 1982), p. 70.
5. Giovanni Boccaccio, *De claris mulieribus* (1355–1359), trans. Guido A. Guarino as *Concerning Famous Women* (New Brunswick, 1963).
6. Jérôme de Lalande, *Astronomie des dames* (1786; Paris, 1820), pp. 5–6.
7. Harless, *Die Verdienste der Frauen,* pp. 1–2.
8. Alphonse Rebière, *Les Femmes dans la science,* 2d ed. (Paris, 1897).
9. Elise Oelsner, *Die Leistungen der deutschen Frau in der letzten vierhundert Jahren auf wissenschaftlichen Gebiete* (Guhrau, 1894).
10. Gino Loria, "Les Femmes mathématiciennes," *Revue scientifique,* 20 (1903): 386.
11. H. J. Mozans, *Woman in Science* (1913; Cambridge, Mass., 1974), pp. 391, 415–416.
12. Robert Merton, *Science, Technology and Society in Seventeenth Century England* (1938; New York, 1970), p. 114.

13. Among leading early sociologists of science, only the French historian of science Alphonse de Candolle (*Histoire des sciences et des savants depuis deux siècles* [Geneva, 1885], pp. 270–271) and Dorothy Stimson (*Scientists and Amateurs: A History of the Royal Society* [New York, 1948] pp. 82–83) touched on the issue of women in science. A few women scientists in this period provided autobiographical accounts of their experiences; see Ida Hyde, "Before Women Were Human Beings: Adventures of an American Fellow in German Universities of the '90s," *Journal of the American Association of University Women,* 31 (1938): 226–236; and Lise Meitner, "The Status of Women in the Professions, *Physics Today,* 13 (1960): 16–21.
14. Kathleen Lonsdale, "Women in Science: Reminiscences and Reflections," *Impact of Science on Society,* 20 (1970): 45–59; and Sara Ruddick and Pamela Daniels, eds., *Working It Out: 23 Women Writers, Artists, Scientists, and Scholars Talk about Their Lives and Work* (New York, 1977). For personal accounts by women scientists in India, Italy, the U.S.S.R., Kenya, and elsewhere, see Derek Richter, ed., *Women Scientists: The Road to Liberation* (London, 1982). See also Naomi Weisstein, "Adventures of a Woman in Science," *Biological Woman—The Convenient Myth,* ed. Ruth Hubbard, Mary Henifin, and Barbara Fried (Cambridge, Mass., 1982), pp. 265–281; *Cecilia Payne-Gaposchkin: An Autobiography and Other Recollections,* ed. Katherine Haramundanis (Cambridge, Eng., 1984).
15. Louis Bucciarelli and Nancy Dworsky, *Sophie Germain: An Essay in the History of the Theory of Elasticity* (Dordrecht, 1980); Patterson, *Mary Somerville and the Cultivation of Science;* Ann Hibner Koblitz, *A Convergence of Lives: Sofia Kovalevskaia—Scientist, Writer, Revolutionary* (Boston, 1983); and Geneviève Fraisse, *Clemence Royer: Philosophe et femme de science* (Paris, 1985).
16. Margaret Rossiter, *Women Scientists in America: Struggles and Strategies to 1940* (Baltimore, 1982); see also Pnina Abir-Am and Dorinda Outram, eds., *Uneasy Careers and Intimate Lives: Women in Science, 1789–1979* (New Brunswick, 1987).
17. Evelyn Fox Keller, *A Feeling for the Organism: The Life and Work of Barbara McClintock* (San Francisco, 1983). See also S. J. Gould's review, "Triumph of a Naturalist," *New York Review of Books* (March, 1984): 3.
18. I have reviewed this literature elsewhere, in "The History and Philosophy of Women in Science: A Review Essay," in *Sex and Scientific Inquiry,* ed. Sandra Harding and Jean O'Barr (Chicago, 1987), pp. 7–34. For bibliographies on women and science, see Patricia Siegel and Kay Finley, *Women in the Scientific Search: An American Bio-bibliography, 1724–1979* (Metuchen, 1985); *Biological Woman—The Convenient Myth,* ed. Hubbard, Henifin, and Fried, pp. 289–376; and Susan Searing and Rima Apple, "The History of Women and Science, Health, and Technology: A Bibliographic Guide," photocopy, University of Wisconsin, Madison, 1988.
19. See, among others, L. H. Fox, L. Brody, and D. Tobin, eds., *Women and the Mathematical Mystique* (Baltimore, 1976); Anne Briscoe and Sheila Pfafflin, eds., *Expanding the Role of Women in the Sciences* (New York, 1979); *Women Scientists in Industry and Government: How Much Progress in the*

1970s? (Washington, D.C., 1980); Sue Berryman, *Who Will Do Science? Minority and Female Attainment of Science and Mathematics Degrees: Trends and Causes* (New York, 1983); *Climbing the Ladder: An Update on the Status of Doctoral Women Scientists and Engineers* (Washington, D.C., 1983); Violet Haas and Carolyn Perrucci, eds., *Women in Scientific and Engineering Professions* (Ann Arbor, 1984).

20. See Donna Haraway, "In the Beginning Was the Word: The Genesis of Biological Theory," *Signs,* 6 (1981): 469–482; Helen Longino and Ruth Doell, "Body, Bias, and Behavior: A Comparative Analysis of Reasoning in Two Areas of Biological Science," *Signs,* 9 (1983): 206–227; Ruth Hubbard and Marian Lowe, eds., *Woman's Nature: Rationalizations of Inequality* (New York, 1983); Ruth Bleier, *Science and Gender: A Critique of Biology and Its Theories on Women* (New York, 1984); Anne Fausto-Sterling, *Myths of Gender: Biological Theories about Women and Men* (New York, 1985); Lynda Birke, *Women, Feminism, and Biology: The Feminist Challenge* (New York, 1986).
21. See Sandra Harding and Merrill Hintikka, eds., *Discovering Reality: Feminist Perspectives on Epistemology, Metaphysics, Methodology, and Philosophy of Science* (Boston, 1983); Evelyn Fox Keller, *Reflections on Gender and Science* (New Haven, 1985); Sandra Harding, *The Science Question in Feminism* (Ithaca, 1986); Ruth Bleier, ed., *Feminist Approaches to Science* (New York, 1986); Karin Hausen and Helga Nowotny, eds., *Wie männlich ist die Wissenschaft?* (Frankfurt, 1986); and Harding and O'Barr, eds., *Sex and Scientific Inquiry.*

1. Institutional Landscapes

1. Archives of the Académie des Sciences, Comité Secret, 27 December 1910. See also, *Le Temps* (4 January 1911); and Marie Curie, "Autobiographical Notes," *Pierre Curie,* trans. Charlotte and Vernon Kellogg (New York, 1923).
2. See Lina Eckenstein, *Woman under Monasticism* (Cambridge, 1896); and Barbara Newman, *Sister of Wisdom: St. Hildegard's Theology of the Feminine* (Berkeley, 1987).
3. John Lawson and Harold Silver, *A Social History of Education in England* (London, 1973), pp. 97–99. Convents on the continent continued to be active centers of learning until the eighteenth and in some cases the nineteenth century.
4. Joan Ferrante, "The Education of Women in the Middle Ages in Theory, Fact, and Fantasy," in *Beyond Their Sex: Learned Women of the European Past,* ed. Patricia Labalme (New York, 1984), p. 17.
5. Lawrence Stone, "The Education Revolution in England, 1560–1640," *Past and Present,* 28 (1964): 41–80.
6. Between 1560 and 1640, 2.5 percent of men over seventeen years of age attended university in England. This level declined in the late seventeenth

century, not to be regained and surpassed until the 1930s (Lawrence Stone, "Literacy and Education in England, 1640–1900," *Past and Present,* 42 [1969]: 74).

7. There is as yet no adequate study of the history of women in Italian universities. These names have been drawn from: Mozans, *Woman in Science,* pp. 55–58; Rebière, *Les Femmes dans la science;* and Paul Kristeller, "Learned Women of Early Modern Italy: Humanists and University Scholars," in *Beyond Their Sex,* ed. Labalme, p. 102.
8. Nicola Fusco, *Elena Lucrezia Cornaro Piscopia, 1646–1684* (Pittsburgh, 1975). See also Maria Tonzig, "Elena Lucrezia Cornaro Piscopia (1646–1684) prima donna laureata," *Quaderni per la storia dell'Università di Padova,* 6 (1973): 183–192.
9. Benedict to Agnesi, September 1750, cited in Rebière, *Les Femmes dans la science,* p. 11.
10. Edna Kramer, "Maria Gaetana Agnesi," *Dictionary of Scientific Biography,* ed. Charles Gillispie (New York, 1973), vol. 1, pp. 75–77.
11. See Richard Rosen, "The Academy of Sciences and the Institute of Bologna, 1690–1804" (Ph.D. diss., Case Western Reserve University, 1971); and Charles Burney, *The Present State of Music in France and Italy* (1773), ed. Percy Scholes (London, 1959), pp. 159–160.
12. Mozans, *Woman in Science,* p. 208. Bassi does not have an entry in the *Dictionary of Scientific Biography.*
13. Kristeller, "Learned Women of Early Modern Italy," p. 116, n. 56.
14. Some of the theses Agnesi defended on logic, ontology, mechanics, hydromechanics, elasticity, celestial mechanics, botany, zoology, chemistry, among other topics, are published in her *Propositiones philosophicae* (Milan, 1738).
15. Dorothea Leporinin [Erxleben], *Gründliche Untersuchung der Ursachen, die das weibliche Geschlecht vom Studieren abhalten* (Berlin, 1742), p. 143.
16. Rosen, "The Academy of Sciences and the Institute of Bologna," pp. 75–76.
17. Cited in J. H. Hexter, "The Education of the Aristocracy in the Renaissance," *Journal of Modern History,* 22 (1950): 1–2. See also Stone, "Literacy and Education in England," p. 74.
18. Henricus Agrippa [Agrippa von Nettesheim], *Female Pre-eminence* (1532; London, 1670), reprinted in *The Feminist Controversy of the Renaissance,* ed. Diane Bornstein (Delmar, N.Y., 1980), p. 36.
19. Baldassare Castiglione, *The Book of the Courtier* (1528), trans. Charles S. Singleton (Garden City, N.Y., 1959), pp. 213–215.
20. Werner L. Gundersheimer, "The Play of Intellect: The *Discorsi* of Annibale Romei" (The Folger Shakespeare Library, 1984, photocopy), pp. 17–21.
21. Castiglione, *Book of the Courtier,* p. 15 and book 3.
22. Ibid., pp. 16–18, 29, 212, and 276.
23. Gundersheimer, "The Play of Intellect," p. 19.
24. Jürgen Voss, "Die Akademien als Organisationsträger der Wissenschaften im 18. Jahrhundert," *Historische Zeitschrift,* 231 (1980): 45, 50–53.

25. Frances Yates, *The French Academies of the Sixteenth Century* (London, 1947), p. 1. Catherine de Medici carried the Medici cultural traditions with her to France when she married Henri II. On women as cultural ambassadors, see Susan Groag Bell, "Medieval Women Book Owners: Arbiters of Lay Piety and Ambassadors of Culture," *Signs,* 7 (1982): 742–768.
26. Edouard Frémy, *L'Académie des derniers Valois* (Paris, 1887), p. 151; and Yates, *French Academies,* p. 32.
27. Pierre de Bourdeille, seigneur de Brantôme, cited in Marjorie Ilsley, *A Daughter of the Renaissance: Marie le Jars de Gournay* (The Hague, 1963), p. 222, n. 22.
28. See Paul Pellisson and P. J. Thoulier d'Olivet, *Histoire de l'Académie française* (Paris, 1858), vol. 1, p. 256.
29. D. Maclaren Robertson, *A History of the French Academy, 1635–1910* (New York, 1910), p. 61. It is doubtful that these discussions were recorded. My search of academy archives in the spring of 1986 turned up nothing.
30. *Supplément manuscrit au Menagiana,* ed. Pierre Le Gouz, Bibliothèque Nationale, MF 23254, no. 184; also reprinted in P.-L. Joly, *Remarques critiques sur le dictionnaire de Bayle* (Paris, 1752), vol. 2, p. 605. I thank Joan DeJean for calling this passage to my attention. See also Gilles Ménage, *Historia mulierum philosopharum* (Lyon, 1690), also translated as *The History of Women Philosophers* by Beatrice Zedler (Lanham, 1984). Ménage's work was known and cited in defense of learned women throughout Europe. See, e.g., *The Works of Mrs. Catharine Cockburn* (London, 1751), vol. 1, p. ii (Ménage's Latin title appears here as it does on his title page.)
31. Mlle Bernard also won academy prizes in 1691, 1693, and 1697. See Monsieur Bosquillon, "Eloge de Mademoiselle de Scudéry," *Journal des Savants,* 29 (1701): 513–525; and Pellisson and d'Olivet, *Histoire de l'Académie française,* vol. 2, p. 15.
32. *Les Registres de l'Académie française, 1672–1793* (Paris, 1895), vol. 1, p. 332.
33. G. Bigourdan, "Les premières sociétés scientifiques de Paris au XVIIe siècle," *Comptes rendus de l'Académie des sciences,* 163 (1916): 937–938.
34. Théophraste Renaudot, ed., *Recueil général des questions traictées és conférences du Bureau d'Adresse, sur toutes sortes de matières; par les plus beaux esprits de ce temps* (Paris, 1656).
35. Claude Clerselier, ed., *Lettres de M. Descartes* (1659; Paris, 1724), vol. 2, preface.
36. See Gustave Reynier, "La Science des dames au temps de Molière," *Revue des deux mondes* (May 1929): 436–464; and Paul Mouy, *Le Développement de la physique cartésienne, 1646–1712* (Paris, 1934), pp. 98–113.
37. Pierre Richelet, *Dictionnaire de la langue françoise, ancienne et moderne* (Lyon, 1759), vol. 1, p. 21. Compare the 1719 (Rouen, vol. 1, p. 12) and 1759 editions of Richelet's dictionary. In the 1719 edition, the editors have left Richelet's remark that the academy at Arles should be praised for its "glorious conduct" with respect to women. The editors of the 1759 edition removed this sentence, noting only that "one no longer speaks of that academy."

38. Harcourt Brown, *Scientific Organizations in Seventeenth-Century France: 1620–1680* (Baltimore, 1934).
39. Charles Gillispie, *Science and Polity in France at the End of the Old Regime* (Princeton, 1980), pp. 81–82.
40. The salary of only 2,000 livres per year was not enough to maintain a bourgeois lifestyle at this time in France. Members had to supplement their income from the academy with private funds. Ibid.
41. Thomas Sprat, *History of the Royal Society of London* (London, 1667), pp. 62–63, 72, 435.
42. Michael Hunter reports the distribution by occupation of the Royal Society for the years 1660–1664: aristocrats 14 percent; courtiers and politicians 24 percent; gentlemen 12 percent; lawyers 6 percent; divines 8 percent; doctors 16 percent; scholars and writers 7 percent; civil servants 5 percent; merchants and tradesmen 4 percent (*The Royal Society and Its Fellows, 1660–1700: The Morphology of an Early Scientific Institution* [Chalfont St. Giles, Bucks., 1982], table 6, p. 116).
43. See Samuel Mintz, "The Duchess of Newcastle's Visit to the Royal Society," *Journal of English and Germanic Philology,* 51 (1952): 168–176.
44. Thomas Birch, *History of the Royal Society* (London, 1756–57), vol. 2, p. 175; and Samuel Pepys, *The Diary of Samuel Pepys,* ed. Robert Latham and William Matthews (London, 1970–1983), vol. 8, p. 243.
45. Pepys, *Diary,* vol. 8, p. 243; and Douglas Grant, *Margaret the First: A Biography of Margaret Cavendish, Duchess of Newcastle, 1623–1673* (London, 1957), p. 26.
46. Mrs. Evelyn quoted in John Evelyn, *The Diary of John Evelyn,* ed. Austin Dobson (London, 1906), vol. 2, p. 271.
47. Kathleen Lonsdale and Marjory Stephenson were elected to the Royal Society in 1945 (*Notes and Records of the Royal Society of London,* 4 [1946]: 39–40).
48. "A Catalogue of the Natural and Artificial Rarities belonging to the Royal Society, and preserved at Gresham College," in H. Curzon, *The Universal Library: Or, Compleat Summary of Science* (London, 1712), vol. 1, p. 439.
49. The Berlin academy, unlike its fraternal counterparts in London and Paris, awarded honorary membership to a few women of high standing. See also Chapter 3 below. Yvonne Choquet-Bruhat was elected to the Académie des Sciences in 1979.
50. Agnesi's *Istituzioni analitiche* was also translated into English, by a Cambridge University professor, John Colson (London, 1801).
51. Archives Nationales, AJ XV 510, no. 331. Basseporte received 800 livres pension plus 300 livres living expenses. Compare this to the 1,200 to 3,000 livres paid to the anatomical demonstrators.
52. *Inventaire général des richesses d'art de la France* (Paris, 1887), vol. 2, p. 117.
53. Ann Sutherland Harris and Linda Nochlin, *Women Artists: 1550–1950* (Los Angeles, 1976), pp. 34–37. Despite this ruling, several women were elected to the academy of painting in the eighteenth century, but their numbers were never allowed to exceed four.

54. See Richard Altick, *The Shows of London* (Cambridge, Mass., 1978); and E. J. Pyke, ed., *A Biographical Dictionary of Wax Modellers* (Oxford, 1973).
55. See Charles Burney's description of the models at the Institute of Bologna in his *Present State of Music in France and Italy,* p. 158.
56. Friedrich Melchior von Grimm, *Correspondance littéraire, philosophique et critique* (Paris, 1829), vol. 7, pp. 221–222. On Biheron, see also Mélanie Lipinska, *Histoire des femmes médecins, depuis l'antiquité jusqu'à nos jours* (Paris, 1900), pp. 187–190; and P. Dorveaux, "Les femmes médecins: Notes sur Mademoiselle Biheron," *La Médecine anecdotique,* 1 (1901): 165–171.
57. *Histoire de l'Académie royale des sciences. Année 1759* (Paris, 1765), p. 94.
58. Louis Prudhomme, *Biographie universelle et historique des femmes* (Paris, 1830), pp. 363–364.
59. *Histoire de l'Académie royale des sciences. Année 1770* (Paris, 1773), p. 49. On the portrayal of women in wax models, see L. J. Jordanova, "Natural Facts: A Historical Perspective on Science and Sexuality," in *Nature, Culture, and Gender,* ed. Carol P. MacCormack and Marilyn Strathern (Cambridge, 1980), pp. 42–69.
60. Mireille Laget, "Childbirth in Seventeenth- and Eighteenth-Century France: Obstetrical Practices and Collective Attitudes," *Medicine and Society in France,* ed. Robert Forster and Orest Ranum (Baltimore, 1980), p. 169.
61. *Académie des sciences, procès-verbaux,* 6 (18 October 1819): 503, and 9 (2 August 1830): 484.
62. *Etat de médecine, chirurgie et pharmacie, en Europe. Pour l'année, 1776* (Paris, 1776), p. 230.
62. *Histoire de l'Académie. 1759,* p. 94. See also Diderot, letter to General Betsky, 15 June 1774, *Oeuvres complètes de Diderot,* ed. J. Assézat (Paris, 1875), vol. 20, pp. 62–63.
64. Marie Biheron, *Anatomie artificielle* (Paris, 1761). See also *Magasin encyclopédique,* 4 (1796): 414–415.
65. Letter from Denis Diderot to Antoine Petit, July 1771, *Oeuvres complètes de Diderot,* vol. 9, pp. 240–241.
66. For a similar analysis of *préciosité,* see Domna Stanton, "The Fiction of *Préciosité* and the Fear of Women," *Yale French Studies,* 62 (1981): 107–134.
67. Gillispie, *Science and Polity in France,* pp. 7, 94.
68. Carolyn Lougee, *Le Paradis des Femmes: Women, Salons, and Social Stratification in Seventeenth Century France* (Princeton, 1976), p. 27. Salons served as a model for educational institutions later in the century. When Jean Pilâtre de Rozier, the first person to ascend in a balloon, founded his Athénée in the 1780s, he explicitly used the system of salons as a model. See Pierre Laffitte, "L'Athénée," *La Revue occidentale philosophique, sociale et politique,* 12 (1889): 6.
69. Lougee, *Le Paradis des Femmes,* pp. 41–53 and 117–118. Seventeenth-century salon women came overwhelmingly from the nobility: of the women for whom such information is available 74 percent were noble (48 percent coming from old noble families and 26 percent from newly noble

families). Lougee finds significant, however, the proportion of women who were newly noble (26 percent) or nonnoble (14 percent); the status of the remaining 12 percent is unknown. In a society that equated elite with noble status, Lougee points to the inclusion of such a large proportion of non- and new nobility as evidence of the extent to which salons were forging new alliances within French society. See also Joan Landes, *Women and the Public Sphere in the Age of the French Revolution* (Ithaca, 1988), pp. 23–31; Alan Kors, *D'Holboch's Coterie* (Princeton, 1976); and Evelyn Bodek, "Salonières and Bluestockings: Educated Obsolescence and Germinating Feminism," *Feminist Studies,* 3 (1976): 186. A full study of the eighteenth-century salons and their role in eighteenth-century intellectual culture remains to be written.

70. *The Achievement of Bernard Le Bovier de Fontenelle,* ed. Leonard Marsak (New York, 1970), pp. 33–36.
71. See Sara Malueg, "Women and the *Encyclopédie,*" in *French Women and the Age of Enlightenment,* ed. Samia Spencer (Bloomington, 1984), p. 260; and Terry Dock, "Woman in the *Encyclopédie,*" (Ph.D. diss., Vanderbilt University, 1979).
72. Dorinda Outram, "Before Objectivity: Wives, Patronage, and Cultural Reproduction in Early Nineteenth-Century French Science," in *Uneasy Careers and Intimate Lives,* ed. Pnina Abir-Am and Dorinda Outram (New Brunswick, 1987), p. 19.
73. Margaret Cavendish, "The Female Academy," in *Playes* (London, 1662), pp. 653–679.
74. Ibid., p. 671.
75. Mary Astell, *A Serious Proposal to the Ladies* (1694; London, 1701), part 1, p. 24. See also Ruth Perry, *The Celebrated Mary Astell: An Early English Feminist* (Chicago, 1986), p. 134.
76. "Der Frau von ***" [Pauline Erdt], *Philotheens Frauenzimmer-Akademie: Für Liebhaberinnen der Gelehrsamkeit,* trans. from the French (Augsburg, 1783), p. v. I quote the German edition throughout; I have been unable to find the original French, if indeed it ever existed.
77. Ibid., pp. xi-xiii.
78. Ibid., pp. vi, ix.

2. Noble Networks

1. Benjamin Martin, *The Young Gentleman's and Lady's Philosophy* (London, 1763), vol. 1, pp. 1–32.
2. Joseph Sigaud de Lafond, *Physique générale* (Paris, 1788).
3. Jacques Roger, *Les Sciences de la vie dans la pensée française du XVIII*[e] *siècle* (Paris, 1963), pp. 165, 181–182.
4. The Bibliothèque Nationale in Paris holds the catalogues of some thirty-five natural history collections that were put up for sale sometime in the eighteenth century. Five of these collections belonged to women. See, e.g.,

P. Remy, *Catalogue d'une collection de très belles coquilles, madrépores, stalactiques, . . . de Madame Bure* (Paris, 1763).

5. In addition to the works cited individually, see John Harris, *Astronomical Dialogues Between a Gentleman and a Lady* (London, 1719); James Ferguson, *Easy Introduction to Astronomy for Gentlemen and Ladies* (London, 1768); and *Le Buffon des demoiselles,* 4 vols. (Paris, 1819). In England, journals were also launched that included instruction on science for ladies—among these were *The Athenian Mercury* (1690–1697); *The Free-Thinker* (1718–1721); and *The Female Spectator* (1744–1746). In Germany, popular science for women flourished toward the end of the eighteenth century. J. C. Gottsched's *Die vernünftigen Tadlerinnen* contained some science for women. See also [Lorenz Suckow], *Briefe an das schöne Geschlecht über verschiedene Gegenstände aus dem Reiche der Natur* (Jena, 1770); Jakob Weber, *Fragmente von der Physik für Frauenzimmer und Kinder* (Tübingen, 1779); Christoph Leppentin, *Naturlehre für Frauenzimmer* (Hamburg, 1781); August Batsch, *Botanik für Frauenzimmer* (Weimar, 1795); and Christian Steinberg, *Naturlehre für Frauenzimmer* (Breslau, 1796). See also Gerald Meyer's excellent *The Scientific Lady in England: 1650–1760* (Berkeley, 1955); and Henry Lowood, "Natural Philosophy for the Ladies: Female Readership and the Diffusion of Science in the German Enlightenment" (Stanford University, 1985, photocopy).
6. Margaret Cavendish, *The Description of a New World, called the Blazing World* (London, 1666).
7. Bernard Le Bovier de Fontenelle, *Entretiens sur la pluralité des mondes* (1686), ed. Robert Shackleton (Oxford, England, 1955), p. 54.
8. Bernard Le Bovier de Fontenelle, *A Discovery of the New Worlds,* trans. Aphra Behn (London, 1688).
9. Francesco Algarotti, *Il Newtonianismo per le dame* (Naples, 1737). Algarotti's work was translated into French as *Le Newtonianisme pour les dames,* trans. M. du Perron de Castera (Paris, 1738) and into English by Elizabeth Carter, *Sir Isaac Newton's Philosophy Explain'd: for the Use of the Ladies* (London, 1739). See also Johanna Charlotte Unzer, *Grundriss einer Weltweisheit für Frauenzimmer* (Altona, 1761); and Leonhard Euler, *Lettres à une princesse d'Allemagne sur divers points de physique et de philosophie* (St. Petersburg, 1768).
10. Letter from Gottfried W. Leibniz to Queen Sophie Charlotte, November 1697, reprinted in *Die Werke von Leibniz,* ed. Onno Klopp (Hanover, 1864–1884), vol. 8, pp. 47–53.
11. John Dunton, *The Athenian Oracle* (London, 1728), vol. 4, p. 7.
12. John Tripper et al., eds., *The Ladies' Diary* (London, 1706), advertisement. I am much indebted to Teri Perl's excellent article, "The Ladies' Diary or Woman's Almanack, 1704–1841," *Historia Mathematica,* 6 (1979): 36–53.
13. *The Ladies' Diary* (1709).
14. Ibid. (1718), letter to the readers.
15. Thomas Leybourn, *The Mathematical Questions proposed in the Ladies' Diary* (London, 1817), vol. 1, pp. vi-vii.

16. Walter Houghton, "The English Virtuoso in the Seventeenth Century," *Journal of the History of Ideas,* 3 (1942): 57.
17. Roger Hahn, *The Anatomy of a Scientific Institution: The Paris Academy of Science, 1666–1803* (Berkeley, 1971), p. 35; Henry Peacham, *Peacham's Compleat Gentleman* (1634), ed. G. S. Gordon (London, 1906), p. 18.
18. Yates, *The French Academies,* p. 285.
19. Bell, "Medieval Women Book Owners, p. 767.
20. *A Collection of Papers which Passed between the late Learned Mr. Leibnitz and Dr. Clarke* (1717), in *The Leibniz-Clarke Correspondence,* ed. H. G. Alexander (Manchester, 1956).
21. Caroline to Leibniz, in *Die Werke von Leibniz,* vol. 11, pp. 52, 71, and 90.
22. Letter from Elizabeth to Descartes, 12 June 1645, in A. Foucher de Careil, *Descartes, la Princesse Elisabeth, et la Reine Christine* (Paris, 1909), pp. 65–66.
23. Letter from Elizabeth to Descartes, 10 June 1643, in ibid., p. 50.
24. A. Foucher de Careil, *Descartes et la princesse palatine ou de l'influence du cartésianisme sur les femmes au XVIIe siècle* (Paris, 1862), p. 16. On the intellectual relationship between Descartes and Elizabeth, see also Samuel Sorbière, *Sorberiana* (Paris, 1691), p. 102.
25. Claude Clerselier, *Lettres de M. Descartes* (Paris, 1724), vol. 1, preface.
26. *Carpentariana ou remarques . . . de M. Charpentier* (Paris, 1724), p. 316.
27. In the 1750s, Marie Ardinghelli of Tuscany charged the Abbé Nollet with the task of keeping her in touch with all recent discoveries in physics. In 1757, Nollet read part of Ardinghelli's description of an erupting volcano before a session of the Académie Royale des Sciences (*Procès-verbaux,* 76 [1757]: 335).
28. Between 1475 and 1700, women in England and America published some six hundred books, or about half of one percent of the total number of books published (Elaine Hobby, "English Women in Print, 1640–1700," paper delivered at the Sixth Berkshire Conference on the History of Women, June 3, 1984). Twenty-one of these books were published by Margaret Cavendish alone; fifteen were original works which then appeared in various editions to equal twenty-one publications in all. For secondary literature on Cavendish see Grant, *Margaret the First;* Henry Ten Eyck Perry, *The First Duchess of Newcastle and Her Husband as Figures in Literary History* (Boston, 1918); R. W. Goulding, *Margaret (Lucas), Duchess of Newcastle* (London, 1925); Virginia Woolf, "The Duchess of Newcastle," in *The Common Reader* (London, 1929), pp. 98–109; and Lisa Sarasohn, "A Science Turned Upside Down: Feminism and the Natural Philosophy of Margaret Cavendish," *Huntington Library Quarterly,* 47 (1984): 289–307.
29. Margaret Cavendish, "A true relation of my Birth, Breeding, and Life," in *Natures Pictures* (London, 1656), p. 370.
30. Margaret Cavendish, *Worlds Olio* (London, 1655), "The Preface to the Reader."
31. Ibid.

32. Robert Kargon, *Atomism in England from Hariot to Newton* (Oxford, 1966), pp. 68–76.
33. Margaret Cavendish, *Philosophical and Physical Opinions* (London, 1655), "An Epilogue to my Philosophical Opinions."
34. Margaret Cavendish, *Sociable Letters* (London, 1664), p. 38.
35. Margaret Cavendish and Anne Conway shared common philosophical beliefs. Like Cavendish, Conway insisted that nature was not dead. See Anne Conway, *The Principles of the most Ancient and Modern Philosophers* (London, 1692), p. 77. See also Chapter 6 below.
36. Cavendish, *Worlds Olio,* "My Lord."
37. Cavendish, *Observations upon Experimental Philosophy* (London, 1666), "To his Grace the Duke of Newcastle." See also Grant, *Margaret the First,* p. 93.
38. Margaret Cavendish, *Poems and Fancies* (London, 1653), preface.
39. Her earlier *Philosophical and Physical Opinions,* written before she had studied much philosophy, was somewhat incoherent, as she admitted in a preface to her *Grounds of Natural Philosophy* (London, 1668).
40. Carolyn Merchant, *The Death of Nature: Women, Ecology and the Scientific Revolution* (San Francisco, 1980), p. 193.
41. Margaret Cavendish, *Philosophical Letters* (London, 1664), "A Preface to the Reader."
42. Cavendish, *Observations upon Experimental Philosophy,* "Of Knowledge and Perception in General," p. 155.
43. Ibid., "To the Reader," part 1, pp. 49–50.
44. Cavendish, *Philosophical and Physical Opinions,* prefaces.
45. Cavendish, *Observations upon Experimental Philosophy,* "All powerful God and Servant of Nature."
46. Cavendish, *Philosophical Letters,* p. 147.
47. Grant, *Margaret the First,* p. 204.
48. Cavendish, *Observations upon Experimental Philosophy,* "Further Observations upon Experimental Philosophy, The Preface to the ensuing Treatise," and pp. 7–8.
49. *Letters and Poems in Honour of the Incomparable Princess, Margaret, Dutchess of Newcastle* (London, 1676), pp. 108–119.
50. Cavendish, *Observations upon Experimental Philosophy,* "To the Most Famous University of Cambridge."
51. Margaret Cavendish, *The Life of . . . William Cavendishe* (London, 1667), "To his Grace."
52. Cavendish also hoped to have her *opera* translated into Latin in order to make her work available to foreign scholars, but the idiosyncrasies of her terminology baffled the translator (Grant, *Margaret the First,* p. 218).
53. *Letters and Poems in Honour of the . . . Dutchess of Newcastle,* pp. 108–119.
54. Cavendish, *Worlds Olio,* p. 84.
55. Ibid., "The Preface to the Reader."
56. Margaret Cavendish, "Femal Orations," in *Orations of Divers Sorts* (London, 1662), p. 225.
57. Ibid., pp. 227–228.

58. Ibid., p. 229. See also Cavendish, *Worlds Olio,* "The Preface to the Reader."
59. Cavendish, "Femal Orations," p. 231.
60. Cavendish, *Sociable Letters,* "The Preface."
61. Cavendish died in 1673, too early to have known François Poullain de la Barre's radical application of Cartesianism to the woman question. See his *De l'égalité des deux sexes: Discours physique et moral* (1673; Paris, 1984), preface.
62. Cavendish, *Grounds of Natural Philosophy,* pp. 14–15.
63. The duke and duchess of Newcastle were both royalists. William Cavendish was general of all royalist forces in the north of England during the civil war. Hilda Smith has pointed out that at the end of the seventeenth century there was a link between feminism and royalism. See her *Reason's Disciples: Seventeenth-Century English Feminists* (Urbana, 1982), pp. 3–17. See also Catherine Gallagher, "Embracing the Absolute: The Politics of the Female Subject in Seventeenth Century England," *Genders,* 1 (1988): 24–39.
64. Cavendish, *Observations upon Experimental Philosophy,* "The Preface to the Ensuing Treatise."
65. Cavendish, *Philosophical and Physical Opinions,* "Dedication."
66. In 1656, she published one of the first secular autobiographies written by an English woman as part of her *Natures Pictures.* When *Natures Pictures* was reprinted fifteen years later, the autobiography had been removed. See Grant, *Margaret the First,* p. 154.
67. Cavendish, *Natures Pictures,* "An Epistle to my Readers."
68. Nicole Lepaute was born Nicole-Reine Etable de Labrière (1723–1788) at the Luxembourg Palace, Paris. At age twenty-five, she married J. A. Lepaute, royal clockmaker, for whom she observed, calculated, and recorded observations. Jérôme de Lalande, director of the Paris observatory, celebrated her as "the only woman in France who has acquired a true knowledge of astronomy." She worked with Clairaut and did calculations for Lalande's *Connoissance des mouvemens célestes.* Lepaute was elected member of the Académie de Béziers in 1788. Her works include "La Table des longueurs des pendules," in J. A. Lepaute's *Traité d'horlogerie* (1755); "Tables du soleil, de la lune et des autres planètes" in Lalande's *Ephémérides du mouvement céleste* (1774), vol. 7.

    Marie-Jeanne-Amélie Le Français de Lalande, born Harlay (1760–1832), niece of Jérôme de Lalande, married astronomer Michael-Jean-Jérôme Le Francais de Lalande in 1788. She calculated the "Tables horaires" published in Lalande's *Abrégé de navigation* (Paris, 1793).
69. Marie-Anne Pierrette Lavoisier (1758–1836) translated Richard Kirwan's *Essay on Phlogiston* (*Essai sur le phlogistique* [Paris, 1788]), and his "Strength of Acids and the Proportion of Ingredients in Neutral Salts," published in *Annales de chimie,* 14 (1792): 152, 211, 238–286. She prepared thirteen copperplate illustrations for Lavoisier's *Traité élémentaire de chimie* (1789). After Lavoisier's death she edited his *Mémoires de chimie* (1805).
70. Sophie Germain (1776–1831) studied as a young girl in her father's library. She gave a theoretical explanation of E. F. Chladni's vibrating plates that

won her the grand prize of the Academy of Sciences in 1816. Her works include "Tables générales de notation," in *Connaissance des temps* (1807): 484; *Recherches sur la théorie des surfaces élastiques* (1821); *Remarques sur la nature, les bornes et l'étendue de la question des surfaces élastiques et équation générale de ces surfaces* (1826); *Considérations générales sur l'état des sciences et des lettres aux différentes époques de leur culture* (1833).

71. René Taton's "Gabrielle-Emilie Le Tonnelier de Breteuil, Marquise du Châtelet," *Dictionary of Scientific Biography,* vol. 3, pp. 215–217, provides primary and secondary bibliography. See also René Taton, "Madame du Châtelet, traductrice de Newton," *Archives internationales d'histoire des sciences,* 22 (1969): 185–210; Carolyn Iltis, "Madame du Châtelet's Metaphysics and Mechanics," *Studies in History and Philosophy of Science,* 8 (1977): 29–48; Ira O. Wade, *Voltaire and Madame du Châtelet: An Essay on the Intellectual Activity at Cirey* (Princeton, 1941); Elizabeth Badinter, *Emilie, Emilie: L'Ambition féminine au XVIII*[e] *siècle* (Paris, 1983); Esther Ehrmann, *Madame du Châtelet: Scientist, Philosopher and Feminist of the Enlightenment* (New York, 1987); Linda Gardiner, "Women in Science," in *French Women,* ed. Spencer, pp. 181–196; and Linda Gardiner, *Emilie du Châtelet* (Wellesley College Center for Research on Women, 1982, photocopy). I am indebted to Linda Gardiner, on whose work I have relied in this section.
72. Theodore Besterman, ed., *Voltaire's Correspondence* (Geneva, 1968–77), D1411, 23 December 1738, to Thieriot.
73. Linda Gardiner Janik, "Searching for the Metaphysics of Science: The Structure and Composition of Madame du Châtelet's *Institutions de physique,* 1737–1740," *Studies on Voltaire and the Eighteenth Century,* 201 (1982): 87.
74. Besterman, *Voltaire's Correspondence,* D1528, 21 June 1738, to Maupertuis.
75. Gardiner, *Emilie du Châtelet,* chap. 5.
76. Gardiner Janik, "Searching for the Metaphysics of Science," pp. 93–94.
77. For a close analysis of the preparation of the *Institutions,* see ibid.
78. Ibid., p. 97.
79. Voltaire, "Préface hi. ·orique," in Isaac Newton, *Principes mathématiques de la philosophie naturelle,* trans. Marquise du Chastellet, with commentary by Clairaut (Paris, 1756).
80. Gardiner, *Emilie du Châtelet.*
81. For an evaluation of du Châtelet's work, see ibid., chap. 3.
82. I. O. Wade, *Studies on Voltaire* (Princeton, 1947), p. 133.
83. Bernard de Mandeville, *Fable des abeilles,* trans. Emilie du Châtelet (1735), preface, in Wade, *Studies on Voltaire,* pp. 131–138.
84. Ibid.
85. Cited in Gardiner, *Emilie du Châtelet,* chap. 1.

3. Scientific Women in the Craft Tradition

1. This estimate is from the following sources: Joachim von Sandrart, *Teutsche Academie der Edlen Bau-, Bild- und Mahlerey-Künste* (Frankfurt, 1675);

Friedrich Luce, *Fürsten Kron oder eigentliche wahrhaffte Beschreibung ober und nieder Schlesiens* (Frankfurt am Main, 1685); and Friedrich Weidler, *Historia astronomiae* (Wittenberg, 1741).

2. See Edgar Zilsel, "The Sociological Roots of Science," *American Journal of Sociology,* 47 (1942): 545–546; and Arthur Clegg, "Craftsmen and the Origin of Science," *Science and Society,* 43 (1979): 186–201.
3. Pizan, *The Book of the City of Ladies,* pp. 70–80.
4. Margret Wensky, *Die Stellung der Frau in der stadtkölnischen Wirtschaft im Spätmittelalter* (Cologne, 1981), pp. 7, 318–319. Wensky traces women's strength in Cologne's economic life to the fact that textile manufacture predominated, and this was a type of manufacture in which women were strong.
5. Jean Quataert has warned against conflating important distinctions between guilds and households (see her "Shaping of Women's Work in Manufacturing: Guilds, Households, and the State in Central Europe, 1648–1870," *American Historical Review,* 90 [1985]: 1122–1148). For the case of astronomy or entomology, however, the larger danger has been to ignore almost entirely both of these forms of production. Here I use the term *craft* to refer to household production and *guild* to refer to regulated crafts. See also Antony Black, *Guilds and Civil Society in European Political Thought from the Twelfth Century to the Present* (Ithaca, 1984). There were, of course, women artisans in France who fit the patterns I discuss below.
6. Literature on the role of women in household economies written between 1880 and 1920 tended to overestimate women's participation in an attempt to bolster contemporary movements seeking to improve economic conditions for women. See Carl Bücher, *Die Frauenfrage im Mittelalter* (Tübingen, 1882); Henri Hauser, *Ouvriers du temps passé* (Paris, 1899); and Alice Clark, *Working Life of Women in the Seventeenth Century* (London, 1919). More recently, much careful work has been done. In addition to the works cited separately, see Louise Tilly and Joan Scott, *Women, Work, and Family* (New York, 1978); and Natalie Zemon Davis, "Women in the Crafts in Sixteenth-Century Lyon," *Feminist Studies,* 8 (1982): 47–80.
7. See Wolf-Dietrich Beer, "The Significance of the Leningrad Book of Notes and Studies," in Maria Sibylla Merian, *Schmetterlinge, Käfer und andere Insekten: Leningrader Studienbuch,* ed. Wolf-Dietrich Beer (Leipzig, 1976), pp. 51–64. See also Margarete Pfister-Burkhalter, *Maria Sibylla Merian, Leben und Werk, 1647–1717* (Basel, 1980).
8. Sandrart, *Teutsche Academie,* cited in Elisabeth Rücker, "Maria Sibylla Merian," *Fränkische Lebensbilder,* 1 (1967): 225.
9. Harris and Nochlin, *Women Artists,* pp. 17–19.
10. Johann Doppelmayr, *Historische Nachricht von den Nürnbergischen Mathematicis und Künstlern* (1730), ed. Karlheinz Goldmann (Hildesheim, 1972), p. 255.
11. Merian also drew at least one student from the Nuremberg patriciate—Clara Imhoff. See Elisabeth Rücker, "Maria Sibylla Merian," *Germanisches Nationalmuseum Nürnberg* (Nuremberg, 1967), pp. 17, 19, and 21.

12. Maria S. Gräffin [Maria Merian], *Der Raupen wunderbare Verwandlung und sonderbare Blumennahrung* (Nuremberg, 1679).
13. Maria Merian, *Metamorphosis insectorum Surinamensium* (1705), ed. Helmut Decker (Leipzig, 1975), p. 26.
14. M. S. Gräffin [Maria Merian], *Neues Blumenbuch* (Nuremberg, 1680), preface.
15. See Maria Merian, *Die schönsten Tafeln aus dem grossen Buch der Schmetterlings und Pflanzen: Metamorphosis insectorum Surinamensium,* ed. Gerhard Nebel (Hamburg, 1964), pp. 1–40.
16. *Neue Zeitungen von Gelehrten Sachen,* 23 (20 March 1717): 179.
17. Merian, *Metamorphosis,* foreword.
18. Cited in Rücker, "Maria Sibylla Merian," *Fränkische Lebensbilder,* p. 225.
19. *Neue Zeitungen von Gelehrten Sachen,* 23 (20 March 1717): 178.
20. Ibid., 95 (November 1717): 767–768. See also Doppelmayr, *Historische Nachricht,* p. 269.
21. Ratsverlass 12 August 1692, Staatsarchiv Nürnberg, Ratsverlässe no. 2936, pp. 2 f., cited in Rücker, "Maria Sibylla Merian," in *Fränkische Lebensbilder,* p. 234. Graff married again in 1694.
22. See Christian Jöcher, *Allgemeines Gelehrten-Lexicon, darinne die Gelehrten aller Stände sowohl männ- als weiblichen Geschlechts* (Leipzig, 1751), s.v. "Merian."
23. Deckert in Merian, *Metamorphosis,* p. 24. By the end of the eighteenth century, divorce had become quite common in Protestant Germany, especially in Prussia. See Ruth Dawson, "And This Shield Is Called–Self-Reliance," in *German Women in the Eighteenth and Nineteenth Centuries,* ed. Ruth-Ellen Joeres and Mary Jo Maynes (Bloomington, 1986), p. 162.
24. Helga Ullmann, "Maria Sibylla Merian: Zeit, Leben und künstlisches Schaffen," in Maria Merian, *Leningrader Aquarelle,* ed. Ernst Ullmann (Leipzig, 1974), vol. 2, p. 42; *Ratssupplikationen,* 7 September 1690, Stadtarchiv Frankfurt am Main, cited by Deckert in Merian, *Metamorphosis,* p. 16; Merian to Scheurling, 29 August 1697, Stadtbibliothek Nürnberg, Autographen no. 167, reprinted in Rücker, "Maria Sibylla Merian," *Germanisches Nationalmuseum Nürnberg,* p. 21.
25. Deckert in Merian, *Metamorphosis,* p. 16. Merian also prepared the illustrations for G. E. Rumphius's *D'Amboinsche Rariteitkamer.*
26. Merian, *Metamorphosis,* p. 37.
27. Merian to Imhoff, 29 August 1697, reprinted in Rücker, "Maria Sibylla Merian," *Germanisches Nationalmuseum Nürnberg,* p. 21.
28. Merian, *Metamorphosis,* commentary to plate no. 20.
29. Ibid., commentary to plate no. 45; see also plate nos. 7, 25, and 13.
30. Merian to Johan Volckamer, 8 October 1702, Trew-Bibliothek, Brief-Sammlung Ms. 1834, Merian no. 1, Universitätsbibliothek Erlangen, reprinted in Rücker, "Maria Sibylla Merian," *Germanisches Nationalmuseum Nürnberg,* p. 22.
31. Deckert in Merian, *Metamorphosis,* p. 23. The term *insect* was used in the early eighteenth century to designate almost any small invertebrate. Réau-

mur's characterization, however, is an anomaly. Merian to Volckamer, October 1702, reprinted in Rücker, "Maria Sibylla Merian," *Germanisches Nationalmuseum Nürnberg,* pp. 23–24.

32. Zacharias von Uffenbach, *Merkwürdige Reisen durch Niedersachsen, Holland und Engelland* (Ulm, 1753), vol. 3, pp. 552–554.
33. Deckert in Merian, *Metamorphosis,* p. 10. See Francesco Redi, *Esperienze intorno alla generazione degli insetti* (Florence, 1668).
34. See, e.g., Merian, *Metamorphosis,* commentary to plate no. 35; and Merian to Volckamer, 8 October 1702, reprinted in Rücker, "Maria Sibylla Merian," *Germanisches Nationalmuseum Nürnberg,* p. 22.
35. Merian, *Metamorphosis,* p. 38.
36. Ibid., commentary to plate nos. 2, 5, and 11.
37. Deckert in Merian, *Metamorphosis,* p. 23.
38. Merian may have received some assistance from the city of Amsterdam; see Ullmann, "Maria Sibylla Merian," in Merian, *Leningrader Aquarelle,* vol. 2, p. 44.
39. Merian, *Metamorphosis,* p. 38. See also Merian to Volckamer, 16 April 1705, reprinted in Rücker, "Maria Sibylla Merian," *Germanisches Nationalmuseum Nürnberg,* p. 25.
40. Merian, *Metamorphosis,* p. 20.
41. *Acta eruditorum,* November 1707, pp. 481–482.
42. Gräffin [Merian], *Der Raupen,* preface.
43. See *Neue Zeitungen von Gelehrten Sachen,* 23 (1717): 177–180; 95 (1717): 767–768; 73 (1719): 580–582; and 71 (1731): 622–624. See also *Bibliothèque ancienne et moderne,* 11 (1718): 237–254. Merian's Surinam book appeared in a number of catalogues of Parisian natural history cabinets; see, e.g., *Catalogue du cabinet d'histoire naturelle de Mlle Clairon* (Paris, 1773), p. 56.
44. Boris Lukin, "On the History of the Collection of the Leningrad Merian Watercolors," in Merian, *Leningrader Aquarelle,* pp. 118, 120, 122, 124, and 130.
45. Goethe praised Merian's work for "satisfying completely the sensual pleasures; blossoms and buds speak to the eyes, and fruits to the palate"; see his "Blumen-Mahlerei," in *Goethes Werke* (Weimar, 1887–1919), vol. 49, p. 380. See also William Swainson, *The Cabinet Cyclopedia* (London, 1840), pp. 272–273.
46. Lansdown Guilding, "Observations on the Work of Maria Sibilla Merian on the Insects, etc., of Surinam," *Magazine of Natural History,* 7 (1834): 356, 362, 369–371. Emphasis added.
47. Professor Burmeister, "*Metamorphosis insectorum Surinamensium* von Maria Sibilla Merian," *Adhandlungen der Naturforschenden Gesellschaft zu Halle,* 2 (1854): 59.
48. Merian, *Leningrader Aquarelle.*
49. See Richard Westfall, "Science and Patronage: Galileo and the Telescope," *Isis,* 76 (1985): 11–30; Robert Westman, "The Astronomer's Role in the Sixteenth Century: A Preliminary Study," *History of Science,* 18 (1980):

124–125; and Ernst Zinner, *Die Geschichte der Sternkunde* (Berlin, 1931), pp. 587–590.

50. See Maria Cunitz, *Urania propitia* (Oels, 1650). For full bibliographies for Cunitz, Maria Eimmart, and Elisabetha Hevelius, see Londa Schiebinger, "Maria Winkelmann at the Berlin Academy: A Turning Point for Women in Science," *Isis,* 78 (1987): notes 72, 74, and 78.
51. Alphonse des Vignoles, "Eloge de Madame Kirch à l'occasion de laquelle on parle de quelques autres femmes et d'un paisan astronomes," *Bibliothèque germanique,* 3 (1721): 167.
52. Cunitz, *Urania propitia,* p. 147.
53. Johann Eberti, *Eröffnetes Cabinet des gelehrten Frauenzimmers* (Frankfurt and Leipzig, 1706).
54. See Weidler, *Historia astronomiae;* Doppelmayr, *Historische Nachricht,* pp. 259–260; and Jöcher, *Allgemeines Gelehrten-Lexicon,* vol. 3, p. 743.
55. See, e.g., J. C. Poggendorff, *Handwörterbuch zur Geschichte der exacten Wissenschaften* (Leipzig, 1863), vol. 1, p. 65. Eighteenth-century lexicons that list Eimmart's works in great detail attribute the *Ichnographia* to her father. See Doppelmayr, *Historische Nachricht,* p. 126; Georg Will, *Nürnbergisches Gelehrten-Lexicon, oder Beschreibung aller Nürnbergischen Gelehrten beyderley Geschlechtes* (Nuremberg, 1755–1758).
56. According to Peter Ketsch, family trades often passed to the daughter (see his *Frauen im Mittelalter* [Düsseldorf, 1983], vol. 1, p. 29).
57. See Margaret Rossiter, "'Women's Work' in Science, 1880–1910," *Isis,* 71 (1980): 381–398. See also her *Women Scientists in America,* pp. 51–72.
58. Merry Wiesner, "Women's Work in the Changing City Economy, 1500–1650," in *Connecting Spheres: Women in the Western World, 1500 to the Present,* ed. Marilyn Boxer and Jean Quataert (New York, 1987), p. 66.
59. Elisabetha Hevelius is portrayed in three illustrations in Johannes Hevelius, *Machina coelestis* (Danzig, 1673), plates following pp. 222, 254, 450. Two depict her working at the sextant with Johannes; a third shows her using a telescope.
60. Johannes Hevelius, *Prodromus astronomiae* (Danzig, 1690).
61. Vignoles, "Eloge de Madame Kirch," pp. 172–173; and Alphonse des Vignoles, "Lebens Umstände und Schicksale des ehemahles berühmten Gottfried Kirchs," *Dresdenische Gelehrte Anzeigen,* 49 (1761): 775.
62. The Berlin Academy first bore a Latin name, the Societas Regia Scientiarum. From its founding, it was also known as the Brandenburgische or Berlin Societät der Wissenschaften. In the 1740s, it took a French name, Académie Royale des Sciences et Belles-Lettres. In the 1780s it became the Königlich Preussische Akademie der Wissenschaften, which it remained until its reorganization after World War II, when it took its present name, the Akademie der Wissenschaften der Deutschen Demokratischen Republik.
63. On the location of the scientific papers of the Kirch family, see Dietrich Wattenberg, "Zur Geschichte der Astronomie in Berlin im 16. bis 18. Jahrhundert II," *Die Sterne,* 49 (1972): 104–116; and Aufgebauer, "Die Astronomenfamilie Kirch," *Die Sterne,* 47 (1971): 241–247. For a full bibli-

ography for Winkelmann, see Schiebinger, "Maria Winkelmann at the Berlin Academy," note 18.

64. Winkelmann to Leibniz, Leibniz Archiv, Niedersächsische Landesbibliothek, Hanover, Kirch, no. 472, page 11.
65. F. Herbert Weiss, "Quellenbeiträge zur Geschichte der Preussischen Akademie der Wissenschaften," *Jahrbuch der Preussischen Akademie der Wissenschaften* (1939): 223–224; from his private collection. A copy of Winkelmann's report can be found in the Kirch papers, Paris Observatory, MS A.B. 3.7, no. 83, 41, B.
66. Kirch papers, Paris Observatory, MS A.B. 3.5, no. 81 B, p. 33.
67. Adolf Harnack, "Berichte des Secretars der brandenburgischen Societät der Wissenschaften J. Th. Jablonski an den Präsidenten G. W. Leibniz," *Philosophisch-historische Abhandlungen der Königlichen Akademie der Wissenschaften zu Berlin,* 3 (1897): no. 22.
68. See Wattenberg, "Zur Geschichte der Astronomie in Berlin II," p. 107.
69. Vignoles, "Eloge de Madame Kirch," p. 174.
70. See Gottfried Kirch, "Observationes cometae novi," *Acta eruditorum,* 21 April 1702, pp. 256–258; Vignoles, "Eloge de Madame Kirch," pp. 175–176; and Gottfried Kirch, "De cometa anno 1702: Berolini observato," *Miscellanea Berolinensia,* 1 (1710): 213–214.
71. Adolf von Harnack, *Geschichte der Königlich Preussischen Akademie der Wissenschaften zu Berlin* (1900; Hildesheim, 1970), vol. 1, pp. 48–49.
72. Winkelmann (as Kirchin) to Leibniz, 4 November 1707, Leibniz Archiv, Kirch, no. 472, pp. 11–12.
73. Winkelmann often sent Leibniz special reports of her observations. She knew him well enough to drop by to see him to announce that her book was finished and would arrive from the publisher in a few hours. See Winkelmann to Leibniz, n.d., Leibniz Archiv, Kirch, no. 472, p. 10.
74. Leibniz to Sophie Charlotte, January 1709, *Die Werke von Leibniz,* vol. 9, pp. 295–296. Vignoles also reported that Leibniz often tested Winkelmann's knowledge of certain subjects. She was, he wrote, a zealous partisan of the Copernican system; see Vignoles, "Eloge de Madame Kirch," p. 182.
75. Leibniz is referring to the attempt to get an exact measurement of the lunar parallax, which failed because Baron von Krosigk's apprentice, Peter Kolb, was irresponsible and only occasionally made observations. Hans Ludendorff, "Zur Frühgeschichte der Astronomie in Berlin," *Vorträge und Schriften der Preussischen Akademie der Wissenschaften,* 9 (1942): 15.
76. Winkelmann to Leibniz, 17 July 1709, Leibniz Archiv, Jablonski, no. 440, pp. 111–112; Maria Margaretha Winkelmann, *Vorstellung des Himmels bey der Zusammenkunfft dreyer Grossmächtigsten Könige* (Potsdam, 1709). This pamphlet was originally housed in the Preussische Staatsbibliothek, where one still finds a card for it in the catalogue; through the vagaries of war the only extant copy is in the Biblioteka Jagiellońska in Krakow.
77. Leibniz's note in the margin of Winkelmann's letter to him, 17 July 1709, Leibniz Archiv, reprinted in Harnack, "Berichte des Secretars Jablonski an den Präsidenten Leibniz," no. 87.
78. Vignoles, "Eloge de Madame Kirch," p. 182.

79. Winkelmann to Leibniz, 4 November 1707, Leibniz Archiv, Kirch, no. 472, pp. 11–12. Though astrology had begun losing ground in Germany in the sixteenth century, it continued to exercise considerable influence even within scientific circles; see Zinner, *Die Geschichte der Sternkunde,* pp. 558–564.
80. Maria Margaretha Winkelmann, *Vorbereitung, zur grossen Opposition, oder merckwürdige Himmels-Gestalt im 1712* (Cölln an der Spree, 1711). To my knowledge, the only extant copy of this pamphlet is at the Paris Observatory, acquired (I assume) along with other papers of the Kirch family by Joseph-Nicolas Delisle. For a review of this work, see "Praeparatio ad oppositionem magnam," *Acta eruditorum,* 1712, pp. 77–79.
81. Since the Gregorian calendar reform of 1582, Catholics and Protestants had used calendars that differed by ten days (Dietrich Wattenberg, "Zur Geschichte der Astronomie in Berlin im 16. bis 18. Jahrhundert I," *Die Sterne,* 48 [1972]: 165). The "improved" Protestant calendar was similar to the Gregorian calendar except that Easter was calculated differently.
82. Harnack, *Geschichte der Akademie zu Berlin,* vol. 1, p. 124.
83. Gottfried Kirch and Maria Winkelmann, *Das älteste Berliner Wetter-Buch: 1700–1701,* ed. G. Hellmann (Berlin, 1893), pp. 12 and 20–21.
84. Harnack, "Berichte des Secretars Jablonski an den Präsidenten Leibniz," no. 112.
85. See Ludendorff, "Frühgeschichte der Astronomie," p. 12.
86. Winkelmann (as Kirchin) to the Berlin Academy, 2 August 1710; original in Kirch papers, DDR Academy Archives, I–III, 1, pp. 46–48; copy in the Leibniz Archiv, Jablonski, no. 440, pp. 154–156.
87. Harnack, "Berichte des Secretars Jablonski an den Präsidenten Leibniz," no. 115. Leibniz's response to Jablonski has not been preserved.
88. Ibid., no. 116.
89. See Harnack, *Geschichte der Akademie zu Berlin,* vol. 1, pp. 155–156.
90. "Protokollum Concilii, Societatis Scientiarum," 15 December 1710, 18 March 1711, and 9 September 1711, DDR Academy Archives I, IV, 6, pt. 1, pp. 54, 65–66, 93. Unfortunately, we do not know why Winkelmann received a medal.
91. Ibid., 3 February 1712, p. 106.
92. Winkelmann to the council of the Berlin Academy, 3 March 1711, DDR Academy Archives, Kirch papers, I, III, 1, p. 50; and Winkelmann, *Vorbereitung,* pp. 3–4.
93. Harnack, "Berichte des Secretars Jablonski an den Präsidenten Leibniz," nos. 112, 133, 143, 144. In this last, Jablonski was probably referring to Winkelmann's 1711 *Vorbereitung.*
94. See Jablonski to Leibniz, 29 October 1712, in Harnack, "Berichte des Secretars Jablonski an den Präsidenten Leibniz," nos. 143, 167.
95. Wolfram Fischer, *Handwerksrecht und Handwerkswirtschaft um 1800* (Berlin, 1955), p. 18; and W. V. Farrar, "Science and the German University System: 1790–1850," in *The Emergence of Science in Western Europe,* ed. Maurice Crosland (London, 1975), p. 181.
96. Hahn, *The Anatomy of a Scientific Institution,* p. 39.

97. Harnack, *Geschichte der Akademie zu Berlin,* vol. 1, p. 370.
98. Though rights of widows were being restricted in different parts of Europe after about 1650, widows held a surprisingly strong position in the guilds. English common law considered that a widow had served an apprenticeship had she worked in her husband's trade for seven years before his death. See *Baron and Feme: A Treatise of the Common Law concerning Husbands and Wives,* 2d ed. (London, 1719), p. 303. This was also true in France. See "Femme (Jurisp.)" *Encyclopédie, ou Dictionnaire raisonné des sciences, des arts et des métiers* (Paris, 1751–1765), vol. 6, p. 476. In Nuremberg in 1561, more than 10 percent of households—with workshops—were run by widows (Rudolf Endres, "Zur Lage der Nürnberger Handwerkerschaft zur Zeit von Hans Sachs," *Jahrbuch für Fränkische Landesforschung,* 37 [1977]: 122). In most parts of France, as late as 1776 the widow of a surgeon retained the right to continue her husband's practice (*Etat de médecine, chirurgie et pharmacie, en Europe. Pour l'année, 1776* [Paris, 1776], p. 108).
99. Wensky, *Die Stellung der Frau,* pp. 58–59.
100. See Ketsch, *Frauen im Mittelalter,* vol. 1, pp. 29, 204, and 210.
101. "Protokollum Concilii, Societatis Scientiarum," 23 September 1716, DDR Academy Archives, I, IV, 6, pt. 2, pp. 230–232.
102. Christfried Kirch, *Teutsche Ephemeris* (Nuremberg, 1715), p. 82; ibid., (1714), pp. 76–77, 80; and ibid (1715), pp. 78–80, 82–84.
103. See Vignoles, "Eloge de Madame Kirch," p. 180.
104. "Protokollum Concilii, Societatis Scientiarum," 8 October 1716 and 6 April 1718, DDR Academy Archives, I, IV, 6, pt. 2, pp. 236, 318. See also Weiss, "Quellenbeiträge zur Geschichte der Preussischen Akademie," pp. 219–222.
105. Vignoles, "Eloge de Madame Kirch," p. 181.
106. "Protokollum Concilii, Societatis Scientiarum," 18 August 1717, DDR Academy Archives, I, IV, 6, pt. 2, p. 269, 272–273 (emphasis added).
107. Vignoles, "Eloge de Madame Kirch," p. 181; "Protokollum Concilii, Societatis Scientiarum," 21 October 1717, DDR Academy Archives, I, IV, 6, pt. 2, pp. 275–276.
108. Vignoles, "Eloge de Madame Kirch," pp. 181–182.
109. Sophie Charlotte was privately tutored by Leibniz from an early age, well read in Latin, well traveled, and a devotee of French culture. See Leibniz to Sophie Charlotte, November 1697, reprinted in *Die Werke von Leibniz,* vol. 8, pp. 47–53.
110. Frederick II, "Mémoire de l'Académie" (1748), reprinted in Jean-Pierre Erman, *Mémoire pour servir à l'histoire de Sophie Charlotte, Reine de Prusse* (Berlin, 1801), p. 382.
111. "Leibnizens Denkschrift in Bezug auf die Einrichtung einer Societas Scientiarum et Artium in Berlin vom 26 März 1700, bestimmt für den Kurfürsten," in Harnack, *Geschichte der Akademie zu Berlin,* vol. 2, p. 80. I thank Gerda Utermöhlen of the Leibniz Archiv, Hanover, for calling this passage to my attention.
112. Alphonse des Vignoles, "Eloge de M. Kirch le Fils, Astronome de Berlin," *Journal littéraire d'Allemagne de Suisse et du Nord,* 1 (1741): 349.

113. Christine Kirch to Delisle, 24 July 1744, Paris Observatory, Delisle papers, MS A.B. 1. IV, no. 12a; and 28 April 1745, no. 42.
114. Harnack, *Geschichte der Akademie zu Berlin,* vol. 1, p. 491.
115. It should be pointed out that Catherine was elected in 1767, when Frederick the Great, as president of the academy, personally oversaw all academy appointments. The following year Frederick decreed that Catherine's membership in the academy should be elevated from honorary status to that of a regular foreign member (Harnack, *Geschichte der Akademie zu Berlin,* vol. 1, pp. 369, 473).
116. Maria Wentzel's endowment of 1894 was in honor of her architect husband and factory-owning father (ibid, vol. 1, p. 1019).
117. In addition to the fourteen women who became academy members over the last three-and-one-half centuries, fifteen other women won academy prizes. See Werner Hartkopf, *Die Akademie der Wissenschaften der DDR: Ein Beitrag zu ihrer Geschichte* (Berlin, 1983); and Erik Amburger, *Die Mitglieder der deutschen Akademie der Wissenschaft zu Berlin, 1700–1950* (Berlin, 1950).

4. *Women's Traditions*

1. François-Marie Arouet de Voltaire, *Dictionnaire philosophique* (1764; Amsterdam, 1789), vol. 5, p. 255.
2. Pizan, *The Book of the City of Ladies,* p. 81.
3. Mlle Archambault, *Dissertation sur la question: Lequel de l'homme ou de la femme est plus capable de constance?* (Paris, 1750), pp. 69–70.
4. William Alexander, *The History of Women* (London, 1779), vol. 2, pp. 36–43.
5. Poullain de la Barre, *De l'égalité des deux sexes,* p. 58.
6. See especially Jean Donnison, *Midwives and Medical Men* (New York, 1977). See also Monica Green, "Women's Medical Practice and Health Care in Medieval Europe," *Signs,* 14 (1989): 434–473.
7. See G. Elmeer, "The Regulation of German Midwifery in the Fourteenth, Fifteenth, and Sixteenth Centuries," M.D. thesis, Yale University School of Medicine, 1963. Women were active in other areas of health care as well; in both Catholic and Protestant countries, they ran hospitals and orphanages and served as surgeons and even doctors. Albrecht von Haller lists a number of women surgeons in his *Bibliotheca chirurgica* of 1774–1775. See also A. L. Wyman, "The Surgeoness: The Female Practitioner of Surgery, 1400–1800," *Medical History,* 28 (1984): 22–41; and *Etat de médecine, chirurgie et pharmacie, en Europe* (Paris, 1776).
8. Laget, "Childbirth in Seventeenth- and Eighteenth-Century France," pp. 137–176.
9. Thomas Forbes, "The Regulation of English Midwives in the Sixteenth and Seventeenth Centuries," *Medical History,* 8 (1964): 235–244; and Merry Wiesner, *Working Women in Renaissance Germany* (New Brunswick, 1986), chap. 2.

10. Erwin Ackerknecht and Esther Fischer-Homberger, "Five Made It—One Not: The Rise of Medical Craftsmen to Academic Status during the Nineteenth Century," *Clio Medica,* 12 (1977): 255–267.
11. Donnison, *Midwives and Medical Men,* p. 22.
12. Ibid., p. 11.
13. Edmund Chapman, *A Reply to Mr. Douglass's Short Account of the State of Midwifery in London and Westminster* (London, 1737), pp. 4–6.
14. John Maubray, *The Femal Physician* (London, 1724), quoted in Barbara Schnorrenberg, "Is Childbirth Any Place for a Woman? The Decline of Midwifery in Eighteenth-Century England," *Studies in Eighteenth-Century Culture,* 10 (1981): 399–400.
15. See, e.g., "By a physician," *The Ladies Physical Directory* (London, 1716 and 1727). See also Donnison, *Midwives and Medical Men,* pp. 10 and 23.
16. Richard Petrelli, "The Regulation of French Midwifery during the *Ancien Régime,*" *Journal of the History of Medicine and Allied Sciences,* 16 (1971): 282–283. Though educated in the United States (in the 1840s), Elizabeth Blackwell had to travel to France to receive clinical training. La Maternité (an old convent converted into a state institution for the training of midwives) was the only place that Blackwell (who was neither Catholic, nor at that time particularly interested in gynecology) could receive clinical training. Like most doctors of her day, she too was contemptuous of the "ignorant" midwife. See Elizabeth Blackwell, *Opening the Medical Profession to Women,* ed. Mary Walsh (New York, 1977).
17. Elizabeth Cellier, *A Scheme for the Foundation of a Royal Hospital* (1687), reprinted in *The Harleian Miscellany,* ed. Thomas Osborne (London, 1745), vol. 4, pp. 136–139.
18. Jane Sharp, *The Midwives Book* (London, 1671), p. 4.
19. Elizabeth Nihell, *A Treatise on the Art of Midwifery* (London, 1760), p. 15.
20. Merry Wiesner, "Early Modern Midwifery: A Case Study," in *Women and Work in Preindustrial Europe,* ed. Barbara Hanawalt (Bloomington, 1986), pp. 94–114.
21. Nihell, *A Treatise on the Art of Midwifery,* p. 50.
22. Ibid., pp. vii–xii.
23. Donnison, *Midwives and Medical Men,* p. 2.
24. Gunnar Heinsohn and Otto Steiger, *Die Vernichtung der weisen Frauen* (Herbstein, 1985).
25. See J. T. Noonan, *Contraception: A History of Its Treatment by the Catholic Theologians and Canonists* (Cambridge, Mass., 1965).
26. Cellier, quoted in Forbes, "Regulation of English Midwives," p. 235.
27. See Barbara Ehrenreich and Deirdre English, *For Her Own Good: 150 Years of the Experts' Advice to Women* (New York, 1978). For a different view, see Roy Porter, "A Touch of Danger: The Man-midwife as Sexual Predator," in *Sexual Underworlds of the Enlightenment,* ed. G. S. Rousseau and Roy Porter (Manchester, England, 1987), pp. 206–232.
28. See Hans Wiswe, *Kulturgeschichte der Kochkunst* (Munich, 1970), pp. 59–72.

29. Franz de Rongier, *Kunstbuch von mancherley Essen* (Wolfenbüttel, 1598), preface.
30. [Eliza Smith], *The Compleat Housewife* (London, 1728), title page.
31. George Hartman, Chymist, *The True Preserver and Restorer of Health* (London, 1682), "To the Reader."
32. Albrecht von Haller, *Bibliotheca botanica* (London, 1771).
33. Maria Edgeworth, *Letters for Literary Ladies* (London, 1795), p. 67.
34. Frederick Accum, *Culinary Chemist, Exhibiting the Scientific Principles of Cookery* (London, 1821), pp. iii–3.
35. Marie Meurdrac, *La Chymie charitable et facile, en faveur des dames* (Paris, 1674), preface. See also Lloyd Bishop and Will DeLoach, "Marie Meurdrac—First Lady of Chemistry?" *Journal of Chemical Education,* 47 (1970): 448–449.
36. Meurdrac, *La Chymie,* preface and p. 255. For a discussion of other cosmetic books, see Virginia Smith, "Popular Medical 'Knowledges': The Case of Cosmetics," paper delivered at the Institute of Historical Research, London, February 14, 1986.
37. Ginnie Smith, "Thomas Tryon's Regimen for Women: Sectarian Health in the Seventeenth Century," in *The Sexual Dynamics of History: Men's Power, Women's Resistance* (London, 1983), pp. 47–65.
38. [Thomas Cocke], *Kitchin-Physick* (London, 1676).
39. William Buchan, *Domestic Medicine; or the Family Physician* (Edinburgh, 1769).
40. John B. Blake, "The Compleat Housewife," *Bulletin of the History of Medicine,* 49 (1975): 35.
41. Wiswe, *Kulturgeschichte der Kochkunst,* pp. 59–60.
42. William Forster, *A Treatise on the Various Kinds and Qualities of Foods* (Newcastle upon Tyne, 1738).
43. John Gerard, *The Herball or General Historie of Plantes,* enlarged by Thomas Johnson (1597; London, 1636), pp. 151–152. Gerard's wife assisted him professionally. Benjamin Jackson, *A Catalogue of Plants* (London, 1876), p. xvi.
44. Louis Lémery, *New Curiosities in Art and Nature: Or, A Collection of the most Valuable Secrets in all Arts and Sciences,* trans. from the 7th ed. (London, 1711).
45. [Cocke], *Kitchin-Physick.* See also P. J. Marperger, *Küch- und Keller-Dictionarium* (Hamburg, 1716).
46. Wiswe, *Kulturgeschichte der Kochkunst,* p. 68. See also Jean-Claude Bonnet, "Le Réseau culinaire dans l'Encyclopédie," *Annales,* 31 (1976): 891–914.
47. Louis Lémery, *A Treatise of Foods, in General* (London, 1706).
48. *Adam's Luxury, and Eve's Cookery* (London, 1747).
49. Anna Weckerin, *Ein köstlich new Kochbuch* (Amberg, 1697), dedication.
50. These figures are drawn from Virginia Maclean, *A Short-title Catalogue of Household and Cookery Books published in the English Tongue, 1701–1800* (London, 1981). Many cookbooks were produced completely by women. Maria Sophia Conring's cookbook of 1697 was published under the pro-

tection of a noblewoman and printed by the widow of a printer. See, e.g., [Maria Sophia Conring], *Die wol unterweisette Köchinn* (Braunschweig, 1697).

51. By a Lady [Hannah Glasse], *The Art of Cookery, Made Plain and Easy* (London, 1747), p. 118.
52. Elizabeth Raffald, *The Experienced English Housekeeper, For the Use and Ease of Ladies, Cooks, etc., Wrote purely from practice* (1769), 2d ed. (London, 1772), "To the Reader." Raffald also wrote a book on midwifery.
53. Cited in Blake, "The Compleat Housewife," p. 41.
54. Home economics, developed by women in the late nineteenth century, preserved part of this earlier intellectual legacy but at a relatively lower status.
55. Stephen Mennell, *All Manners of Food: Eating and Taste in England and France from the Middle Ages to the Present* (London, 1985), pp. 200–204.
56. *Madam Johnson's Present: Or, Every Young Woman's Companion in Useful and Universal Knowledge* (Dublin, 1770). The most popular cookbooks of the eighteenth century were written by Hannah Glasse, Elizabeth Raffald, and Eliza Smith. See Arnold Oxford, *English Cookery Books to the Year 1850* (London, 1913); and Maclean, *A Short-Title Catalogue of Household and Cookery Books.*
57. William Smellie, *A Treatise on the Theory and Practice of Midwifery* (London, 1752–64), p. 72.
58. Albrecht von Haller, *Physiology,* trans. Samuel Mihlis (London, 1754), p. xvi.
59. *Der aus dem Parnasso ehmals entlaufenen vortrefflichen Köchin welche bey denen Göttinnen Ceres, Diana und Pomona viel Jahre gedienet* (Nuremberg, 1691).
60. Conring, *Die wol unterweisette Köchinn,* dedication poem.
61. Nihell, *A Treatise on the Art of Midwifery,* p. vi.

5. Battles over Scholarly Style

1. See the explanation of the frontispiece by its designer, Charles-Nicolas Cochin ("Explication, Frontispice de l'Encyclopédie," *Encyclopédie,* facing frontispiece. I rely on I. Bernard Cohen's description of the scientific equipment in his *Album of Science: From Leonardo to Lavoisier* (New York, 1980), no. 362. Though the frontispiece is often bound with the first edition (1751), it was first drawn in 1764 and engraved in 1776 (by B. L. Prevost). The frontispiece reproduced here is a variant of the original published in the Geneva edition. I thank Richard N. Schwab for this information.

   In 1665 Jacquette Guillaume suggested that, compared to the solitary Apollo, there were *nine* muses each credited with the cultivation of a particular part of knowledge (*Des dames illustres, ou par bonnes et fortes raisons, il se prouve, que le sexe féminin surpasse en toute sorte de genres le sexe masculin* [Paris, 1665], p. 209). It is interesting to note that the cartouche of successive volumes of the *Encyclopédie* alternate between illustrations of a female and male figure, whom Sara Malueg identifies as Minerva and Apollo (see her "Women and the *Encyclopédie,*" in *French Women,* ed. Spencer, p. 264).

2. Charles Cochin and Hubert Gravelot, *Iconologie par figures; ou Traité complet des allégories, emblèmes, &c.* (1791; Geneva, 1972), vol. 1, pp. vii–x. Iconography—from *icon* for "picture" and *graphy* for "writing"—is literally picture writing. Erwin Panofsky distinguishes between iconography (the classification and description of symbols) and iconology (the symbolic value of symbols or synthetic meaning of a work of art). See his *Meaning in the Visual Arts* (Garden City, N.Y., 1955), chap. 1.
3. Lynn Hunt has described a similar iconographic struggle in French political life (*Politics, Culture, and Class in the French Revolution* [Berkeley and Los Angeles, 1984], pp. 87–119).
4. See the images of nature in Cesare Ripa, *Iconologia* (Rome, 1593), first illustrated in 1603; Jean Baudoin, *Iconologie, ou Explication nouvelle de plusieurs images* (Paris, 1644); Cochin, *Iconologie;* and Merchant, *The Death of Nature.*
5. Ripa, *Iconologia.*
6. Morton Bloomfield, "A Grammatical Approach to Personification Allegory," *Modern Philology,* 60 (1963): 163.
7. Cochin, *Iconologie,* vol. 4, p. 79.
8. The identity of "the queen of the sciences" shifted; for some it was "most holy theology" (Giovanni Pico della Mirandola, *Oration on the Dignity of Man,* trans. A. Robert Caponigri [Chicago, 1956], p. 21); for others it was philosophy.
9. Johannes Hevelius, *Firmamentum Sobiescianum sive Uranographia* (Danzig, 1687). See also Ivan Volkoff, *Johannes Hevelius and His Catalogue of Stars* (Provo, Utah, 1971). The designer of Hevelius's frontispiece may have been influenced by the frontispiece to Johannes Kepler's *Rudolphine Tables* (1627). Here Kepler illustrated the social structure of seventeenth-century astronomy. At the very top, the imperial eagle spreads its wings of protection over the whole temple of astronomy. The talers dropping from its beak symbolize Emperor Rudolph II's financial support. On the dome are seven goddesses or muses (among them Urania), each providing inspiration and holding reminders of Kepler's important scientific innovations. The temple rests on the "shoulders of giants"—astronomers Hipparchus, Copernicus, Tycho Brahe, and Ptolemy. Kepler, the practioner, is shown seated in an inset in the base of the temple (left). Another inset (right) shows apprentices at work. See Cohen, *Album of Science,* p. 53, no. 68.
10. Nicolas Lémery, *A Course of Chymistry,* 4th English ed. (London, 1720). On the tradition of nudity and truth, see Cesare Ripa, *Baroque and Rococo Pictorial Imagery: The 1758–60 Hertel Edition of Ripa's 'Iconologia,' with 200 Engraved Illustrations,* ed. and trans. Edward Maser (New York, 1971), no. 50, "Veritas." See also Marina Warner, *Monuments and Maidens: The Allegory of the Female Form* (New York, 1985), pp. 294–328.
11. Other examples of the use of feminine imagery include: Meurdrac's *La Chymie charitable et facile;* Christian Wolff's *Mathematiches Lexicon* (Leipzig, 1716); and Georges-Louis Leclerc, comte de Buffon's multi-volume *Histoire naturelle, générale et particulière* (Paris, 1749–1804).

12. I have assumed here that the author had something to do with the selection of frontispieces for his or her work. The publishing history of these works, however, is very difficult to come by, and I have been unable to investigate what role authors actually played in choosing frontispieces.
13. The Amsterdam edition of du Châtelet's *Institutions physiques* (published two years after the Paris edition) bears both her name and portrait. Though women rarely appended their portraits to works of what we might call "high science," they more often revealed themselves in medical cookbooks and the works of popular science. A rare portrait of a woman is the lovely one of Margaret Bryan and her daughters in her *Compendious System of Astronomy* (London, 1797).
14. Cochin, *Iconologie,* p. v.
15. Ripa, quoted in Warner, *Monuments and Maidens,* p. 65.
16. François-Marie Arouet de Voltaire, *An Essay upon the Civil Wars of France . . . And also upon the Epic Poetry of the European Nations* (London, 1727), p. 115.
17. See Ripa, *Baroque and Rococo Pictorial Imagery,* no. 79, "Pax."
18. See A. Meillet, *Linguistique historique et linquistique générale* (Paris, 1982), pp. 211–229. In modern Romance languages, action nouns generally became masculine while the actions themselves became feminine. Thus in French, *le juge* (a judge) is masculine while *la justice* (justice) is feminine (Warner, *Monuments and Maidens,* pp. 66–68).
19. Cochin, *Iconologie,* pp. v–xv.
20. Henry More, *Conjectura cabbalistica* (London, 1653), pp. 40 and 70. See Marjorie Hope Nicolson, "Milton and the Conjectura Cabbalistica," *Philological Quarterly,* 6 (1927): 15.
21. Plato, "Timaeus," trans. Benjamin Jowett, in *The Collected Dialogues of Plato, Including the Letters,* ed. Edith Hamilton and Huntington Cairns (Princeton, 1961), pp. 1165–1167.
22. Pico della Mirandola, *Oration on the Dignity of Man,* pp. 22–23.
23. Joan M. Ferrante, *Woman as Image in Medieval Literature* (New York, 1975), p. 50.
24. Ibid., pp. 46–49. See also Robert Worth Frank, Jr., "The Art of Reading Medieval Personification-Allegory," *ELH: A Journal of English Literary History,* 20 (1953): 239.
25. See Gregor Reisch, *Margarita philosophica* (Basel, 1517).
26. T. H. Gent [Thomas Heywood], *The Generall History of Women* (London, 1657), pp. 104–105. Diderot's *Encyclopédie* also defined *muse* as deriving from the Greek and signifying "the explaining of mysteries, because the muses taught to men very curious and very important things" (s.v. "muse"). The ancient world envisioned nine muses; by the sixteenth century, however, muses represented nearly all the arts and sciences (Warner, *Monuments and Maidens,* p. 206). Only rarely is the scientist portrayed leading the muse to truth; see, e.g., Galileo trying to teach three rather bored muses the finer details of his sytem (*Opere di Galileo Galilei* [Bologna, 1655–56], vol. 2).
27. Cochin, *Iconologie,* vol. 2, s.v. "Géométrie."

28. Cunitz, *Urania propitia*. See also Ingrid Guentherodt, "Maria Cunitz und Maria Sibylla Merian: Pionierinnen der deutschen Wissenschaftssprache im 17. Jahrhundert," Zeitschrift für germanistische Linguistik, 14 (1986): 29–31.
29. Cochin, *Iconologie,* vol. 4, p. 79.
30. Abraham Cowley, "To the Royal Society," in Sprat, *History of the Royal Society,*" dedication poem.
31. Thomas Sprat, "An Account of the Life and Writings of Mr. Abraham Cowley," in *The Works of Mr. Abraham Cowley,* 8th ed. (London, 1693), facing p. b2.
32. Francis Bacon, *The Works of Francis Bacon,* ed. James Spedding, Robert Ellis, and Douglas Heath (London, 1857–1874), vol. 3, pp. 524–539. According to the editors, this essay was written in about 1608. See also Evelyn Fox Keller, "Baconian Science: A Hermaphroditic Birth," *The Philosophical Forum,* 11 (1980): 299–308.
33. Benjamin Farrington, "Temporis Partus Masculus: An Untranslated Writing of Francis Bacon," *Centaurus,* 1 (1951): 194, 200.
34. Sprat, *History of the Royal Society,* p. 129.
35. Henry Oldenburg, "The Publisher to the Reader," in Robert Boyle, *Experiments and Considerations Touching Colours* (London, 1664), p. iv.
36. William Harvey, *The Circulation of the Blood* (1628) trans. Robert Willis (London, 1952), p. 7; Francis Bacon, "The Advancement of Learning," in *Selected Works of Francis Bacon,* ed. Hugh Dick (New York, 1955), p. 225.
37. Andreas Vesalius, *De humani corporis fabrica* (Basel, 1543). Similarly, Tycho Brahe portrayed himself actively measuring the heavens at his observatory at Uraniborg. See Brahe, *Astronomiae instauratae mechanica* (1598), reprinted in Cohen, *Album of Science,* no. 54.
38. Oldenburg, "The Publisher to the Reader," in Boyle, *Experiments and Considerations Touching Colours.*
39. Sir Isaac Newton, *Method of Fluxions and Infinite Series,* trans. John Colson (London, 1736). I thank Iain Boal for translating this Greek motto.
40. *Essayes of Natural Experiments, Made in the Academy del Cimento,* trans. Richard Waller (London, 1684). See also Elizabeth Eisenstein, *The Printing Revolution in Early Modern Europe* (Cambridge, 1983), pp. 244–245.
41. Marie Sybille de Merian, *Histoire générale des insectes de Surinam et de toute l'Europe,* ed. Pierre Buch'oz, trans. Jean de Missy and Jean Marret (Paris, 1771).
42. Pizan, *The Book of the City of Ladies,* pp. 73–76. Francis Bacon also spoke of Ceres as a real person. In his *Novum organum* (1620) he called attention "in things mechanical [to] the works of Bacchus and Ceres—that is, of the arts of preparing wine and beer, and of making bread . . ." (*The New Organon,* ed. Fulton Anderson [New York, 1960], p. 82).
43. Gent [Heywood], "To the Reader," *The Generall History of Women,* facing p. A4.
44. Wetenhall Wilkes, *An Essay on the Pleasures and Advantages of Female Literature* (London, 1741), p. 17.

45. Johann Zedler, ed., *Grosses vollständiges Universal-Lexicon aller Wissenschafften und Künste* (Leipzig and Halle, 1733–50), s.v. "Weib."
46. For a discussion of women as muses in eighteenth-century France, see Roseann Rute, "Women as Muse," in *French Women,* ed. Spencer, pp. 143–154. On neo-Platonism in the seventeenth-century French salon, see Lougee, *Le Paradis des Femmes.*
47. Cavendish, *Observations upon Experimental Philosophy,* p. 2. In 1683 James Norris also observed that the ancients had a better opinion of women, for "they ascribed all Sciences to the Muses" (*Haec & Hic; or the Feminine Gender More Worthy than the Masculine* [London, 1683], p. 17).
48. Immanuel Kant, *Critik der reinen Vernunft* (Riga, 1781), p. viii.
49. Immanuel Kant, *Beobachtungen über das Gefühl des Schönen und Erhabenen,* in *Kants Werke,* ed. Wilhelm Dilthey (Berlin, 1900–1919), vol. 2, pp. 229–230. The beard was a common symbol of virility; see Gabriel Jouard, *Nouvel Essai sur la femme considérée comparativement à l'homme* (Paris, 1804), p. 8.
50. Harless, *Die Verdienste der Frauen um Naturwissenschaft,* p. 7.
51. See Ernest Tuveson, *The Imagination as a Means of Grace* (Berkeley, 1960), pp. 5–41, esp. 20.
52. Quoted in Ferrante, *Woman as Image in Medieval Literature,* p. 43.
53. D. J. Gordon, "Ripa's Fate," in *The Renaissance Imagination: Essays and Lectures by D. J. Gordon,* ed. Stephen Orgel (Berkeley, 1975), p. 54.
54. *Allegory* is used here to refer to social meanings encoded in an image or text; it is not used in the older sense of reflecting a higher spiritual reality.
55. *The Scientist,* ed. Henry Margenau, David Bergamini, and the editors of *Life* (New York, 1964), p. 185. On the implications of unveiling, see Caroline Merchant, "Isis' Consciousness Raised," *Isis,* 73 (1982): 404.
56. Novalis, cited in Edgar Zilsel, "Die Gesellschaftlichen Würzeln der romantischen Ideologie," *Der Kampf,* 26 (1933): 154.
57. In France, it was during the period from the 1750s to the 1790s that scientists first tried to dissociate themselves from the literati. See Wolf Lepenies, "Der Wissenschaftler als Autor, Buffons prekärer Nachruhm," in his *Das Ende der Naturgeschichte: Wandel kultureller Selbstverständlichkeiten in den Wissenschaften des 18. und 19. Jahrhunderts* (Frankfurt, 1978), pp. 131–168.
58. Philippe de Prétot, *Le Triomphe des dames, ou Le Nouvel Empire littéraire* (Paris, 1755), p. 3 and 18.
59. Alexander, *The History of Women,* vol. 1, p. 329.
60. Mary Wollstonecraft, *Vindication of the Rights of Woman* (1792; Harmondsworth, England, 1982), p. 155. Roman Stoic philosopher C. Musonius Rufus was one of the few to argue against the idea that marriage disrupted philosophical pursuit ("Is Marriage a Handicap for the Pursuit of Philosophy," in *Visions of Women,* ed. Linda A. Bell [Clifton, N.J., 1983], pp. 75–77).
61. Karl Joël, *Die Frauen in der Philosophie* (Hamburg, 1896), pp. 44 and 48.
62. David Hume, "Of Essay Writing," in *The Essays Moral, Political and Literary* (1741–1742; London, 1963), p. 570.
63. Denis Diderot, "Sur les femmes" (1772) in *Oeuvres complètes de Diderot,* ed.

J. Assézat (Paris, 1875), vol. 2, p. 262. For a different interpretation, see Landes, *Women and the Public Sphere,* p. 45.

64. Anne-Thérèse de Marguenat de Courcelles, marquise de Lambert, *Réflexions nouvelles sur les femmes* (1727; London, 1820), p. 132.
65. Lougee, *Le Paradis des Femmes,* p. 53; Lambert, *Réflexions nouvelles sur les femme,* pp. 110–111. For an analysis of the *précieuses* and *honnête homme,* see Domna Stanton, *The Aristocrat as Art* (New York, 1980), esp. 13–30.
66. Lepenies, "Der Wissenschaftler als Autor," p. 145.
67. For a similar analysis of *préciosité,* see Domna Stanton, "The Fiction of *Préciosité* and the Fear of Women," *Yale French Studies,* 62 (1981): 107–134.
68. Lepenies, "Der Wissenschaftler als Author," p. 141.
69. Voltaire, *An Essay upon the Civil Wars of France,* pp. 121–122.
70. Wren, cited in Katharine Rogers, "The View from England," in *French Women,* ed. Spencer, pp. 358 and 361.
71. Alexander, *The History of Women,* vol. 1, p. 321.
72. [Judith Drake], *An Essay in Defence of the Female Sex* (London, 1696), p. xxi. This essay has also been attributed to Mary Astell and H. Wyatt.
73. François Du Soucy, quoted in Lougee, *Le Paradis des Femmes,* p. 29.
74. J.-J. Rousseau, *Lettre à M. d'Alembert sur les spectacles* (1758), ed. L. Brunel (Paris, 1896). Rousseau's was a common attitude. In 1787 Marie-Armande-Jeanne de Humières, Mme Gacon-Dufour, answered the Chevalier de Feucher's charge that the sciences and the arts were in a state of decay because of the influence of women (*Mémoire pour le sexe féminin contre le sexe masculin* [Paris and London, 1787], p. 4). Though Rousseau's attack on the influence of the salon on scholarship was new, the attack on the salon was not (see Chapter 8 below).
75. Rousseau, *Lettre à M. d'Alembert,* pp. 156–157.
76. Ibid., p. 157.
77. Janice Moulton, "A Paradigm of Philosophy: The Adversary Method, in *Discovering Reality: Feminist Perspectives on Epistemology, Metaphysics, Methodology, and the Philosophy of Science,* ed. S. Harding and M. Hintikka (Dordrecht, Holland, 1983), pp. 149–164.
78. Rousseau, *Lettre à M. d'Alembert,* p. 154.
79. J.-J. Rousseau, *Emile* (1762) in *Oeuvres complètes,* ed. Bernard Gagnebin and Marcel Raymond (Paris, 1959–1969), vol. 4, p. 737.
80. Marie-Jean-Antoine-Nicolas Caritat, marquis de Condorcet, *Esquisse d'un tableau historique des progrès de l'esprit humain* (Paris, 1795).
81. Buffon, *Histoire naturelle,* vol. 2, p. 554.
82. Rousseau, *Lettre à M. d'Alembert,* p. 150.
83. See d'Alembert, *Preliminary Discourse to the Encyclopedia of Diderot,* trans. Richard N. Schwab (Indianapolis, 1963), pp. 93–97.
84. Antoine Lavoisier, *Elements of Chemistry,* trans. Robert Kerr, in *Lavoisier, Fourier, Faraday,* ed. Robert Maynard Hutchins (Chicago, 1952), p. 1. See also Wilda Anderson, *Between the Library and the Laboratory: The Language of Chemistry in Eighteenth Century France* (Baltimore, 1984). I thank Timothy Reiss for calling these sources to my attention.

85. Lepenies, "Der Wissenschaftler als Autor," p. 141.
86. Hume, "Of Essay Writing," p. 570.
87. Lambert, *Réflexiones nouvelles sur les femmes,* pp. 110 and 113.

6. Competing Cosmologies: Locating Sex and Gender in the Natural Order

1. Alexandre Koyré, *From the Closed World to the Infinite Universe* (Baltimore, 1957), pp. vii–viii. See also, e.g., Robert Merton, *Science, Technology and Society in Seventeenth Century England* (1938; New York, 1970); Boris Hessen, "The Social and Economic Roots of Newton's 'Principia,'" in *Science at the Cross Roads* (London, 1931); Paul Feyerabend, *Against Method* (London, 1975); and A. R. Hall, *From Galileo to Newton* (New York, 1963).
2. See Helkiah Crooke, *Mikrokosmographia, A Description of the Body of Man* (London, 1615).
3. See Aristotle, *Generation of Animals,* trans. A. L. Peck (London, 1943), I, i–ii, pp. 11–15, and IV, vi, p. 459. See also Sarah Pomeroy, *Goddesses, Whores, Wives, and Slaves* (New York, 1975); M. C. Horowitz, "Aristotle and Woman," *Journal of the History of Biology,* 9 (1976): 183–213; and G. E. R. Lloyd, *Science, Folklore and Ideology* (Cambridge, 1983).
4. Crooke, *Mikrokosmographia,* p. 271. On the continuity in medical views from the ancient to the medieval world, see Vern Bullough, "Medieval Medical and Scientific Views of Women," *Viator,* 4 (1973): 487; and Lester S. King, "The Transformation of Galenism," *Medicine in Seventeenth-Century England,* ed. Allen Debus (Berkeley, 1974), pp. 7–32.
5. Galen, *On the Usefulness of the Parts of the Body,* trans. Margaret May (Ithaca, 1968), vol. 2, pp. 628–629.
6. Ibid., vol. 2, p. 630. See also Esther Fischer-Homberger, *Krankheit Frau und andere Arbeiten zur Medizingeschichte der Frau* (Bern, 1979), pp. 14–15.
7. Crooke, *Mikrokosmographia,* p. 249. It was a common notion in eighteenth-century Europe that blacks also changed spontaneously into whites. See also Marvin Harris, *The Rise of Anthropological Theory* (New York, 1968); William Stanton, *The Leopard's Spots* (Chicago, 1960). Bert Hansen has pointed out that the ancients may have developed this view to explain the very real phenomenon of a prolapsed uterus (private communication, 1986).
8. This according to Crooke; see his *Mikrokosmographia,* pp. 45, 249.
9. Georges Ascoli, "Essai sur l'histoire des idées féministes en France du XVIe siècle à la révolution," *Revue de synthèse historique,* 13 (1906): 25–183.
10. For differing views on whether women benefited from the Renaissance, see Joan Kelly-Gadol, "Did Women Have a Renaissance," in *Becoming Visible: Women in European History,* ed. Claudia Koonz and Renate Bridenthal (Boston, 1977), pp. 137–164; Ruth Kelso, *Doctrine for the Lady of the Renaissance* (Urbana, 1956); and Constance Jordan, "Feminism and the Humanists: The Case of Sir Thomas Elyot's *Defense of Good Women,*" in *Re-*

writing the Renaissance: The Discourses of Sexual Difference in Early Modern Europe, ed. Margaret Ferguson, Maureen Quilligan, and Nancy Vickers (Chicago, 1986), pp. 227–241.

11. Agrippa von Nettesheim, *Female Pre-eminence,* p. 9. Jacquette Guillaume made a similar argument in the preface to her *Des dames illustres.*
12. This was a common argument; see *Wonders of the Female World* (London, 1683), pp. 104–193.
13. Anna van Schurman, *The Learned Maid, or Whether a Maid may be a Scholar?* (London, 1659), pp. 1–2. See also Una Pope-Hennessy, *Anna van Schurman: Artist, Scholar, Saint* (New York, 1909).
14. Marie le Jars de Gournay, *Egalité des hommes et des femmes* (Paris, 1622), pp. 8 and 18.
15. Jacques Du Bosc, *L'Honneste femme* (1632; Paris, 1658), pp. 117–119. This was a common opinion in Paris at this time; see Renaudot, ed., *Recueil,* pp. 90–96.
16. Samuel Sorbière, *Lettres et discourse de M. de Sorbière sur diverses matières curieuses* (Paris, 1660), p. 71.
17. *The Guardian,* 165 (8 September 1713).
18. Agrippa von Nettesheim, *Female Pre-eminence,* pp. 1–2. The term *female worthies* is Natalie Davis's. On German lexicons of learned ladies in this period, see Jean M. Woods and Maria Faurstenwald, *Schriftstellerinnen, Künstlerinnen und gelehrte Frauen des deutschen Barock: Ein Lexikon* (Stuttgart, 1984).
19. Jean de La Forge, *Le Cercle des femmes sçavantes* (Paris, 1663).
20. Marguerite Buffet, *Nouvelles Observations sur la langue françoise* (Paris, 1668), pp. 200, 221–223.
21. Ibid., pp. 228–229.
22. Augustine made this argument using the analogy of superior and inferior angels. See Eleanor Mclaughlin, "Equality of Souls, Inequality of Sexes: Woman in Medieval Theology," in *Religion and Sexism: Images of Women in the Jewish and Christian Traditions,* ed. Rosemary Ruether (New York, 1974), p. 218.
23. Ibid., pp. 219–221. The notion of a hermaphroditic God also justified the notion that souls have no sex. The eighteenth-century *Encyclopédiste* Louis le Chevalier de Jaucourt reported that this "monstrous" opinion was widely accepted, despite the fact that God, having no "corporeal organs" whatsoever, certainly had no "organs of generation" and thus no sex; "Femme (Anthropologie)" *Encyclopédie,* vol. 6, p. 470). See also Stuart Schneiderman, *An Angel Passes: How the Sexes Became Undivided* (New York, 1988).
24. Pierre Le Moyne, *La Galerie des femmes fortes* (Paris, 1647), p. 251, quoted in Lougee, *Le Paradis des Femmes,* p. 63.
25. *The Ladies Calling,* [attributed to Richard Allestree] (1673; Oxford, 1720), preface.
26. Crooke, *Mikrokosmographia* pp. 45–46. As a French medical doctor wrote: "the soul is to the body, that what the prince is to his estates, the prelate to his dioceses, the liege to his siege, the architect to his buildings, the laborer

to his land, the artisan to his shop . . ." (Louis Couvay, *L'Honneste maîtresse* [Paris, 1654], p. 93).

27. Letter to Abbé Claude Picot, reprinted as a preface to René Descartes, *Principles of Philosophy,* in *The Philosophical Works of Descartes,* trans. Elizabeth Haldane and G. R. Ross (Cambridge, 1911), vol. 1, pp. 205–208.
28. Ibid., p. 210.
29. René Descartes, *Discours de la méthode pour bien conduire sa raison, et chercher la verité dans les sciences,* in *Oeuvres de Descartes,* ed. Charles Adam and Paul Tannery (Paris, 1897–1910), vol. 6, p. 2.
30. Cavendish, *Philosophical and Physical Opinions,* "Letter to the Reader."
31. Descartes, *Principles of Philosophy,* p. 218.
32. Stéphanie-Félicité du Crest, comtesse de Genlis, *Le Club des dames, ou le Retour de Descartes* (Paris, 1784), preface.
33. See René Descartes, "Primae cogitationes circa generationem animalium," in *Oeuvres de Descartes,* ed. Adam and Tannery, vol. 11, pp. 516–528; Descartes, "Traité de l'homme," in ibid., pp. 119–215. See also Michèle Le Doeuff, "Women and Philosophy," *Radical Philosophy,* 17 (1977): 5.
34. On this important point see Melissa Butler, "Early Liberal Roots of Feminism: John Locke and the Attack on Patriarchy," *American Political Science Review,* 72 (1978): 135–150; Susan Moller Okin, *Women in Western Political Thought* (Princeton, 1979); Carole Pateman and Teresa Brennan, "'Mere Auxiliaries to the Commonwealth': Women and the Origins of Liberalism," *Political Studies,* 27 (1979): 183–200.
35. John Locke, "The Reasonableness of Christianity, As Delivered in the Scriptures," in *The Works,* 6th ed. (London, 1759), vol. 2, pp. 585–586.
36. John Locke, *The Educational Writings,* ed. James Axtell (Cambridge, England, 1968), p. 117, 344–346.
37. Cockburn published her work anonymously. According to Thomas Birch, she felt that "the name of a woman would be a prejudice" against a philosophical work. See *The Works of Mrs. Catharine Cockburn* (London, 1751), vol. 1, preface.
38. Paul Hoffmann, *La Femme dans la pensée des lumières* (Paris, 1977), pp. 45–52. See also Lieselotte Steinbrügg, "Vom Aufstieg und Fall der gelehrten Frau: Einige Aspekte der 'Querelle des Femmes' im XVIII. Jahrhundert," *Lendemains,* 25/26 (1982): 158–159; and Hilda Smith, *Reason's Disciples* (Urbana, 1982).
39. Though Descartes detached the mind from the body, his vision of the sex organs remained traditional. Like Galen, Descartes considered the female vagina an inverted version of the male penis. Sex was determined in the fetus by its position in the mother's uterus. In another passage in his "Primae cogitationes," the male sex is determined by a predominance of solid and dry elements.
40. Ruth Perry, "Radical Doubt and the Liberation of Women," *Eighteenth-Century Studies,* 18 (1985): 479.
41. Genevieve Lloyd, *The Man of Reason: "Male" and "Female" in Western Phi-*

losophy (Minneapolis, 1984), pp. 38–50. See also Susan Bordo, *The Flight to Objectivity: Essays in Cartesianism and Culture* (Albany, 1987).

42. See Merchant, *The Death of Nature;* Brian Easlea, *Witch-Hunting, Magic, and the New Philosophy* (Brighton, 1980); Sally Allen and Joanna Hubbs, "Outrunning Atalanta: Feminine Destiny in Alchemical Transmutation," *Signs,* 6 (1980): 210–229; Evelyn Fox Keller, *Reflections on Gender and Science* (New Haven, 1985), chap. 3.

43. As late as the nineteenth century, there was still talk of the "hermaphroditic Adam." The medical doctor Moreau de la Sarthe cited Mirabeau's work showing that the first human was neither man nor woman, but an "androgyne." The hermaphroditic Adam—like many plants and animals—had the ability to reproduce himself as witnessed by God's command to the solitary Adam to "be fruitful and multiply, and replenish the earth." According to Mirabeau and Moreau de la Sarthe, individuals of distinctive sex did not exist until the creation of woman, during the seventh day or epoch. Mirabeau, quoted in Jacques-Louis Moreau de la Sarthe, *Histoire naturelle de la femme* (Paris, 1803), vol. 1, pp. 63–66.

44. Conway, *Principles,* pp. 77, 98, 146–147. See also Marjorie Nicolson, *Conway Letters* (New Haven, 1930); and Carolyn Merchant, "The Vitalism of Anne Conway: Its Impact on Leibniz's Concept of the Monad," *Journal of the History of Philosophy,* 17 (1979): 255–269.

45. Henri Piéron, "De l'influence sociale des principes cartésiens. Un précurseur inconnu du féminisme et de la révolution: Poulain de la Barre," *Revue de synthèse historique,* 5 (1902): 153, 171. See also Marie Louise Stock, "Poullain de la Barre: A Seventeenth-Century Feminist" (Ph.D. diss., Columbia University, 1961); and Bernard Magne, "Le Féminisme de Poullain de la Barre: Origine et signification," Thèse pour le Doctorat de 3e cycle, Université de Toulouse, 1964.

46. François Poullain de la Barre, *De l'éducation des dames pour la conduite de l'esprit dans les sciences et dans les moeurs* (Paris, 1674), pp. 327, 331–334.

47. Poullain de la Barre, *De l'égalité des deux sexes,* preface. This work was translated into English as *The Woman as Good as the Man, or the Equality of Both Sexes* (London, 1677).

48. Poullain, *De l'égalité des deux sexes,* pp. 59–62.

49. Ibid., pp. 28 and 25.

50. Poullain, *De l'éducation des dames,* advertisement and pp. 307–310.

51. [Drake], *An Essay in Defence of the Female Sex,* p. 32.

52. Ibid., pp. 12–13.

53. In addition to Poullain and Drake, see Hannah Woolley, *The Gentlewomans Companion* (London, 1675), p. 29; *Biographium Faemineum: The Female Worthies* (London, 1766), p. vii; and Martin, *Young Gentleman's and Lady's Philosophy,* vol. 1, p. vi.

54. J. P. Lotichium summarized this literature in his *Gynaecologia* (Frankfurt, 1645). See also Ian Maclean, *The Renaissance Notion of Woman* (Cambridge, England, 1980).

55. Andreas Vesalius, *Tabulae anatomicae sex* (1538), plate 87, figures 2–4, in J. B. Saunders and C. D. O'Malley, *Illustrations from Works of Andreas Vesalius* (New York, 1950). Also Jakob Ackermann, *De discrimine sexuum praeter genitalia* (Mainz, 1788); translated as *Über die körperliche Verschiedenheit des Mannes vom Weibe ausser Geschlechtstheilen,* by Joseph Wenzel (Koblenz, 1788), appendix. The German edition is used throughout. See also Plato, *Timaeus,* p. 91c. In the ancient world the uterus was compared to a variety of animals—a tortoise, a newt, a crocodile; Harold Speert, *Iconographia Gyniatrica: A Pictorial History of Gynecology and Obstetrics* (Philadelphia, 1973), p. 8.
56. "Woman conserves, nourishes, protects the child, not only for nine months in her own body, but during the entire period of childhood. The man serves in his way, but not in the same fashion as the woman." Couvay, *L'Honneste maîtresse,* pp. 288–296. See also Crooke, *Mikrokosmographia,* p. 271.
57. For a fuller treatment of this topic, see Maryanne Horowitz, "The 'Science' of Embryology before the Discovery of the Ovum," in *Connecting Spheres,* ed. Boxer and Quataert, chap. 4.
58. James Keill, *The Anatomy of the Humane Body Abridged* (London, 1698), p. 93. See also Francis J. Cole, *Early Theories of Sexual Generation* (Oxford, 1930).
59. James Drake, *Anthropologia Nova; or, a New System of Anatomy,* ed. Judith Drake (London, 1707), vol. 1, p. 352 (note p. 352 follows p. 335).
60. Buffon, *Histoire naturelle,* vol. 3, p. 264. See also Carlo Castellani, "The Problem of Generation in Bonnet and in Buffon: A Critical Comparison," in *Science, Medicine, and Society in the Renaissance,* ed. Allen Debus (New York, 1972), vol. 2, pp. 265–288.
61. Buffon, *Histoire naturelle,* vol. 2, p. 329.
62. J.B. Saunders and C.D. O'Malley, eds., *The Illustrations from the Works of Andreas Vesalius* (New York, 1950), pp. 222–223.
63. See Fritz Weindler, *Geschichte der gynökologisch-anatomischen Abbildung* (Dresden, 1908), p. 41, figure 37. See also G. Wolf-Heidegger and Anna Maria Cetto, *Die anatomische Sektion in bildlicher Darstellung* (Basel, 1967).
64. Mary Niven Alston, "The Attitude of the Church towards Dissection before 1500," *Bulletin of the History of Medicine,* 16 (1944): 227–228.
65. Kate Campbell Hurd-Mead, *A History of Women in Medicine* (Haddam, Connecticut, 1938), pp. 358–359.
66. Vesalius, *De humani corporis fabrica,* frontispiece. I thank I. B. Cohen for pointing out to me the presence of the nun.
67. Saunders and O'Malley, eds., *Illustrations from the Works of Andreas Vesalius,* p. 170. Female bodies were hard to come by, but then so were male bodies. The story is told that Vesalius saw a body of a male who had been executed swinging from a tree. Waiting until nightfall, he returned and cut it down.
68. William Cheselden, *Anatomy of the Bones* (1713; Boston, 1795), p. 276.
69. "A Methodical Catalogue of all the Chiefest Rarities in the Publick Theatre and Anatomical-Hall in the University of Leyden," in *A Compleat Volume of the Memoirs for the Curious* (November, 1707), pp. 389–391.

70. William Cowper, *The Anatomy of Humane Bodies* (London, 1737). Cowper pirated his illustrations from Godfried Bidloo's *Anatomia humani corporis* (Amsterdam, 1685).
71. Cowper, *The Anatomy of Humane Bodies,* commentary to plate 2.
72. William Harvey, *Lectures on the Whole of Anatomy on the Male and Female Body,* trans. Gweneth Whitteridge (1616; London, 1964), pp. 53, 313–315.
73. Crooke, *Mikrokosmographia,* pp. 272–274.
74. Ibid., p. 250.
75. Ambroise Paré, *The Works of that Famous Chirurgion Ambrose Parey,* trans. Thomas Johnson (1585; London, 1634), p. 27.

7. More Than Skin Deep: The Scientific Search for Sexual Difference

1. Pierre Roussel, *Système physique et moral de la femme, ou Tableau philosophique de la constitution, de l'état organique, du tempérament, des moeurs, et des fonctions propres au sexe* (Paris, 1775), p. 2. Carl Klose also argued that it is not the uterus that makes woman what she is. Even women from whom the uterus has been removed, he stressed, retain feminine characteristics. See his *Über den Einfluss des Geschlechts-Unterschiedes auf Ausbildung und Heilung von Krankheiten* (Stendal, 1829), pp. 28–30.
2. Ackermann, *Über die körperliche Verschiedenheit,* pp. 2–5.
3. Thomas Laqueur, "Organism, Generation, and the Politics of Reproductive Biology," in *The Making of the Modern Body: Sexuality and Society in the Nineteenth Century,* ed. Catherine Gallagher and Thomas Laqueur (Berkeley, 1987), pp. 1–41.
4. It is important to recall, however, that in the nineteenth century the uterus would increasingly be considered a source of female weakness, especially hysteria.
5. Moreau de la Sarthe, *Histoire naturelle de la femme,* vol. 1, pp. 68–69.
6. Klose, *Über den Einfluss des Geschlechts-Unterschiedes,* pp. 28–33.
7. Jouard, *Nouvel Essai sur la femme,* pp. 51–55.
8. See, e.g., Edmond-Thomas Moreau, *Quaestio medica: An praeter genitalia sexus inter se discrepent?* (Paris, 1750).
9. Bernard Albinus, *Table of the Skeleton and Muscles of the Human Body* (London, 1749), "Account of the Work."
10. Albinus quoted in Ludwig Chouland, *History and Bibliography of Anatomic Illustration,* trans. Mortimer Frank (Chicago, 1920), p. 277.
11. Albinus gave a description of a female skeleton in his *De sceleto humano* (Leiden, 1762, chap. 126) but did not provide an illustration.
12. Gaspard Bauhin, *Theatrum anatomicum* (Frankfurt, 1605), plate 4, p. 247. This skeleton was reprinted in Crooke's *Mikrokosmographia* and the English edition of Ambroise Paré's works (*The Works of . . . Ambrose Parey,* chap. 41), though it did not appear in Paré's work itself (see his *Oeuvres* [Paris, 1585]). It then dropped from sight.

13. In addition to the skeletons discussed separately, see Pierre Tarin, *Ostéographie, ou Description des os de l'adulte* (Paris, 1753), plate 23.
14. Alexander Monro, *The Anatomy of the Humane Bones* (Edinburgh, 1726), appendix, p. 341.
15. Drake, *Anthropologia Nova,* plates 21 and 22.
16. William Cheselden, *Osteographia, or the Anatomy of the Bones* (London, 1733), figures 34 and 35.
17. Cheselden did not comment on sex differences even in the pelvis. See his *Anatomy,* 3d ed. (London, 1726), p. 29.
18. John Douglas, *Animadversions on a late Pompous Book, intitled Osteographia: Or, The Anatomy of the Bones by William Cheselden esq.* (London, 1735), p. 37.
19. Cheselden, *Anatomy,* p. 266.
20. Jean-J. Sue, *Traité d'ostéologie, traduit de l'Anglois de M. Monro* (Paris, 1759), plate 4.
21. Since at least the late sixteenth century, a large pelvis had been a mark of femininity. See also Antoine Portal, *Histoire de l'anatomie et de la chirurgie* (Paris, 1773), vol. 6, part 1, p. 90.
22. Sue, *Traité d'ostéologie,* text to plate 4.
23. J.-B. Winslow, "Sur les mauvais effets de l'usage des corps à baleine," *Mémoires de l'Académie des sciences* (Paris, 1741).
24. Samuel Thomas von Soemmerring, *Tabula sceleti feminini juncta descriptione* ([Utrecht], 1796). See also Gunter Mann, "'Die schöne Mainzerin' Samuel Thomas Soemmerrings," *Medizin historisches Journal,* 12 (1977): 172–174.
25. *Journal der Empfindungen, Theorien und Widersprüche in der Natur- und Arztneiwissenschaft,* 6 (1797): 17–18.
26. It was thought that a woman did not reach maturity with the onset of menstruation but only at age eighteen or twenty, after the birth of her first child. See Johann Jörg, *Handbuch der Krankheiten des Weibes, nebst einer Einleitung in die Physiologie und Psychologie des weiblichen Organismus* (Leipzig, 1831), p. 6 ff.
27. Choulant, *History of Anatomic Illustration,* pp. 306–307.
28. John Barclay, *The Anatomy of the Bones of the Human Body* (Edinburgh, 1829), commentary to plate 32.
29. Quoted in Choulant, *History of Anatomic Illustrations,* p. 277.
30. Albinus, *Table of the Skeleton and Muscles of the Human Body,* "An Account of the Work."
31. Hendrik Punt, "Bernard Siegfried Albinus (1697–1770) und die anatomische Perfektion," *Medizin historisches Journal,* 12 (1977): 325–345.
32. H. Punt, *Bernard Siegfried Albinus, 1697–1770, on "Human Nature": Anatomical and Physiological Ideas in Eighteenth Century Leiden* (Amsterdam, 1983), p. 13.
33. Ibid., pp. 17–18.
34. Wenzel, in Ackermann, *Über die körperliche Verschiedenheit,* p. 5.
35. According to Johann von Döllinger, there is in women a preponderance of femininity and in men a preponderance of masculinity. The preponderance

of masculinity in the male genitalia lies in the testes (the prostate is feminine); the preponderance of femininity in the woman lies in the ovaries (the uterus is masculine). See his "Versuch einer Geschichte der menschlichen Zeugung," reprinted in Arthur Meyer's *Human Generation: Conclusions of Burdach, Döllinger and von Baer* (Stanford, 1956), p. 42.

36. Quoted in Ackermann, *Über die körperliche Verschiedenheit,* p. 5–7.
37. Choulant, *History of Anatomic Illustrations,* p. 302.
38. Soemmerring, *Tabula sceleti feminini,* commentary to plate.
39. Albinus, *Table of the Skeleton and Muscles of the Human Body,* "An Account of the Work."
40. See, e.g., [Pierre Maupertuis], *Venus physique* (Paris, 1745); and [Anonymous], *The Ladies Dispensatory* (London, 1739), p. 171.
41. See Stephen Jay Gould, *Ontogeny and Phylogeny* (Cambridge, Mass., 1977), chap. 5. See also Ruth Hubbard, "Have Only Men Evolved?" *Biological Woman,* ed. Hubbard, Hennifin, and Fried, pp. 17–46.
42. G. W. F. Hegel, *Phänomenologie des Geistes* (1807), in his *Werke,* ed. Eva Moldenhauer and Karl Michel (Frankfurt, 1969–1971), vol. 3, p. 248.
43. On women and craniology, see Elizabeth Fee, "Nineteenth-Century Craniology: The Study of the Female Skull," *Bulletin of the History of Medicine,* 53 (1979): 415–433; and Stephen Jay Gould, *The Panda's Thumb* (Boston, 1980), chap. 14.
44. Samuel Thomas von Soemmerring, *Vom Baue des menschlichen Körpers* (1796; Frankfurt am Main, 1800), vol. 1, p. 82. Male and female brains continued to be compared throughout the nineteenth century. In the 1890s, the brain of the Russian-born mathematician, Sofia Kovalevskaia, was among those of eminent scientists investigated for the correlation between achievement and brain size. After having soaked in alcohol for four years, Kovalevskaia's brain weighed in at 1,385 grams. The brain of her male contemporary, Hermann von Helmholtz, was heavier, weighing 1,440 grams. Compared to body-size, however, Kovalevskaia's brain was much larger. Gustaf Retzius, *Das Gehirn des Mathematikers Sonja Kovalewski in biologische Untersuchungen* (Stockholm, 1900), cited in Mozans, *Woman in Science,* pp. 124–125, note 2.
45. Ackermann, *Über die körperliche Verschiedenheit,* p. 146.
46. Barclay, *The Anatomy of the Bones of the Human Body,* text to plate 32. An earlier edition of this book was published under a slightly different title in 1819.
47. Ibid., commentary to plate 32.
48. E. W. Posner, *Das Weib und das Kind* (Glogau, 1847), p. 9–10.
49. Kelso, *The Doctrine of the Renaissance Lady,* p. 213.
50. "For the free rules the slave, the male the female, and the man the child [each] in a different way." Aristotle justified the subordination of slaves, women, and children in terms of the supposedly lesser degree of rationality each posesses: ". . . all possess the various parts of the soul, but possess them in different ways; for the slave has not yet got the deliberative part at all, the female has it, but without full authority, while the child has it, but

in an undeveloped form." See Aristotle, *The Politics,* trans. H. Rackham (London, 1932), p. 63.

51. Catherine Macaulay, *Letters on Education,* (London, 1790), p. 209.
52. Heidi Rosenbaum, *Formen der Familien* (Frankfurt, 1982), p. 288–289.
53. Alexander Monro (Primus), "Essay on female Conduct written by a Father to his Daughter," manuscript held by Dr. P. A. G. Monro at St. John's College, Cambridge, p. 185.
54. Johann Ziegenbein, *Aussprüche über weibliche Natur, weibliche Bestimmung, Erziehung und Bildung* (Blankenburg, 1808), pp. 16–17.
55. Samuel Thomas von Soemmerring, *Über die körperliche Verschiedenheit des Negers vom Europäer* (Frankfurt, 1785), p. 2. On metaphors of race and gender in science, see Nancy Leys Stepan, "Race and Gender: The Role of Analogy in Science," *Isis,* 77 (June 1986): 261–277. See also Sander L. Gilman, *On Blackness without Blacks: Essays on the Image of the Black in Germany* (Boston, 1982).
56. G. Hervé cited in Stephen Jay Gould, *The Mismeasure of Man* (New York, 1981), p. 103; and F. Pruner and James Hunt cited in Fee, "Nineteenth-Century Craniology," p. 424. Racial and sexual rankings by craniologists were much more elaborate than Table 2 suggests; Carl Vogt, for example, ranked the males of thirty-three races (including "Australians, Polynesians, Hottentots, Peruvians, Oceanic Negroes, Mexicans, Wild Indians, Parisians, Germans, and English") according to cranial capacity. See his *Lectures on Man* (London, 1864), p. 88.
57. Theodor von Bischoff, *Das Studium und die Ausübung der Medicin für Frauen* (Munich, 1872), p. 16.
58. Vogt, *Lectures on Man,* pp. 81–82.
59. Carl Vogt, *Vorlesungen über den Menschen: seine Stellung in der Schöpfung und in der Geschichte der Erde* (Giessen, 1863), p. 151. See also Lucile Hoyme, "The Earliest Use of Indices for Sexing Pelves," *American Journal of Physical Anthropology,* 15 (1957): 537–546.
60. George Humphry, *A Treatise on the Human Skeleton* (London, 1858), pp. 446–447.
61. Soemmerring, *Über die körperliche Verschiedenheit des Negers vom Europäer,* preface; and Frank Dougherty, "Eine Auseinandersetzung im anthropologischer Hinsicht," in *Samuel Thomas Soemmerring und die Gelehrten der Goethezeit,* ed. Gunter Mann (Stuttgart, 1985), p. 39. Kant, in his *Beobachtungen,* where he explained why women were inherently incapable of scientific endeavor, also cited with approval Hume's belief that "African negroes have by nature no feeling that rises above the simple-minded"; *Kants Werke,* ed. Wilhelm Dilthey (Berlin, 1900–1919), vol. 2, p. 253).

8. The Triumph of Complementarity

1. Ackermann, *Über die körperliche Verschiedenheit,* "Krankheitslehre der Frauenzimmer."
2. Exupère-Joseph Bertin, *Traité d'ostéologie* (Paris, 1754), vol. 1, pp. 22–23.

3. See Adalbert von Hanstein, *Die Frauen in der Zeit des Aufschwunges des deutschen Geistesleben* (Leipzig, 1899); David Williams, "Political Feminism in the French Enlightenment," in *The Varied Pattern: Studies in the 18th Century,* ed. Peter Hughes and David Williams (Toronto, 1971), pp. 333–351; Maïte Albistur and Daniel Armogathe, *Histoire du féminisme français* (Paris, 1977); Katherine Rogers, *Feminism in Eighteenth-Century England* (Urbana, 1982).
4. [Louis le Chevalier de Jaucourt], "Femme (droit nat.)," *Encyclopédie,* vol. 6, p. 470. It is well known that both Voltaire and Diderot thought the articles on women in the *Encyclopédie* lacked subtlety, yet they were published unrevised.
5. Marie-Jean-Antoine-Nicolas de Caritat, marquis de Condorcet, "Sur l'admission des femmes au droit de cité," in *Oeuvres* (Stuttgart, 1968), vol. 10, p. 129. See Maurice Bloch and Jean Bloch, "Women and the Dialectics of Nature in Eighteenth-Century French Thought," in *Nature, Culture and Gender,* ed. Carol P. MacCormack and Marilyn Strathern (Cambridge, England, 1980), pp. 25–41; and Steven Rose, Leon Kamin, and Richard Lewontin, *Not in Our Genes: Biology, Ideology, and Human Nature* (London, 1984), pp. 63–81.
6. Christoph Meiners, *Geschichte des weiblichen Geschlechts* (Hanovers, 1788–1800), vol. 4, p. 314–315.
7. See Elizabeth Blochman, "Das Frauenzimmer und die Gelehrsamkeit," *Anthropologie und Erziehung,* 17 (1966): 10–75; Karin Hausen, "Die Polarisierung der 'Geschlechtscharakter,'" in *Sozialgeschichte der Familie in der Neuzeit Europas,* ed. Werner Conze (Stuttgart, 1976); Susan Okin, "Women and the Making of the Sentimental Family," *Philosophy and Public Affairs,* 11 (1982): 65–88.
8. See Marlene LeGates, "The Cult of Womanhood in Eighteenth-Century Thought," *Eighteenth-Century Studies,* 10 (1976): 21–39; Paul Fritz and Richard Morton, eds., *Woman in the Eighteenth Century and Other Essays* (Toronto, 1976); Ruth Bloch, "Untangling the Roots of Modern Sex Roles: A Survey of Four Centuries of Change," *Signs,* 4 (1978): 237–257; Susanne Risse-Stumbries, *Erziehung und Bildung der Frau in der zweiten Hälfte des 18. Jahrhunderts* (Tübingen, 1980); and Elizabeth Fox-Genovese, "Introduction," in *French Women,* ed. Spencer, pp. 1–32.
9. Carolyn Lougee, "*Noblesse,* Domesticity, and Social Reform: The Education of Girls by Fénelon and Saint-Cyr," *History of Education Quarterly,* 14 (1974): 88–89.
10. See Antoine Thomas, *Essai sur le caractère, les moeurs, et l'esprit des femmes dans les différens siècles* (Paris, 1772), p. 130.
11. Jean-Baptiste Molière, *Les Femmes savantes* (1672), ed. Jean Cordier (Paris, 1959), pp. 36–37.
12. Samuel Chapuzeau, *L'Académie des femmes* (Paris, 1661).
13. Molière, *Les Femmes savantes,* p. 37.
14. Louis de Lesclache, *Les Avantages que les femmes peuvent recevoir de la philosophie et principalement de la morale* (Paris, 1667), pp. 4, 11–12.

15. By the seventeenth century, the use of wet nurses had spread to the bourgeoisie; by the eighteenth, to most segments of urban society, including silk manufacturers and merchants and the artisanal classes. See Elisabeth Badinter, *Mother Love: Myth and Reality,* trans. from the French (New York, 1981), part I.
16. Cited in Badinter, *Mother Love,* pp. 122–123, 125.
17. These include: John Gregory, *A Father's Legacy to his Daughters* (London, 1774); Alexander, *The History of Women;* [Anonymous], *Der Arzt der Frauenzimmer, oder die Kunst, dieselben gesund zu erhalten,* trans. from the French (Leipzig, 1773); and J. J. Sachs, *Ärztliches Gemälde des weiblichen Lebens im gesunden und krankhaften Zustande aus physiologischen, intellektuellem und moralischem Standpunkte* (Berlin, 1830).
18. Alexander Monro (Primus), "Essay on female Conduct written by a Father to his Daughter," manuscript held by Dr. P. A. G. Monro at St. John's College, Cambridge.
19. Badinter, *Mother Love,* p. 154.
20. *Allgemeines Landrecht* (1794), part II, title II, art. 67, in Susan Bell and Karen Offen, eds. *Women, the Family and Freedom: The Debate in Documents, 1750–1880* (Stanford, 1983), vol. 1, p. 39.
21. See Timothy Reiss, "Revolution in Bounds: Wollstonecraft, Women and Reason," photocopy, Department of Comparative Literature, New York University, 1985, p. 2. The ideal of domestic motherhood was never realized, for example, by working-class women.
22. Rousseau, *Emile,* p. 692. On Rousseau's views on women, see Silvia Bovenschen, *Die imaginierte Weiblichkeit* (Frankfurt, 1979); Zillah Eisenstein, *The Radical Future of Liberal Feminism* (New York, 1981), chap. 4; and Joel Schwartz, *The Sexual Politics of Jean-Jacques Rousseau* (Chicago, 1984).
23. Rousseau, *Emile,* pp. 693–697.
24. Ibid., pp. 357–361.
25. D. J. [Louis le Chevalier de Jaucourt], "Squelette," *Encyclopédie, ou Dictionnaire raisonné des sciences, des arts, et des métiers* (Neuchâtel, 1765), vol. 15, pp. 482–483.
26. Roussel, *Système physique et moral de la femme,* pp. xvi, 12, and 22–23. Roussel's book was answered by an anonymous book with a similar title, which argued that it was women's natural right to practice the arts and sciences (*De l'éducation physique et morale des femmes avec une notice alphabétique de celles qui se sont distinguées dans les différentes carrières des sciences* [Brussels and Paris, 1779]). Roussel's book was translated into German (by C. F. Michaelis in 1786). On Roussel, see Yvonne Kniehler, "Les Médecins et la 'nature féminine' au temps du code civil," *Annales: Economies, sociétés, civilisations,* 31 (1976): 824–845; L. J. Jordanova, "Natural Facts: A Historical Perspective on Science and Sexuality," in *Nature, Culture and Gender,* ed. Carol P. MacCormack and Marilyn Strathern (Cambridge, England, 1980), pp. 42–69; and Michèle Le Doeuff, "Pierre Roussel's Chiasmas: From Imaginary Knowledge to the Learned Imagination," *I&C,* 9 (Winter 1981/82): 39–63.

27. Soemmerring, *Über die körperliche Verschiedenheit,* p. ix.
28. Ziegenbein, *Aussprüche über weibliche Natur,* p. 1.
29. See Karl Pockels, *Versuch einer Charakteristik des weiblichen Geschlechts* (Hanover, 1799–1802), vol. 1, pp. 6 and 8.
30. Voltaire, *Dictionnaire philosophique,* vol. 4, pp. 375, 378–379; and vol. 5, p. 264. Voltaire reported one happy consequence following from the fact that women are excluded from corrupt and barbarous professions: women commit fewer crimes, as witnessed by the fact that fifty men are executed for every woman executed.
31. Rousseau, *Emile,* p. 693.
32. Karoline von Woltmann, *Über Natur, Bestimmung, Tugend und Bildung der Frauen* (Vienna, 1826).
33. [Jakob Mauvillon], *Mann und Weib nach ihren gegenseitigen Verhältnissen geschildert. Ein Gegenstück zu der Schrift: Über die Weiber* (Leipzig, 1791), p. 9. See also Ernst Brandes, *Über die Weiber* (Leipzig, 1787); and Carl Haase, *Ernst Brandes, 1758–1810* (Hildesheim, 1973).
34. Mauvillon, *Mann and Weib,* p. 234. In Plato's version, narrated by Aristophanes in the *Symposium,* there were three types of beings—the double male, the double female, and the hermaphrodite, which is both man and women. After being severed by the gods, each half searches for its mate. The woman who is a slice of the original female searches for a mate among women rather than men; the man who is a slice of the hermaphrodite searches for a woman; and the man who is a slice of the original male searches for another man. Whereas Aristophanes had denied the importance of the hermaphrodite in order to praise the virtues of male love, writers after the Renaissance almost unanimously exalted the hermaphrodite and thus heterosexual love.
35. Gregory, *A Father's Legacy to his Daughters,* pp. 22 and 42.
36. Philippine von Knigge, *Versuch einer Logic für Frauenzimmer* (Hanover, 1789), pp. 48–50.
37. Cited in Darline Levy, Harriet Applewhite, and Mary Johnson, eds., *Women in Revolutionary Paris, 1789–1795* (Chicago, 1979), pp. 215–216.
38. Condorcet, "Sur l'admission des femmes," p. 125.
39. Wollstonecraft, *Vindication of the Rights of Woman,* p. 82.
40. Condorcet, *Esquisse d'un tableau historique des progrès de l'esprit humain.*
41. Olympe de Gouges, "Les Droits de la Femme," in Levy, Applewhite, and Johnson, *Women in Revolutionary Paris,* pp. 87–96. De Gouges also speaks for the liberty of "colored men" (p. 96). Women's demands were far from met. French women did not receive the vote until after World War II.
42. Theodor von Hippel sharply criticized the French constitution for having neglected the rights of "an entire half of a nation" (Über die bürgerliche Verbesserung der Weiber [1792], in *Sämtliche Werke* [Berlin 1828], vol. 6, p. 120).
43. Mary Wollstonecraft, for example, held that "that society is formed in the wisest manner whose constitution is founded on the nature of man" (*Vindication of the Rights of Woman,* p. 92).

44. Bertin, *Traité d'ostéologie,* vol. 3, pp. 211–212.
45. Eliza Haywood, *The Female Spectator,* 2 (1744): 240–241.
46. *Magasin encyclopédique,* 5 (1796): 539–548. According to Yvonne Knibiehler, this letter was written by Constance de Theis ("Les Médecins et la 'nature féminine' au temps du code civil," *Annales: Economies, sociétés, civilisations,* 31 (1976): 830.
47. Hippel, *Über die bürgerliche Verbesserung der Weiber,* p. 21.
48. Immanuel Kant, *Anthropologie in pragmatischer Hinsicht* (1798; Frankfurt and Leipzig, 1799), p. 309.
49. Meiners, *Geschichte des weiblichen Geschlechts,* vol. 2, p. 166.
50. For a discussion of how Wollstonecraft's work is bounded by the terms of individualism and liberalism, see Mary Poovey, *The Proper Lady and Woman Writer* (Chicago, 1984).
51. Jean Le Rond d'Alembert, *Lettre de M. d'Alembert à M. J.-J. Rousseau* (Amsterdam, 1759), p. 130.
52. Wollstonecraft, *Vindication of the Rights of Woman,* pp. 80, 83, 124. One can, as many have, find lapses of essentialism, e.g., on the naturalness of motherhood, in the works of Wollstonecraft and other Enlightenment revolutionaries. This is not, however, the dominant theme in her work.
53. Ibid., p. 82.
54. Condorcet, "Sur l'admission des femmes," pp. 121–130. See also Barbara Brookes, "The Feminism of Condorcet and Sophie de Grouchy," *Studies on Voltaire and the Eighteenth Century,* 189 (1980): 297–361.
55. *Female Rights Vindicated; or the Equality of the Sexes Morally and Physically proved,* by a Lady (London, 1763), preface.
56. Wollstonecraft, *Vindication of the Rights of Woman,* p. 84.
57. Hippel, *Über die bürgerliche Verbesserung der Weiber,* p. 223.
58. *Magasin encyclopédique,* 5 (1796): 539–548. Wollstonecraft, in contrast, thought child rearing a "natural" duty of women (*Vindication of the Rights of Woman,* p. 265).
59. See Jean Elshtain, *Public Man, Private Woman: Women in Social and Political Thought* (Princeton, 1981); Lloyd, *The Man of Reason,* esp. p. 3; "Naturwissenschaftlerinnen: Einmischung statt Ausgrenzung," *Feministische Studien,* 4 (1985); and Prudence Allen, *The Concept of Woman: The Aristotelian Revolution, 750 B.C. to A.D. 1250* (Montreal, 1985).
60. Woltmann, *Über Natur, Bestimmung . . . der Frauen,* p. 219.
61. Hegel, *Phänomenologie,* p. 319. See also Joan B. Landes, "Hegel's Conception of the Family," *Polity,* 14 (1981): 5–28.
62. Wollstonecraft, *Vindication of the Rights of Woman,* p. 142.
63. In addition to works cited separately, see also Thomas, *Essai sur le caractère;* Johann Ewald, *Die Kunst ein gutes Mädchen, eine gute Gattin, Mutter und Hausfrau zu werden* (Leipzig, 1798); Jouard, *Nouvel Essai sur la femme;* and W. von Humboldt, "Über den geschlechtsunterschied und dessen Einfluss auf die organische Natur," and "Über die männliche und weibliche Form," reprinted in *Neudrucke zur Psychologie,* 1 (1917).
64. Rousseau, *Lettre à M. d'Alembert,* pp. 152–155.

65. D. J. [Louis le Chevalier de Jaucourt], "Le Sexe (Morale)" *Encyclopédie,* vol. 15, p. 138.
66. Cited in Amalia Holst, *Über die Bestimmung der Weibes zur höhern Geistesbildung* (Berlin, 1802; rpt. Zürich, 1983), p. 143.
67. See Leybourn, *Mathematical Questions in the Ladies' Diary,* vol. 4, pp. 414–436.
68. J. C. Gottsched, ed., *Die vernünftigen Tadlerinnen* (Leipzig, 1727). See also Hanstein, *Die Frauen in der Zeit der Aufschwunges des deutschen Geisteslebens,* vol. 1, pp. 75–90.
69. Letter from Sophie Germain to C. F. Gauss, 20 February 1807, in Sophie Germain, *Oeuvres philosophiques de Sophie Germain,* ed. H. Stupuy (Paris, 1896), p. 271.
70. John Tripper, in *The Ladies' Diary* (1709).
71. Leybourn, *Mathematical Questions in The Ladies' Diary,* vol. 1, pp. vi-viii. See also Ruth and Peter Wallis, "Female Philomaths," *Historia Mathematica,* 7 (1980): 58–59. Of the estimated 913 contributors to the journal from 1704 to 1816, there were 32 women (Perl, "The Ladies' Diary," p. 45).
72. Noël Antoine Pluche, *Spectacle de la nature, ou Entretiens sur les particularités de l'histoire naturelle* (1732–1748; Paris, 1782), vol. 6, p. 88.
73. Jean Formey, *La Belle Wolfienne* (The Hague, 1741–1753), reprinted in Christian Wolff, *Gesammelte Werke: Materialien und Dokumente,* ed. Jean Ecole (Hildesheim, 1983), pt. III, vol. 16.1, preface and avertissement; and preface to volume four.
74. Jérôme de Lalande's *Astronomie des dames* (1786; Paris, 1841), pp. 1–2.
75. Ibid., 1820 ed., pp. 4 and 117.
76. Pierre-J. Boudier de Villemert, *L'Ami des femmes* (Paris, 1758), p. 31–33. Charlotte Lennox translated and published long passages from Boudier de Villemert's work in her *Lady's Museum,* which appeared in England in 1760. See also David Williams, "The Fate of French Feminism: Boudier de Villemert's *Amis des Femmes,*" *Eighteenth-Century Studies,* 14 (1980): 37–55.
77. Edgeworth, *Letters for Literary Ladies,* pp. 52 and 66.
78. Hegel compared the male mind to an animal that acquires knowledge only through much struggle and technical exertion. The female mind, by contrast, does not (cannot) rise above its plantlike existence and remains rooted in its *an sich* existence (*Grundlinien der Philosophie des Rechts* [1821] in his *Werke,* vol. 7, pp. 319–320). See also J. F. A. Adams, "Is Botany a Suitable Study for Young Men?" *Science,* 9 (1887): 117–118; and Emmanuel Rudolph, "How It Developed That Botany Was the Science Thought Most Suitable for Victorian Young Ladies," *Children's Literature,* 2 (1973): 92–97.
79. J.-J. Rousseau, "Lettres sur la botanique," *Oeuvres complètes,* ed. Bernard Gagnebin and Marcel Raymond (Paris, 1959–1969), vol. 4, p. 1151. See also Ann B. Shteir, "Linnaeus's Daughters: Women and British Botany," in *Women and the Structure of Society,* ed. Barbara Harris and JoAnn McNamara (Durham, 1984).

80. See François Delaporte, *Nature's Second Kingdom: Explorations of Vegetality in the Eighteenth Century,* trans. Arthur Goldhammer (Cambridge, Mass., 1982), p. 145.
81. See Carolus Linnaeus, *Systema naturae* (Leiden, 1735).
82. Rousseau, "Lettres sur la botanique," pp. 1151–1152, 1188.
83. J.-J. Rousseau, *Letters on the Elements of Botany, addressed to a Lady,* trans. Thomas Martyn (1785; London, 1796), preface, and pp. xi, 13.
84. Priscilla Wakefield, *Introduction to Botany* (London, 1796), preface.
85. Judith Brody, "Women as Popular Science Authors in the Early-Nineteenth Century," *Proceedings of the International Conference on the Role of Women in the History of Science, Technology and Medicine in the Nineteenth and Twentieth Centuries* (Veszprém, Czechoslovakia, 1983), pp. 16–25.
86. [Antoine Mongez], *Algèbre* (Paris, 1789), preface. The *Bibliothèque des dames* also published a geometry for women in 1790 and a trigonometry in 1791.

9. The Public Route Barred

1. On changes in family structure see, Jean Louis Flandrin, *Families in Former Times: Kinship, Household, and Sexuality,* trans. Richard Southern (1975; Cambridge, England, 1979); Lawrence Stone, *The Family, Sex, and Marriage in England, 1500–1800* (London, 1977); Richard Evans and W. R. Lee, *The German Family: Essays on the Social History of the Family in Nineteenth- and Twentieth-Century Germany* (London, 1981); and Leonore Davidoff and Catherine Hall, *Family Fortunes: Men and Women of the English Middle Class, 1780–1850* (Chicago, 1987).
2. See Abir-Am and Outram, eds., *Uneasy Careers.*
3. See Charles McClelland, *State, Society, and University in Germany, 1700–1914* (Cambridge, England, 1980), chap. 2; and F. Paulsen, *Geschichte des gelehrten Unterrichts* (Berlin, 1896).
4. Though Sophie Germain tried, in the 1790s, to pursue her studies at the new *Ecole Polytechnique,* her attempts were short-lived.
5. Stéphanie-Félicité du Crest, comtesse de Genlis, *De l'influence des femmes sur la littérature française* (Paris, 1811), p. xxv.
6. *Biographie universelle* (Paris, 1843), vol. 41, pp. 381–382. Though it is said that Thiroux d'Arconville left thirteen volumes of manuscripts at her death, I have been unable to locate them.
7. In the 1750s, Thiroux d'Arconville did a number of translations from English, including George Saville, the marquis of Halifax's *Avis d'un père à sa fille* (1756), Peter Shaw's *Leçons de chymie,* and Monro's *Traité de ostéologie* (1759). Her own work followed in the 1760s, including *Pensées et réflexions morales sur divers sujets* (1760), *De l'amitié* (1761) *Des passions* (1764), *Essai pour servir à l'histoire de la putréfaction* (1766), *Vie de Marie de Médicis, . . . Reine de France et de Navarre* (1744).
8. Jean-Paul Contant, *L'Enseignement de la chimie au Jardin Royal des Plantes*

de Paris (Coueslant, 1952), pp. 26–27. Women were also admitted to courses at the Collège Royal, but these were probably too technical.

9. Roger, *Les sciences de la vie,* p. 175.
10. A typical entry reads: "on the first of April 1762, the day was cool and cloudy, and I placed an ordinary piece of beef in two ounces of mineral water. Four days later, the sky was still cloudy and I found a rose colored and cloudy liquid. One could see a film of grease on the surface; the odor was horribly putrid, the meat white and soft. I washed it, but that did not diminish its stench. I threw it away." See [Thiroux d'Arconville], *Essai pour servir à l'histoire de la putréfaction* (Paris, 1766), p. 31. See also Alain Corbin, *The Foul and the Fragrant: Odor and the French Social Imagination* (Cambridge, Mass., 1986), p. 19.
11. Madame de Blot, cited in Diderot's review of Thiroux d'Arconville's "Vie du Cardinal d'Ossat," *Oeuvres complètes de Diderot,* vol. 9, p. 455.
12. "Essai sur l'amour-propre envisagé comme principe de morale," *Discours prononcé à l'Assemblée Ordinaire de l'Académie royale des Sciences et Belles-Lettres de Prusse* (January 11, 1770).
13. Though d'Arconville's *Ostéologie* was published under Sue's protection, both the translation and illustrations are hers, as is made clear in the preface republished among her collected works in 1775. Here she described how she both oversaw the drawing of the illustrations and added numerous observations to Monro's text. See [Marie Thiroux d'Arconville], "Sur l'ostèologie," in *Mêlanges de littérature, de morale et de physique,* ed. Rossel (her secretary), (Amsterdam, 1775), vol. 3, pp. 186–216.
14. Ibid., pp. 210–216.
15. Ibid., pp. 195–196.
16. [Marie Thiroux d'Arconville], "Sur les femmes," in *Mêlanges,* vol. 1, pp. 368–387.
17. Pierre-Henri-Hippolyte Bodard, *Cours de botanique médicale comparée* (Paris, 1810), vol. 1, p. xxvi-xxx.
18. I have chosen to use Erxleben's married name. Though she published her first book under her maiden name Leporinin (the feminine form of her family name), she used her married name for her later publications. Her university degree also bears her married name, and she was known as Frau Doctorin Erxleben. For a full bibliography of works on and by Erxleben, see Heinz Böhm, *Dorothea Christiane Erxleben: Ihr Leben und Wirken* (Quedlinburg, 1965), pp. 55–61.
19. See Erxleben's "Lebenslauf" in Dorothea Erxleben, *Academische Abhandlung von der gar zu geschwinden und angenehmen aber deswegen öfters unsichern Heilung der Krankheiten* (Halle, 1755), p. 123.
20. Leporinin, *Gründliche Untersuchung,* introduction, section 8; see also "Lebenslauf," in Erxleben, *Academische Abhandlung,* pp. 124–125.
21. Böhm, *Dorothea Christiane Erxleben,* pp. 5 and 8.
22. See, e.g., Johann Junker, "Reflexion über das Studieren und die academischen Würden des Frauenzimmers," *Wöchentliche Hallische Anzeigen,* 26 (July 1754): 450–458.

23. Leporinin, *Gründliche Untersuchung,* sections 66 and 67. Leporinin's book was pirated in 1749 and printed without her name or knowledge as *Vernünftige Gedanken vom Studieren des schönen Geschlechts.* I have been unable to find a copy of this edition.
24. Leporinin, *Gründliche Untersuchung,* introduction, and sections 111–113.
25. Michaelis's petition was published in rhymed verse, and he published it anonymously. See [Johann Michaelis], "Allerunterthänigste Bittschrifft an seine Königliche Majestät in Preussen, eine Anlegung einer Universität für das schöne Geschlecht," 1747. In 1946, Ida Hakemeyer found the petition at the university library in Göttingen and traced authorship to Michaelis (*Bemühungen um Frauenbildung in Göttingen 1747* [Göttingen, 1949], p. 4).
26. Leporinin, *Gründliche Untersuchung,* sections 125, 126, and 130.
27. Tobias Eckhard to Dorothea Leporinin, June 21, 1732, reprinted in Böhm, *Dorothea Christiane Erxleben,* p. 4. See also Ménage, *Historia mulierum philosopharum;* and Christian Paullini, *Hoch- und Wohlgelahrtes teutsches Frauenzimmer* (Frankfurt and Leipzig, 1712).
28. Leporinin, *Gründliche Untersuchung,* sections 125, 46, 86–104, 307.
29. Böhm, *Dorothea Christiane Erxleben,* p. 10.
30. See also Wiesner, *Working Women in Renaissance Germany,* pp. 50–55; and Wyman, "The Surgeoness," p. 25.
31. Doctors Herweg, Grasshoff, and Zeitz to Stiftshauptmann Paul von Schellersheim, February 5, 1753, original in Archiv des Rates der Stadt Quedlinburg, IV R 263 Stiftshauptmanney—"Acta betreffend die Medicinische Pfuscherey," reprinted in Werner Fischer-Defoy, "Die Promotion der ersten deutschen Ärztin, Dorothea Christiana Erleben, und ihre Vorgeschichte," *Archiv für Geschichte der Medizin,* 4 (1911): 444–445.
32. Schellersheim to Erxleben, February 16, 1753, reprinted in ibid., p. 445.
33. Dorothea Erxleben to Stiftshauptmann von Schellersheim, February 21, 1753, reprinted in full in Fischer-Defoy, "Die Promotion der ersten deutschen Ärztin," pp. 446–451.
34. "Lebenslauf," in Erxleben, *Academische Abhandlung,* p. 131.
35. Doctors Herweg, Grasshoff, and Zeitz to Stiftshauptmann Paul von Schellersheim, in Fischer-Defoy, "Die Promotion der ersten deutschen Ärztin," pp. 451–545.
36. Johann Junker, "Programma, mit welchem die Inauguraldissertation der . . . Frauen Dorothea Christiana Erxlebin . . ." reprinted in Erxleben, *Academische Abhandlung,* p. 139; and Junker, "Reflexion über das Studieren," p. 453.
37. Junker, "Programma," in Erxleben, *Academische Abhandlung,* p. 138.
38. This according to Böhm, *Dorothea Christiane Erxleben,* p. 23.
39. Dorothea Erxleben, *Dissertatio inauguralis medica, exponens, quod nimis cito ac jucunde curare saepius fiat caussa minus tutae curationis* (Halle, 1754); translated into German as *Academische Abhandlung von der gar zu geschwinden und angenehmen aber deswegen öfters unsichern Heilung der Krankheiten* (Halle, 1755).

40. Johann Junker, "Beschluss der Reflexion über des Studieren und die academischen Würden des Frauenzimmers," *Wöchentliche Hallische Anzeigen,* 27 (July 1754): 468; and Junker, "Programma," in Erxleben, *Academische Abhandlung,* p. 141.
41. Dorothea Erxleben, "Viri per Singulos Ordines Honoratissimi!" reprinted in Junker, "Beschluss der Reflexion," pp. 469–470.
42. See Fischer-Defoy, "Die Promotion der ersten deutschen Ärztin," p. 461. Women were not formally admitted to European universities until the 1860s in Switzerland, the 1870s in England, the 1880s in France, and the 1900s in Germany. See Rita McWilliams-Tullberg, "Women and Degrees at Cambridge University, 1862–1897," in *A Widening Sphere: Changing Roles of Victorian Women,* ed. Martha Vincinus (Bloomington, 1977), pp. 117–146; and Laetitia Böhm, "Von dem Anfängen des akademischen Frauenstudiums in Deutschland," *Historisches Jahrbuch,* 77 (1958): 2,298–2,327.
43. Johann Basedow, *Das Methodenbuch für Väter und Mütter der Familien und Völker* (Altona, 1770).
44. Cited in Leopold von Schlözer, *Dorothea von Schlözer* (Berlin, 1923), p. 31.
45. Dorothea von Schlözer, "Lebenslauf," August 17, 1787, University of Göttingen Archive, Philosophische Fakultät, no. 71, 1787, L. For bibliography on Dorothea Schlözer, see Martha Küssner, *Dorothea Schlözer: Ein Göttinger Gedenken* (Göttingen, 1976).
46. Cited in Schlözer, *Dorothea von Schlözer,* p. 104.
47. Ibid., pp. 122–123.
48. Ibid., pp. 125 and 134.
49. *Annalen der Braunschweig-Lüneburgischen Churlande,* 2 (1787): 120.
50. Dorothea Schlözer, *Nützliches Buch für die Küche bey Zubereitung der Speisen von dem Koch August Erdmann Lehmann* (Dresden, 1818). See Birgit Panke-Kochinke, "Göttinger Professorenfamilien im 18. und im ersten Drittel des 19. Jahrhunderts. Strukturmerkmale weiblichen Lebenszusammenhanges," Inaugural-Diss., Technische Universität Berlin, 1984, p. 330, n. 101.
51. Meiners, *Geschichte des weiblichen Geschlechts,* vol. 4. See also Brandes, *Über die Weiber;* Johann Ewald, *Die Kunst ein gutes Mädchen, eine gute Gattin, Mutter und Hausfrau zu werden* (Leipzig, 1798); Pockels, *Versuch einer Charakteristik des weiblichen Geschlechts;* and Karl Pockels, *Der Mann, Ein anthropologisches Charaktergemälde seines Geschlechts: Ein Gegenstück zur Charakteristik des weiblichen Geschlechts* (Hanover, 1805–1808).
52. Caroline Herschel, *Memoir and Correspondence of Caroline Herschel,* ed. Mrs. John Herschel (New York, 1876), p. 52. See also Mary Clerke, *The Herschels and Modern Astronomy* (New York, 1895); Constance Lubbock, *The Herschel Chronicle: The Life-Story of William Herschel and His Sister, Caroline Herschel* (Cambridge, England, 1933); Marilyn Ogilvie, "Caroline Herschel's Contributions to Astronomy," *Annals of Science,* 32 (1975): 149–161; and Michael Hoskin and Brian Warner, "Carolyn Herschel's Comet Sweepers," *Journal of the History of Astronomy,* 12 (1981): 27–34.

53. Entry from May 1, 1795, Herschel, *Memoir,* p. 147.
54. Ibid., p. 20.
55. "An Account of a New Comet," by Miss Caroline Herschel, read at the Royal Society, November 9, 1786, London, printed in the *Philosophical Transactions of the Royal Society of London,* 77 (1787): 1–3.
56. Caroline Herschel, *Catalogue of Stars, taken from Mr. Flamsteed's Observations* (London, 1798).
57. Letter of 12 July 1790, in Herschel, *Memoir,* pp. 90 and 99.
58. Caroline Herschel was elected to the Royal Astronomical Society along with Mary Somerville. Herschel and Somerville exchanged courteous letters at the time of their election, but had no other encounters (ibid., pp. 274–276).

10. The Exclusion of Women and the Structure of Knowledge

1. Sophia, a Person of Quality, *Woman not Inferior to Man: or, a short and modest Vindication of the Natural Right of the Fair-Sex to a perfect Equality of Power, Dignity, and Esteem with the Men* (London, 1739), p. 8 (emphasis in original). Cora Rosenkrantz has suggested that "Sophia" was Lady Sophia Fermor. See *A Dictionary of British and American Women Writers, 1660–1800,* ed. Janet Todd (London, 1984), p. 292. Entire sections of *Woman not Inferior to Man* were taken from Poullain de la Barre's *De l'égalité des deux sexes.*
2. [Mauvillon], *Mann und Weib,* pp. 8 and 14. Like Soemmerring, Mauvillon considered himself impartial (*unparteyisch*) in this matter.
3. Pockels, *Versuch einer Charakteristik des weiblichen Geschlechts,* vol. 1, pp. viii–xviii, and vol. 2, p. 331; and *Der Mann,* vol. 4, p. xi.
4. Holst, *Über die Bestimmung des Weibes,* p. x.
5. On the origin of value-neutrality in science, see Robert N. Proctor, "The Politics of Purity: Origins of the Ideal of Neutral Science" (Ph.D. diss., Harvard University, 1984).
6. *Female Rights Vindicated; or the Equality of the Sexes Morally and Physically proved,* by a Lady (London, 1763), p. 71.
7. Auguste Comte to J. S. Mill, 5 October 1843, in *Lettres inédites de J. S. Mill à A. Comte avec les réponses de Comte,* ed. L. Lévy-Bruhl (Paris, 1899), p. 250. I thank Mary Pickering for calling this correspondence to my attention.
8. Theodor von Bischoff, *Das Studium und die Ausübung der Medicin durch Frauen* (Munich, 1872), pp. 14–15, 20, and 47–48. For similar developments in England and the United States, see Joan Burstyn, "Education and Sex: The Medical Case against Higher Education for Women," *Proceedings of the American Philosophical Society,* 117 (1973): 79–89; Janet Sayer, *Biological Politics: Feminist and Anti-Feminist Perspectives* (London, 1982); Louise Newman, ed., *Men's Ideas/Women's Realities: Popular Science, 1870–1915* (New York, 1985).

9. Auguste Comte, *Cours de philosophie positive* (Paris, 1839), vol. 4, pp. 569–570.
10. Comte to Mill, 5 October 1843, *Lettres inédites,* ed. Lévy-Bruhl, p. 256.
11. Mill to Comte, 30 October 1843, ibid., p. 269.
12. Ibid., p. 270; 13 July, 1843, p. 223; and 30 August 1843, pp. 239–240.
13. Hedwig Dohm, *Die wissenschaftliche Emancipation der Frauen* (Berlin, 1874). On Dohm see Renate Duelli-Klein, "Hedwig Dohm: Passionate Theorist," in *Feminist Theorists: Three Centuries of Women's Intellectual Traditions,* ed. Dale Spender (London, 1983), pp. 165–183.
14. Jenny d'Héricourt, *A Woman's Philosophy of Woman; or Woman Affranchised* (New York, 1864), trans. from the French.
15. Friedrich W. T. Ravoth, "Über die Ziele und Aufgaben der Krankenpflege," quoted in Helga Rehse, "Die Rolle der Frau auf den Naturforscherversammlungen des 19. Jahrhunderts," in *Die Versammlung Deutscher Naturforscher und Ärzte im 19. Jahrhundert,* ed. Heinrich Schipperges, in *Schriftenreihe der Bezirksärztekammer Nordwürttemberg* 12 (1968), p. 126. At a recent meeting of the National Association of Scholars, conservative scholars similarly complained that political objectives, many of them flowing from affirmative action programs, had contaminated "objectivity" on decisions about curriculum, promotion, and academic discourse. *New York Times,* November 15, 1988, p. A22.
16. Leporinin, *Gründliche Untersuchung,* section 9.
17. Holst, *Über die Bestimmung des Weibes,* pp. 80–83.
18. See L. W. Weissenborn, *Briefe über die bürgerliche Selbstständigkeit der Weiber* (Gotha, 1806).
19. Recent attempts to replace Western civilization with world civilization in college curricula have addressed this problem. We must be wary, however, of being satisfied with the addition of a few books by women and minorities to these courses. The larger goal will be to rethink entirely the theoretical framework, purposes, and goals of such courses.
20. The problem goes beyond that of ignoring gender issues. Given the emphasis traditionally put on analytical philosophy, many students were left with the impression that philosophers such as Hume and Locke wrote only about epistemology (their political, historical, and religious essays went unanalyzed).
21. Kant, *Beobachtungen,* in *Kants Werke,* ed. Dilthey, vol. 2, p. 230.
22. Dieter Krallmann and Hans Martin, eds., *Wortindex zu Kants gesammelten Schriften* (Berlin, 1967), s.v. "Frau" and "Weib."
23. On Kant's views on women, see Lawrence Blum, "Kant's and Hegel's Moral Rationalism: A Feminist Perspective," *Canadian Journal of Philosophy,* 12 (1982): 287–302; see also Robin Schott, *Cognition and Eros: A Critique of the Kantian Paradigm* (Boston, 1988).
24. [Mauvillon], *Mann und Weib,* pp. 20–21.
25. Timothy Sellner in Hippel, *On Improving the Status of Women* (1792), trans. Timothy Sellner (Detroit, 1979), p. 29.

26. Hippel, *Über die bürgerliche Verbesserung der Weiber,* p. 35.
27. Weissenborn, *Briefe über die bürgerliche Selbstständigkeit der Weiber,* p. 22.
28. [Henriette], *Philosophie der Weiber* (Leipzig, 1802), p. 111.
29. Holst, *Über die Bestimmung des Weibes,* p. 95. Weissenborn held that only a businessman could fully appreciate the value of a learned wife (p. 27).
30. Carol Gilligan, *In a Different Voice: Psychological Theory and Women's Development* (Cambridge, Mass., 1982). This has been a best-seller for its publisher, Harvard University Press.
31. See, e.g., E. O. Wilson, *Sociobiology: The New Synthesis* (Cambridge, Mass., 1975).
32. Karl Joël, *Die Frauen in der Philosophie* (Hamburg, 1896), p. 32.
33. Otto Weininger, *Geschlecht und Charakter* (1903; Vienna and Leipzig, 1905), pp. 81–84.
34. Robert Merton, "The Normative Structure of Science" (1942), in *The Sociology of Science* (Chicago, 1973), pp. 267–278.
35. Oelsner, *Die Leistungen der deutschen Frau,* pp. 3–5.
36. This idea is found in the work of Hélène Cixous, for example.
37. See, among others, Nancy Hartsock, "The Feminist Standpoint: Developing the Ground for a Specifically Feminist Historical Materialism," in *Discovering Reality; Feminist Perspectives on Epistemology, Metaphysics, Methodology, and Philosophy of Science,* ed. Sandra Harding and Merrill Hintikka (Boston, 1983).
38. Nel Noddings, *Caring, A Feminine Approach to Ethics and Moral Education* (Berkeley, 1984); Hilary Rose, "Hand, Brain, and Heart: A Feminist Epistemology for the Natural Sciences," *Signs,* 9 (1983): 73–90; and Sarah Ruddick, "Maternal Thinking," paper delivered at the Boston Colloquium for Feminist Theory, Spring 1984. On the problem of other "others," see Harding, *The Science Question in Feminism,* chap. 7.
39. Martha Minnow cited in Joan W. Scott, "Deconstructing Equality-versus-Difference: Or, the Uses of Poststructuralist Theory for Feminism," *Feminist Studies,* 14 (Spring 1988): 39.

Selected Bibliography

Abir-Am, Pnina, and Dorinda Outram, eds. *Uneasy Careers and Intimate Lives: Women in Science, 1789–1979*. New Brunswick, 1987.

Accum, Frederick. *Culinary Chemist, Exhibiting the Scientific Principles of Cookery.* London, 1821.

Ackerknecht, Erwin, and Esther Fischer-Homberger. "Five Made It—One Not: The Rise of Medical Craftsmen to Academic Status during the Nineteenth Century." *Clio Medica,* 12 (1977): 255–267.

Ackermann, Jakob. *Über die körperliche Verschiedenheit des Mannes vom Weibe ausser Geschlechtstheilen,* trans. Joseph Wenzel. Koblenz, 1788.

Adam's Luxury, and Eve's Cookery. London, 1747.

Agnesi, Maria. *Istituzioni analitiche.* Milan, 1748.

——— *Propositiones philosophicae.* Milan, 1738.

Agrippa von Nettesheim, Henricus. *Female Pre-eminence or the Dignity and Excellency of that Sex, above the Male* (1532). London, 1670. Reprinted in *The Feminist Controversy of the Renaissance,* ed. Diane Bornstein. Delmar, N.Y., 1980.

Albinus, Bernard. *Table of the Skeleton and Muscles of the Human Body.* London, 1749.

Albistur, Maïté, and Daniel Armogathe. *Histoire du féminisme français.* 2 vols. Paris, 1977.

Alexander, William. *The History of Women.* 2 vols. London, 1779.

Algarotti, Francesco. *Il Newtonianismo per le dame.* Naples, 1737.

——— *Le Newtonianisme pour les dames,* trans. M. du Perron de Castera. Paris, 1738.

——— *Sir Isaac Newton's Philosophy Explain'd: for the Use of the Ladies,* trans. Elizabeth Carter. London, 1739.

[Allestree, Richard.] *The Ladies Calling* (1673). Oxford. 1720.

Archambault, Mlle. *Dissertation sur la question: Lequel de l'homme ou de la femme est plus capable de constance?* Paris, 1750.

Astell, Mary. *A Serious Proposal to the Ladies, for the Advancement of their True and Greatest Interest by a Lover of her Sex* (1694). London, 1701.

Aufgebauer, P. "Die Astronomenfamilie Kirch." *Die Sterne,* 47 (1971): 241–247.

Bacon, Francis. *The New Organon* (1620), ed. Fulton Anderson. Indianapolis, 1960.

——— *The Works of Francis Bacon,* ed. James Spedding, Robert Ellis, and Douglas Heath. 14 vols. London, 1857–1874.

Badinter, Elisabeth. *Emilie, Emilie: L'Ambition féminine au XVIII*[e] *siècle.* Paris, 1983.

——— *Mother Love: Myth and Reality.* New York, 1981.

Barclay, John. *The Anatomy of the Bones of the Human Body,* ed. Edward Mitchell and R. Knox. Edinburgh, 1829.

Basedow, Johann. *Das Methodenbuch für Väter und Mütter der Familien und Völker.* Altona, 1770.

Batsch, August. *Botanik für Frauenzimmer.* Weimar, 1795.

Bauhin, Gaspard. *Theatrum anatomicum.* Frankfurt, 1605.

Bell, Susan Groag. "Medieval Women Book Owners: Arbiters of Lay Piety and Ambassadors of Culture." *Signs: Journal of Women in Culture and Society,* 7 (1982): 742–768.

Bertin, Exupère-Joseph. *Traité d'ostéologie.* 4 vols. Paris, 1754.

Besterman, Theodore, ed. *Voltaire's Correspondence.* Geneva, 1968–1977.

Biheron, Marie. *Anatomie artificielle.* Paris, 1761.

Biographium Faemineum: The Female Worthies. London, 1766.

Blake, John B. "The Compleat Housewife." *Bulletin of the History of Medicine,* 49 (1975): 30–42.

Bleier, Ruth, ed. *Feminist Approaches to Science.* Elmsford, N.Y., 1986.

Blum, Lawrence. "Kant's and Hegel's Moral Rationalism: A Feminist Perspective." *Canadian Journal of Philosophy,* 12 (1982): 287–302.

Bodek, Evelyn. "Salonières and Bluestockings: Educated Obsolence and Germinating Feminism." *Feminist Studies,* 3 (1976): 185–199.

Böhm, Heinz. *Dorothea Christiane Erxleben: Ihr Leben und Wirken.* Quedlinburg, 1965.

Bordo, Susan. *The Flight to Objectivity: Essays in Cartesianism and Culture.* Albany, 1987.

Bovenschen, Silvia. *Die imaginierte Weiblichkeit.* Frankfurt, 1979.

Boxer, Marilyn, and Jean Quataert, eds. *Connecting Spheres: Women in the Western World, 1500 to the Present.* New York, 1987.

Boyle, Robert. *Experiments and Considerations Touching Colours*. London, 1664.

Brandes, Ernst. *Über die Weiber*. Leipzig, 1792.

Bucciarelli, Louis, and Nancy Dworsky. *Sophie Germain: An Essay in the History of the Theory of Elasticity*. Dordrecht, Holland, 1980.

Buffon, Georges-Louis Leclerc, comte de. *Histoire naturelle, générale et particulière*. 44 vols. Paris, 1749–1804.

Bullough, Vern L. "Medieval Medical and Scientific Views of Women." *Viator*, 4 (1973): 485–510.

Burney, Charles. *The Present State of Music in France and Italy* (1773), ed. Percy Scholes. London, 1959.

Candolle, Alphonse de. *Histoire des sciences et des savants depuis deux siècles*. Geneva, 1885.

Castiglione, Baldassare. *The Book of the Courtier* (1528), trans. Charles S. Singleton. Garden City, N.Y., 1959.

Cavendish, Margaret. *The Description of a New World, called the Blazing World*. London, 1666.

——— "The Female Academy." In *Playes*. London, 1662.

——— "Femal Orations." In *Orations of Divers Sorts*. London, 1662.

——— *Grounds of Natural Philosophy*. London, 1668.

——— *The Life of the Thrice Noble, High and Puissant Prince William Cavendishe, Duke, Marquess, and Earl of Newcastle*. London, 1667.

——— *Natures Pictures Drawn by Fancies Pencil to the Life*. London, 1656.

——— *Observations upon Experimental Philosophy*. London, 1666.

——— *The Philosophical and Physical Opinions*. London, 1655.

——— *Philosophical Letters*. London, 1664.

——— *Poems, and Fancies*. London, 1653.

——— *Sociable Letters*. London, 1664.

——— *The Worlds Olio*. London, 1655.

Cellier, Elizabeth. *A Scheme for the Foundation of a Royal Hospital* (1687). Reprinted in *The Harleian Miscellany*, ed. Thomas Osborne, vol. 4, 136–139. London, 1745.

Chapuzeau, Samuel. *L'Académie des femmes*. Paris, 1661.

Cheselden, William. *Anatomy of the Bones* (1713). 3d ed. London, 1726.

——— *Osteographia or the Anatomy of the Bones*. London, 1733.

Choulant, Ludwig. *History and Bibliography of Anatomic Illustration* (1852), trans. Mortimer Frank. New York, 1945.

Cochin, Charles, and Hubert-François Gravelot. *Iconologie par figures; ou Traité complet des allégories, emblèmes, &c.* (1791). Geneva, 1972.

Cockburn, Catharine. *The Works of Mrs. Catharine Cockburn*. London, 1751.

Cohen, I. Bernard. *Album of Science: From Leonardo to Lavoisier*. New York, 1980.

Condorcet, Marie-Jean-Antoine-Nicolas Caritat, marquis de. *Esquisse d'un tableau historique des progrès de l'esprit humain* (1795), ed. O. H. Prior. Paris, 1933.

——— "Sur l'admission des femmes au droit de cité." In *Oeuvres,* vol. 10. Stuttgart, 1968.

[Conring, Maria Sophia.] *Die wol unterweisette Köchinn.* Braunschweig, 1697.

Conway, Anne. *The Principles of the most Ancient and Modern Philosophers.* London, 1692.

Conze, Werner, ed. *Sozialgeschichte der Familie in der Neuzeit Europas.* Stuttgart, 1976.

Couvay, Louis. *L'Honneste maîtresse.* Paris, 1654.

Cowper, William. *The Anatomy of Humane Bodies* (1697). London, 1737.

Crooke, Helkiah. *Mikrokosmographia, A Description of the Body of Man.* London. 1615.

Cunitz, Maria. *Urania propitia.* Oels, 1650.

Davidoff, Leonore, and Catherine Hall. *Family Fortunes: Men and Women of the English Middle Class, 1780–1850.* Chicago, 1987.

Der aus dem Parnasso ehmals entlaufenen vortrefflichen Köchin welche bey denen Göttinnen Ceres, Diana und Pomona viel Jahre gedienet. Nuremberg, 1691.

Descartes, René. *Oeuvres de Descartes,* ed. Charles Adam and Paul Tannery. Paris, 1897–1910.

——— *Principles of Philosophy* (1644). In *The Philosophical Works of Descartes,* trans. Elizabeth S. Haldane and G. R. T. Ross. 2 vols. Cambridge, England, 1911.

Diderot, Denis. *Oeuvres complètes de Diderot,* ed. J. Assézat. 20 vols. Paris, 1875.

Dijk, Suzanne van. *Traces de femmes: Présence féminine dans le journalisme français du XVIIIe siècle.* Amsterdam, 1988.

Dock, Terry. "Woman in the *Encyclopédie.*" Ph.D. diss., Vanderbilt University, 1979.

Doeuff, Michèle Le. "Women and Philosophy." *Radical Philosophy,* 17 (1977): 2–11.

Dohm, Hedwig. *Die wissenschaftliche Emancipation der Frauen.* Berlin, 1874.

Donnison, Jean. *Midwives and Medical Men.* New York, 1977.

Doppelmayr, Johann. *Historische Nachricht von den Nürnbergischen Mathematicis und Künstlern* (1730), ed. Karlheinz Goldmann. Hildesheim, 1972.

Drake, James. *Anthropologia Nova; or, a New System of Anatomy,* ed. Judith Drake. 2 vols. London, 1707.

[Drake, Judith.] *An Essay in Defence of the Female Sex*. London, 1696.
Du Bosc, Jacques. *L'Honneste femme* (1632). Paris, 1658.
Easlea, Brian. *Witch-hunting, Magic and the New Philosophy.* Brighton, Sussex, 1980.
Eberti, J. C. *Eröffnetes Cabinet des gelehrten Frauenzimmers*. Frankfurt and Leipzig, 1706.
Edgeworth, Maria. *Letters for Literary Ladies*. London, 1795.
Ehrmann, Esther. *Madame du Châtelet: Scientist, Philosopher and Feminist of the Enlightenment*. New York, 1987.
Elshtain, Jean Bethke. *Public Man, Private Woman: Women in Social and Political Thought*. Princeton, 1981.
Encyclopédie, ou Dictionnaire raisonné des sciences, des arts et des métiers. Paris, 1751–1765.
[Erdt, Pauline.] *Philotheens Frauenzimmer-Akademie: Für Liebhaberinnen der Gelehrsamkeit,* trans. from the French. Augsburg, 1783.
Erxleben, Dorothea. *Academische Abhandlung von der gar zu geschwinden und angenehmen aber deswegen öfters unsichern Heilung der Krankheiten*. Halle, 1755.
Etat de médicine, chirurgie et pharmacie, en Europe. Pour l'année, 1776. Paris, 1776.
Euler, Leonhard. *Lettres à une princesse d'Allemagne sur divers points de physique et de philosophie*. St. Petersburg, 1768.
Evelyn, John. *The Diary of John Evelyn,* ed. Austin Dobson. London, 1906.
Fahy, Conor. "Three Early Renaissance Treatises on Women." *Italian Studies,* 11 (1956): 30–55.
Farrington, Benjamin. "Temporis Partus Masculus: An Untranslated Writing of Francis Bacon." *Centaurus,* 1 (1951): 193–205.
Fausto-Sterling, Anne. *Myths of Gender.* New York, 1985.
Fee, Elizabeth. "Nineteenth-Century Craniology: The Study of the Female Skull." *Bulletin of the History of Medicine,* 53 (1979): 415–433.
Female Rights Vindicated; or the Equality of the Sexes Morally and Physically proved. By a Lady. London, 1763.
Ferguson, James. *Easy Introduction to Astronomy for Gentlemen and Ladies*. London, 1768.
Ferguson, Margaret, Maureen Quilligan, and Nancy Vickers, eds. *Rewriting the Renaissance: The Discourses of Sexual Difference in Early Modern Europe*. Chicago, 1986.
Ferrante, Joan M. *Woman as Image in Medieval Literature*. New York, 1975.
Fischer-Defoy, Werner. "Die Promotion der ersten deutschen Ärztin, Do-

rothea Christiana Erxleben, und ihre Vorgeschichte." *Archiv für Geschichte der Medizin,* 4 (1911): 440–461.

Fischer-Homberger, Esther. *Krankheit Frau und andere Arbeiten zur Medizingeschichte der Frau.* Bern, 1979.

Fontanus, Nicholas. *The Womans Docteur: or, an exact and distinct Explanation of all such Diseases as are peculiar to that Sex.* London, 1652.

Fontenelle, Bernard Le Bovier de. *A Discovery of the New Worlds,* trans. Aphra Behn. London, 1688.

——— *Entretiens sur la pluralité des mondes* (1686), ed. Robert Shackleton. Oxford, 1955.

Forbes, Thomas R. "Regulation of English Midwives in the Sixteenth and Seventeenth Centuries." *Medical History,* 8 (1964): 235–244.

[Fores, S. W.] *Man-Midwifery Dissected.* London, 1793.

Fraisse, Geneviève. *Clemence Royer: Philosophe et femme de science.* Paris, 1985.

[Gacon-Dufour, Marie-Armande-Jeanne de Humières, Mme.] *Mémoire pour le sexe féminin contre le sexe masculin.* Paris and London, 1787.

Gallagher, Catherine, and Thomas Laqueur, eds. *The Making of the Modern Body: Sexuality and Society in the Nineteenth Century.* Berkeley, 1987.

Gardiner, Linda. *Emilie du Châtelet.* Wellesley College Center for Research on Women. Photocopy. 1982.

——— [Linda Gardiner Janik]. "Searching for the Metaphysics of Science: The Structure and Composition of Madame du Châtelet's *Institutions de physique,* 1737–1740." *Studies on Voltaire and the Eighteenth Century,* 201 (1982): 85–113.

Genlis, Stéphanie-Félicité du Crest, comtesse de. *Le Club des dames, ou le Retour de Descartes.* Paris, 1784.

——— *De l'influence des femmes sur la littérature française.* Paris, 1811.

The Gentleman's Diary, or the Mathematical Depository; an Almanack. London, 1741.

Gerhard, Ute. *Verhältnisse und Verhinderungen: Frauenarbeit, Familie und Rechte der Frauen im 19. Jahrhundert.* Frankfurt, 1978.

Germain, Sophie. *Oeuvres philosophiques de Sophie Germain,* ed. H. Stupuy. Paris, 1896.

Gillispie, Charles. *Science and Polity in France at the End of the Old Regime.* Princeton, 1980.

[Glasse, Hannah.] *The Art of Cookery, Made Plain and Easy.* London, 1747.

Gould, Stephen Jay. *The Panda's Thumb.* Boston, 1980.

Gournay, Marie le Jars de. *Egalité des hommes et des femmes.* Paris, 1622.

Grant, Douglas. *Margaret the First: A Biography of Margaret Cavendish, Duchess of Newcastle, 1623–1673.* London, 1957.

Gregory, John. *A Father's Legacy to his Daughters*. London, 1774.

Guillaume, Jacquette. *Les Dames illustres, où par bonnes et fortes raisons, il se prouve, que le sexe féminin surpasse en toute sorte de genres le sexe masculin*. Paris, 1665.

Gundersheimer, Werner L. "The Play of Intellect: The *Discorsi* of Annibale Romei." The Folger Shakespeare Library. Photocopy. 1984.

Hahn, Roger. *The Anatomy of a Scientific Institution: The Paris Academy of Science, 1666–1803*. Berkeley, 1971.

Hanstein, Adalbert von. *Die Frauen in der Zeit des Aufschwunges des deutschen Geistesleben*. Leipzig, 1899.

Harding, Sandra. *The Science Question in Feminism*. Ithaca, 1986.

Harding, Sandra, and Jean O'Barr, eds. *Sex and Scientific Inquiry*. Chicago, 1987.

Harless, Christian. *Die Verdienste der Frauen um Naturwissenschaft, Gesundheits- und Heilkunde*. Göttingen, 1830.

Harnack, Adolf von. "Berichte des Secretars der brandenburgischen Societät der Wissenschaften J. Th. Jablonski an den Präsidenten G. W. Leibniz." *Philosophisch-historische Abhandlungen der Königlichen Akademie der Wissenschaften zu Berlin,* 3 (1897).

——— *Geschichte der Königlich Preussischen Akademie der Wissenschften zu Berlin*. 3 vols. (1900). Hildesheim, 1970.

Harris, Ann Sutherland, and Linda Nochlin. *Women Artists: 1550–1950*. Los Angeles, 1976.

Hartman, George. *The True Preserver and Restorer of Health*. London, 1682.

Hausen, Karin, and Helga Nowotny, eds. *Wie männlich ist die Wissenschaft?* Frankfurt, 1986.

Haywood, Eliza. *The Female Spectator*. London, 1744–1746.

Hegel, Georg Wilhelm Friedrich. *Phänomenologie des Geistes* (1807). Vol. 3 in *Werke,* ed. Eva Moldenhauer and Karl Michel. Frankfurt, 1969–1971.

——— *Grundlinien der Philosophie des Rechts* (1821). Vol. 7 in *Werke,* ed. Eva Moldenhauer and Karl Michel. Frankfurt, 1969–1971.

Heinsohn, Gunnar, and Otto Steiger. *Die Vernichtung der weisen Frauen*. Herbstein, 1985.

[Henriette.] *Philosophie der Weiber*. Leipzig, 1802.

Herschel, Caroline. *Memoir and Correspondence of Caroline Herschel,* ed. Mrs. John Herschel. New York, 1876.

Hessen, Boris. "The Social and Economic Roots of Newton's 'Principia.'" In *Science at the Cross Roads*. London, 1931.

Hevelius, Johannes. *Firmamentum Sobiescianum sive Uranographie*. Danzig, 1687.

——— *Machina coelestis*. Danzig, 1673.

[Heywood, Thomas.] *The Generall History of Women*. London, 1657.

[Hippel, Theodor von.] *Über die bürgerliche Verbesserung der Weiber* (1792). Vol. 6 in *Sämmtliche Werke*. Berlin, 1828.

Hoffmann, Paul. *La Femme dans la pensée des lumières*. Paris, 1977.

Holst, Amalia. *Über die Bestimmung des Weibes zur höhern Geistesbildung*. Berlin, 1802.

Hubbard, Ruth, Mary Henifin, and Barbara Fried, eds. *Biological Woman—The Convenient Myth*. Cambridge, Mass., 1982.

Humboldt, Wilhelm von. "Über den Geschlechtsunterschied und dessen Einfluss auf die organische Natur," and "Über die männliche und weibliche Form," *Neudrucke zur Psychologie*, ed. Fritz Giese, 1 (1917): 1–231.

Hume, David. "Of Essay Writing" (1741). In *Essays Moral, Political and Literary*, 568–572. London, 1963.

Hunter, Michael. *The Royal Society and Its Fellows, 1660–1700: The Morphology of an Early Scientific Institution*. Chalfont St. Giles, Bucks., 1982.

Hurd-Mead, Kate Campbell. *A History of Women in Medicine*. Haddam, Conn., 1938.

Jöcher, Christian. *Allgemeines Gelehrten-Lexicon, darinne die Gelehrten aller Stände sowohl männ- als weiblichen Geschlechts*. Leipzig, 1751.

Joël, Karl. *Die Frauen in der Philosophie*. Hamburg, 1896.

Joeres, Ruth-Ellen, and Mary Jo Maynes, eds. *German Women in the Eighteenth and Nineteenth Centuries*. Bloomington, 1986.

Jörg, Johann. *Handbuch der Krankheiten des Weibes, nebst einer Einleitung in die Physiologie und Psychologie des weiblichen Organismus*. Leipzig, 1831.

Jouard, Gabriel. *Nouvel Essai sur la femme considérée comparativement à l'homme*. Paris, 1804.

Junker, Johann. "Reflexion über das Studieren und die academischen Würden des Frauenzimmers." *Wöchentliche Hallische Anzeigen*, 26 (July 1754): 450–458.

——— "Beschluss der Reflexion über das Studieren und die academischen Würden des Frauenzimmers," *Wöchentliche Hallische Anzeigen*, 27 (July 1754): 466–470.

Kant, Immanuel. *Anthropologie in pragmatischer Hinsicht* (1798). Frankfurt and Leipzig, 1799.

——— *Beobachtungen über das Gefühl des Schönen und Erhabenen* (1766). Vol. 2 in *Kants Werke*, ed. Wilhelm Dilthey. 24 vols. Berlin, 1900–1919.

Keller, Evelyn Fox. *Reflections on Gender and Science*. New Haven, 1985.

Kelly-Gadol, Joan. "Did Women Have a Renaissance?" In *Becoming Visible: Women in European History,* ed. Claudia Koonz and Renate Bridenthal, 137–164. Boston, 1977.

Kelso, Ruth. *The Doctrine of the Renaissance Lady.* Chicago, 1978.

Ketsch, Peter. *Frauen im Mittelalter.* 2 vols. Düsseldorf, 1983.

Kirch, Gottfried, and Maria Winkelmann. *Das älteste Berliner Wetter-Buch: 1700–1701,* ed. G. Hellmann. Berlin, 1893.

Klose, Carl. *Über den Einfluss des Geschlechts-Unterschiedes auf die Ausbildung und Heilung von Krankheiten.* Stendal, 1829.

Knibiehler, Yvonne, and Catherine Fouquet. *La Femme et les médecins.* Paris, 1983.

Knigge, Philippine von. *Versuch einer Logic für Frauenzimmer.* Hanover, 1789.

Koblitz, Ann Hibner. *A Convergence of Lives: Sofia Kovalevskaia—Scientist, Writer, Revolutionary.* Boston, 1983.

Koyré, Alexandre. *From the Closed World to the Infinite Universe.* Baltimore, 1957.

Labalme, Patricia, ed. *Beyond Their Sex: Learned Women of the European Past.* New York, 1984.

The Ladies' Diary, ed. John Tripper, Henry Beighton, Caelia Beighton, et al. London, 1704–1841.

The Ladies Dispensatory. London, 1739.

The Ladies Physical Directory. By a physician. London, 1716.

Laget, Mireille. "Childbirth in Seventeenth- and Eighteenth-Century France: Obstetrical Practices and Collective Attitudes." In *Medicine and Society in France,* ed. Robert Forster and Orest Ranum, 137–176. Baltimore, 1980.

Lagrange, E. "Les Femmes-Astronomes," *Ciel et terre,* 5 (1885): 513–527.

Lalande, Jérôme de. *Astronomie des dames.* Paris, 1786.

Lambert, Anne Thérèse de Marguenat de Courcelles, marquise de. *Réflexions nouvelles sur les femmes* (1727). London, 1820.

LeGates, Marlene. "The Cult of Womanhood in Eighteenth-Century Thought." *Eighteenth-Century Studies,* 10 (1976): 21–39.

Leibniz, Gottfried Wilhelm. *Die Werke von Leibniz,* ed. Onno Klopp. 11 vols. Hanover, 1864–1884.

Lémery, Nicolas. *A Course of Chymistry.* 4th English ed. London, 1720.

Lennox, Charlotte, ed. *The Lady's Museum.* London, 1760–1761.

Lepenies, Wolf. "Der Wissenschaftler als Autor, Buffons prekärer Nachruhm." In *Das Ende der Naturgeschichte: Wandel Kultureller Selbstverständlichkeiten in den Wissenschaften des 18. und 19. Jahrhunderts.* Frankfurt, 1978.

Leporinin, Dorothea [Dorothea Erxleben]. *Gründliche Untersuchung der Ursachen, die das weibliche Geschlecht vom Studieren abhalten.* Berlin, 1742.

Leppentin, Christoph. *Naturlehre für Frauenzimmer.* Hamburg, 1781.

Lesclache, Louis de. *Les Avantages que les femmes peuvent recevoir de la philosophie et principalement de la morale.* Paris, 1667.

Lévy-Bruhl, L., ed. *Lettres inédites de J. S. Mill à A. Comte avec les réponses de Comte.* Paris, 1899.

Leybourn, Thomas. *The Mathematical Questions proposed in the Ladies' Diary.* 4 vols. London, 1817.

La Liberté des dames. Paris, 1685.

Lipinska, Mélanie. *Histoire des femmes médecins, depuis l'antiquité jusqu'à nos jours.* Paris, 1900.

Lloyd, Genevieve. *The Man of Reason: "Male" and "Female" in Western Philosophy.* Minneapolis, 1984.

Locke, John. *The Works.* 3 vols. London, 1759.

Lonsdale, Kathleen. "Women in Science: Reminiscences and Reflections." *Impact of Science on Society,* 20 (1970): 45–59.

Lotichium, J. P. *Gynaecologia.* Frankfurt, 1645.

Lougee, Carolyn. *Le Paradis des Femmes: Women, Salons, and Social Stratification in Seventeenth Century France.* Princeton, 1976.

Ludendorff, Hans. "Zur Frühgeschichte der Astronomie in Berlin." *Vorträge und Schriften der Preussischen Akademie der Wissenschaften,* 9 (1942): 3–23.

MacCormack, Carol P., and Marilyn Strathern, eds. *Nature, Culture and Gender.* Cambridge, England, 1980.

Maclean, Ian. *The Renaissance Notion of Woman: A Study in the Fortunes of Scholasticism and Medical Science in European Intellectual Life.* Cambridge, England, 1980.

Maclean, Virginia. *A Short-Title Catalogue of Household and Cookery Books Published in the English Tongue, 1701–1800.* London, 1981.

Madame Johnson's Present: Or, Every Young Woman's Companion in Useful and Universal Knowledge. Dublin, 1770.

Martin, Benjamin. *Young Gentleman's and Lady's Philosophy* (1763). 2 vols. London, 1772.

[Mauvillon, Jakob.] *Mann und Weib nach ihren gegenseitigen Verhältnissen geschildert. Ein Gegenstück zu der Schrift: Über die Weiber.* Leipzig, 1791.

Meiners, Christoph. *Geschichte des weiblichen Geschlechts.* 4 vols. Hanover, 1788–1800.

Ménage, Gilles. *Historia mulierum philosopharum.* Lyon, 1690.

Merchant, Carolyn. *The Death of Nature: Women, Ecology and the Scientific Revolution.* San Francisco, 1980.

——— "Isis' Consciousness Raised." *Isis,* 73 (1982): 398–409.

Merian, Maria Sibylla [Maria S. Gräffin]. *Leningrader Aquarelle,* ed. Ernst Ullmann. 2 vols. Leipzig, 1972.

——— *Metamorphosis insectorum Surinamensium* (1705), ed. Helmut Decker. Leipzig, 1975.

——— *Der Raupen wunderbare Verwandlung und sonderbare Blumennahrung.* Nuremberg, 1679.

——— *Schmetterlinge, Käfer und andere Insekten: Leningrader Studienbuch,* ed. Wolf-Dietrich Beer. 2 vols. Leipzig, 1976.

——— *Die schönsten Tafeln aus dem grossen Buch der Schmetterlinge und Pflanzen: Metamorphosis insectorum Surinamensium,* ed. Gerhard Nebel. Hamburg, 1964.

Merton, Robert. *Science, Technology and Society in Seventeenth Century England* (1938). New York, 1970.

Meurdrac, Marie. *La Chymie charitable et facile, en faveur des dames.* Paris, 1665.

Meyer, Gerald. *The Scientific Lady in England: 1650–1760.* Berkeley, 1955.

Mintz, Samuel. "The Duchess of Newcastle's Visit to the Royal Society." *Journal of English and Germanic Philology,* 51 (1952): 168–176.

Molière, Jean-Baptiste. *Les Femmes savantes* (1672), ed. Jean Cordier. Paris, 1959.

Monro, Alexander. *The Anatomy of the Humane Bones.* Edinburgh, 1726.

——— "Essay on Female Conduct written by a Father to his Daughter." Manuscript held by Dr. P. A. G. Monro at St. John's College, Cambridge.

——— "Traité d'ostéologie," trans. Marie Thiroux d'Arconville. Paris, 1759.

Moreau, Edmond Thomas. *Quaestio medica: An praeter genitalia sexus inter se discrepent?* Paris, 1750.

Moreau de la Sarthe, Jacques-Louis. *Histoire naturelle de la femme.* Paris, 1803.

Moulton, Janice. "A Paradigm of Philosophy: The Adversary Method." In *Discovering Reality: Feminist Perspectives on Epistemology, Metaphysics, Methodology, and the Philosophy of Science,* ed. S. Harding and M. Hintikka, 149–164. Dordrecht, Holland, 1983.

Mozans, H. J. [John Zahm]. *Woman in Science: With an Introductory Chapter on Woman's Long Struggle for Things of the Mind* (1913). Cambridge, Mass., 1974.

Newman, Barbara. *Sister of Wisdom: St. Hildegard's Theology of the Feminine.* Berkeley, 1987.

Nihell, Elizabeth. *A Treatise on the Art of Midwifery.* London, 1760.

Nollet, Jean-Antoine. *Essai sur l'électricité des corps.* Paris, 1746.

——— *Leçons de physique experimentale.* 6 vols. Paris, 1743–1748.

Oelsner, Elise. *Die Leistungen der deutschen Frau in der letzten vierhundert Jahren auf wissenschaftlichen Gebiete.* Guhrau, 1894.

Okin, Susan Moller. "Women and the Making of the Sentimental Family." *Philosophy & Public Affairs,* 11 (1982): 65–88.

——— *Women in Western Political Thought.* Princeton, 1979.

Paré, Ambroise. *The Works of that Famous Chirurgion Ambrose Parey,* trans. Thomas Johnson (1585). London, 1634.

Pateman, Carole, and Teresa Brennan. "'Mere Auxiliaries to the Commonwealth': Women and the Origins of Liberalism." *Political Studies,* 27 (1979): 183–200.

Patterson, Elizabeth. *Mary Somerville and the Cultivation of Science, 1815–1840.* The Hague, 1983.

Paullini, Christian. *Hoch- und Wohlgelahrtes teutsches Frauenzimmer.* Frankfurt, 1712.

Pellisson, Paul, and P.-J. Thoulier d'Olivet. *Histoire de l'Académie française.* Paris, 1858.

Perl, Teri. "The Ladies' Diary or Woman's Almanack, 1704–1841." *Historia Mathematica,* 6 (1979): 36–53.

Perry, Ruth. *The Celebrated Mary Astell: An Early English Feminist.* Chicago, 1986.

Pfister-Burkhalter, Margarete. *Maria Sibylla Merian, Leben und Werk, 1647–1717.* Basel, 1980.

Pico della Mirandola, Giovanni. *Oration on the Dignity of Man,* trans. A. Robert Caponigri. Chicago, 1956.

Pizan, Christine de. *The Book of the City of Ladies* (1405), trans. Earl Jeffrey Richards. New York, 1982.

Pluche, Noel. *Spectacle de la nature.* Paris, 1732–1748.

Pockels, Karl. *Der Mann, ein anthropologisches Charaktergemälde seines Geschlechts: Ein Gegenstück zu der Charakteristik des weiblichen Geschlechts.* 4 vols. Hanover, 1805–1808.

——— *Versuch einer Charakteristik des weiblichen Geschlechts.* 5 vols. Hanover, 1799–1802.

Pope, Barbara Corrado. "Revolution and Retreat: Upper-Class French Women after 1789." In *Women, War and Revolution,* ed. Carol Berkin and Clara Lovett, 215–236. New York, 1980.

Posner, E. W. *Das Weib und das Kind.* Glogau, 1847.

Poullain de la Barre, François. *De l'éducation des dames pour la conduite de l'esprit dans les sciences et dans les moeurs.* Paris, 1674.

——— *De l'égalité des deux sexes: Discours physique et moral.* Paris, 1673.

Prétot, Philippe de. *Le Triomphe des dames, ou Le Nouvel Empire littéraire.* Paris, 1755.

Proctor, Robert N. "The Politics of Purity: Origins of the Ideal of Neutral Science." Ph.D. diss., Harvard University, 1984.

Prudhomme, Louis. *Biographie universelle et historique des femmes.* Paris, 1830.

Quataert, Jean. "Shaping of Women's Work in Manufacturing: Guilds, Households, and the State in Central Europe, 1648–1870." *American Historical Review,* 90 (1985): 1122–1148.

Raffald, Elizabeth. *The Experienced English Housekeeper, For the Use and Ease of Ladies, Cooks, etc., Wrote purely from practice* (1769). 2d ed. London, 1772.

Rebière, Alphonse. *Les Femmes dans la science.* 2d ed. Paris, 1897.

Remy, P. *Catalogue d'une collection de très belles coquilles, madrépores, stalactiques, . . . de Madame Bure.* Paris, 1763.

Renaudot, Théophraste, ed. *Recueil général des questions traictées ès conférences du Bureau d'Adresse, sur toutes sortes de matières; par les plus beaux esprits de ce temps.* Paris, 1656.

Reynier, Gustave. "La Science des dames au temps de Molière." *Revue des deux mondes,* May 1929, 436–464.

Ripa, Cesare. *Baroque and Rococo Pictorial Imagery: The 1758–60 Hertel Edition of Ripa's 'Iconologia,' with 200 Engraved Illustrations,* ed. and trans. Edward Maser. New York, 1971.

——— *Iconologia.* Rome, 1593.

Risse-Stumbries, Susanne. *Erziehung und Bildung der Frau in der zweiten Hälfte des 18. Jahrhunderts.* Tübingen, 1980.

Roger, Jacques. *Les Sciences de la vie dans la pensée française du XVIIIe siècle.* Paris, 1963.

Rosen, Richard. "The Academy of Sciences and the Institute of Bologna, 1690–1804." Ph.D. diss., Case Western Reserve University, 1971.

Rosenbaum, Heidi. *Formen der Familie.* Frankfurt, 1982.

Rossiter, Margaret. *Women Scientists in America: Struggles and Strategies to 1940.* Baltimore, 1982.

Rousseau, Jean-Jacques. *Emile.* (1762). In *Oeuvres complètes,* ed. Bernard Gagnebin and Marcel Raymond, vol. 4. Paris, 1959–1969.

——— *Letters on the Elements of Botany, addressed to a Lady,* trans. Thomas Martyn. 6th ed. London, 1802.

——— *Lettre à M. d'Alembert sur les spectacles* (1758), ed. L. Brunel. Paris, 1896.

——— "Lettres sur la botanique." In *Oeuvres complètes,* ed. Bernard Gagnebin and Marcel Raymond, vol. 4. Paris, 1959–1969.

Roussel, Pierre. *Système physique et moral de la femme, ou Tableau philosophique de la constitution, de l'état organique, du tempérament, des moeurs, et des fonctions propres au sexe.* Paris, 1775.

Rudolph, Emmanuel. "How It Happened That Botany Was the Science Thought Most Suitable for Victorian Young Ladies." *Children's Literature,* 2 (1973): 92–99.

Rücker, Elisabeth. "Maria Sibylla Merian." *Fränkische Lebensbilder,* 1 (1967): 221–247.

——— "Maria Sibylla Merian." In *Germanisches Nationalmuseum Nürnberg.* Nuremberg, 1967.

Sachs, J. J. *Ärztliches Gemälde des weiblichen Lebens im gesunden und krankhaften Zustande aus physiologischem, intellektuellem und moralischem Standpunkte.* Berlin, 1830.

Sandrart, Joachim von. *Teutsche Academie der Edlen Bau-, Bild- und Mahlerey-Künste.* Frankfurt, 1675.

Schiebinger, Londa. "The History and Philosophy of Women in Science: A Review Essay," *Signs,* 12 (1987): 305–332.

——— "Maria Winkelmann at the Berlin Academy: A Turning Point for Women in Science." *Isis,* 78 (1987): 174–200.

Schlözer, Leopold von. *Dorothea von Schlözer: der Philosophie Doctor.* Berlin, 1923.

Schurman, Anna van. *The Learned Maid, or Whether a Maid may be a Scholar? A Logic Exercise.* London, 1659.

[Smith, Eliza.] *The Compleat Housewife.* London, 1728.

Smith, Hilda. *Reason's Disciples: Seventeenth-Century English Feminists.* Chicago, 1982.

Soemmerring, Samuel Thomas von. *Tabula sceleti feminini juncta descriptione.* [Utrecht], 1796.

——— *Über die körperliche Verschiedenheit des Negers vom Europäer.* Frankfurt and Mainz, 1785.

——— *Über die Wirkungen der Schnürbruste.* Berlin, 1793.

Sophia, a Person of Quality. *Woman not Inferior to Man: or, a short and modest Vindication of the Natural Right of the Fair-Sex to a perfect Equality of Power, Dignity, and Esteem with the Men.* London, 1739.

Sorbière, Samuel. *Sorberiana.* Paris, 1691.

Speert, Harold. *Iconographia Gyniatrica: A Pictorial History of Gynecology and Obstetrics.* Philadelphia, 1973.

Spencer, Samia, ed. *French Women and the Age of Enlightenment.* Bloomington, 1984.

Sprat, Thomas. *History of the Royal Society of London, For the Improving of Natural Knowledge.* London, 1667.

Steinberg, Christian. *Naturlehre für Frauenzimmer.* Breslau, 1796.

Steinbrügg, Lieselotte. "Vom Aufstieg und Fall der gelehrten Frau: eine Aspekte der 'Querelle des Femmes' im XVIII. Jahrhundert." *Lendemains,* 25/26 (1982): 157–167.

Stone, Lawrence. *The Family, Sex, and Marriage in England, 1500–1800.* London, 1977.

——— "Literacy and Education in England, 1640–1900." *Past and Present,* 42 (1969): 69–139.

[Suckow, Lorenz.] *Briefe an das schöne Geschlecht über verschiedene Gegenstände aus dem Reiche der Nature*. Jena, 1770.

Sue, Jean-J. [Marie Thiroux d'Arconville]. *Traité d'ostéologie, traduit de l'Anglois de M. Monro*. Paris, 1759.

"Tabula sceleti feminini." *Journal der Empfindungen, Theorien und Widersprüche in der Natur- und Arztneiwissenschaft*, 6 (1797): Intelligenzblatt.

Taton, René. "Mme du Châtelet, traductrice de Newton." *Archives internationales d'histoire des sciences*, 22 (1969): 185–210.

Thiroux d'Arconville, Marie. *Essai pour servir à l'histoire de la putréfaction*. Paris, 1766.

——— *Mêlanges de littérature, de morale et de physique*, ed. Rossel. 7 vols. Amsterdam, 1775.

Thomas, Antoine. *Essai sur le caractère, les moeurs, et l'esprit des femmes dans les differens siècles*. Paris, 1772.

Tilly, Louise, and Joan Scott. *Women, Work, and Family*. New York, 1978.

Todd, Janet. *Dictionary of British and American Women Writers, 1660–1800*. London, 1984.

Tonzig, Maria. "Elena Lucrezia Cornaro Piscopia (1646–1684), prima donna laureata." *Quaderni per la storia dell'Università di Padova*, 6 (1973): 183–192.

Unzer, Johanna. *Grundriss einer Weltweisheit für Frauenzimmer*. Altona, 1761.

Vesalius, Andreas. *De humani corporis fabrica*. Basel, 1543.

Vignoles, Alphonse des. "Eloge de Madame Kirch à l'occasion de laquelle on parle de quelques autres femmes et d'un paisan astronomes," *Bibliothèque germanique*, 3 (1721): 115–183.

Villemert, Pierre-J. Boudier de. *L'Ami des femmes*. Paris, 1758.

Vogt, Carl. *Lectures on Man*, ed. James Hunt. London, 1864.

——— *Vorlesungen über den Menschen*. Giessen, 1863.

Voltaire, François-Marie Arouet de. *Dictionnaire philosophique* (1764). Amsterdam, 1789.

——— *An Essay upon the Civil Wars of France . . . And also upon the Epic Poetry of the European Nations*. London, 1727.

Wade, I. O. *Studies on Voltaire*. Princeton, 1947.

Wallis, Ruth and Peter. "Female Philomaths." *Historia Mathematica*, 7 (1980): 57–64.

Warner, Marina. *Monuments and Maidens: The Allegory of the Female Form*. New York, 1985.

Wattenberg, Dietrich. "Zur Geschichte der Astronomie in Berlin im 16. bis 18. Jahrhundert I." *Die Sterne*, 48 (1972): 161–172.

——— "Zur Geschichte der Astronomie in Berlin im 16. bis 18. Jahrhundert II." *Die Sterne*, 49 (1972): 104–116.

Weber, Jakob. *Fragmente von der Physik für Frauenzimmer und Kinder.* Tübingen, 1779.

Weckerin, Anna. *Ein köstlich new Kochbuch.* Amberg, 1697.

Weidler, Frederick. *Historia astronomiae.* Wittenberg, 1741.

Weindler, Fritz. *Geschichte der gynäkologisch-anatomischen Abbildung.* Dresden, 1908.

Weininger, Otto. *Geschlecht und Charakter* (1903). Vienna and Leipzig, 1905.

Weiss, F. Herbert. "Quellenbeiträge zur Geschichte der Preussischen Akademie der Wissenschaften." *Jahrbuch der Preussischen Akademie der Wissenschaften,* 1939: 214–224.

Weissenborn, L. W. *Briefe über die bürgerliche Selbstständigkeit der Weiber.* Gotha, 1806.

Wensky, Margret. *Die Stellung der Frau in der stadtkölnischen Wirtschaft im Spätmittelalter.* Cologne, 1981.

Wiesner, Merry. *Working Women in Renaissance Germany.* New Brunswick, 1986.

Wilkes, Wetenhall. *An Essay on the Pleasures and Advantages of Female Literature.* London, 1741.

Will, Georg. *Nürnbergisches Gelehrten-Lexicon, oder Beschreibung aller Nürnbergischen Gelehrten beyderley Geschlechtes.* Nuremberg, 1755–1758.

Williams, David. "Political Feminism in the French Enlightenment." In *The Varied Pattern: Studies in the 18th Century,* ed. Peter Hughes and David Williams, vol. 1, 333–351. Toronto, 1971.

Winkelmann, Maria. *Vorbereitung, zur grossen Opposition, oder merckwürdige Himmels-Gestalt im 1712.* Cölln an der Spree, 1711.

——— *Vorstellung des Himmels bey der Zusammenkunfft dreyer Grossmächtigsten Könige.* Potsdam, 1709.

Winslow, J.-B. "Sur les mauvais effets de l'usage des corps à baleine." *Mémoires de l'Académie des sciences.* Paris, 1741.

Wiswe, Hans. *Kulturgeschichte der Kochkunst.* Munich, 1970.

Wolf-Heidegger, G., and Anna Maria Cetto. *Die anatomische Sektion in bildlicher Darstellung.* Basel, 1967.

Wollstonecraft, Mary. *Vindication of the Rights of Woman* (1792), ed. Miriam Brody Kramnick. Harmondsworth, England, 1982.

Woltmann, Karoline von. *Über Natur, Bestimmung, Tugend und Bildung der Frauen.* Wien, 1826.

Woolley, Hannah. *The Gentlewomans Companion.* London, 1675.

Wyman, A. L. "The Surgeoness: The Female Practitioner of Surgery, 1400–1800." *Medical History,* 28 (1984): 22–41.

Yates, Frances. *The French Academies of the Sixteenth Century.* London, 1947.

Ziegenbein, Johann. *Aussprüche über weibliche Natur, weibliche Bestimmung, Erziehung und Bildung*. Blankenburg, 1808.
Zilsel, Edgar. "The Sociological Roots of Modern Science." *American Journal of Sociology*, 47 (1942): 245–279.
Zinner, Ernst. *Die Geschichte der Sternkunde*. Berlin, 1931.

Index